Tropical Climatology

This book is dedicated to
my co-author Simon Nieuwolt

Tropical Climatology

An Introduction to the Climates of the Low Latitudes

Second Edition

Glenn R. McGregor
The University of Birmingham, UK

Simon Nieuwolt
University of Guelph, Canada

JOHN WILEY & SONS
Chichester · New York · Weinheim · Brisbane · Singapore · Toronto

Other Wiley Editorial Offices

John Wiley & Sons, Inc., 605 Third Avenue,
New York NY 10158-0012, USA

WILEY-VCH Verlags GmbH, Pappelallee 3,
D-69469 Weinheim, Germany

Jacaranda Wiley Ltd, 33 Park Road, Milton,
Queensland 4064, Australia

John Wiley & Sons (Asia) Pte Ltd, 2 Clementi Loop #02-01,
Jin Xing Distripark, Singapore 129809

John Wiley & Sons (Canada) Ltd, 22 Worcester Road,
Rexdale, Ontario M9W 1L1, Canada

Library of Congress Cataloging-in-Publication Data

McGregor, G. R. (Glenn R.)
    Tropical climatology: an introduction to the climates of the low
latitudes / G. R. McGregor and S. Nieuwolt.
      p. cm.
    Includes bibliographical references and index.
    ISBN 0-471- 96610-X (alk. paper). — ISBN 0-471-96611-8 (alk.
paper)
    1. Tropics—Climate. I. Nieuwolt, S. II. Title.
QC993.5.M34   1998
551.6913—dc21   97-41190
                                            CIP

British Library Cataloguing in Publications Data

A catalogue record for this book is available from the British Library

ISBN 0-471-96610-X (hardback)
    0-471-96611-8 (paperback)
Typeset in 10/12pt Times from the author's disks by Vision Typesetting, Manchester

Contents

Preface

Since 1977, when *Tropical Climatology: An Introduction to the Climates of the Low Latitudes* was first published (Nieuwolt, 1977), there have been many advances in the understanding of the workings of the tropical atmosphere and the nature of tropical climates at a variety of space and time scales. At the global tropical scale much has been learnt about the intertropical convergence zone, the quasi-biennial oscillation, the 40–50 day tropical oscillation and the El Nino southern oscillation phenomena, all of which play a major role in climatic variability and therefore life in the tropics. Central to the sustainability of life and economic production systems in large parts of the tropics are the monsoons. Over the last two decades there has also been rapid knowledge increases concerning these seasonal circulation systems. For example, the Asian monsoon is now known to be composed of two component systems, namely the Indian and East Asian systems. Furthermore, it appears that the general Asian monsoon system may be part of a larger interactive system involving Eurasian snow cover and sea surface temperature anomalies in the Pacific Basin. Over the eastern Pacific satellite-based studies of the annual cycle of the ocean and atmosphere have also revealed the possible existence of a further monsoon circulation while increased monitoring of the monsoon systems in West and East Africa have revealed these to be quite different in nature. At smaller space and time scales our knowledge concerning the meteorology and climatology of tropical disturbances such as tropical cyclones, easterly waves and squall lines has been increased markedly, mainly due to the information available from satellites.

In addition to considerable progress having been made concerning tropical circulation features at a variety of space and time scales, a scientific issue which has not only provided the scientific community with one of its greatest challenges but also has caught the imagination of the public over the last two decades is that of climate change. This is of particular relevance to the tropics as the economies of many low latitude countries are heavily dependent on the climate resource. Changes in the nature and availability of that resource may bring climate advantages to some low latitude countries while others may experience climate disadvantages as a consequence of an anthropogenically enhanced greenhouse effect.

Given the above, a second updated edition of *Tropical Climatology* seemed appropriate. This has been accomplished with a co-author. The fundamental aims of the second edition remain unchanged from those of the first. These are to provide a geographical viewpoint on the physical processes in the tropical atmosphere; to offer explanations of how a location's climate is a product of these processes; and to highlight the implications of tropical atmospheric behaviour for those living in the tropics. To partly meet these aims and because contemporary climatology places emphasis on explanation of the behaviour of the atmosphere and climate, a new chapter (Chapter 2) on the fundamental laws of weather and climate has been added. Covering the very basic physics and dynamics of the atmosphere, this chapter attempts to provide the reader with a non-mathematical overview of the physical laws that describe the relationship between those physical atmospheric variables that make up a location's climate.

The addition of a new chapter (Chapter 6) on non-seasonal variations of the tropical circulation, although including some material from the first edition, partly reflects the various advances that have been made concerning our understanding of the quasi-biennial oscillation, the 40–50 day tropical oscillation and the El Nino southern oscillation phenomena.

Climate change and its implications for the tropics is the topic of the remaining new chapter (Chapter 13). Following an introduction to the general causes of climate change, material on the potential effects of climate change in the tropics is introduced. The often perceived, tropics related issues of deforestation and desertification are also addressed in this new chapter.

To reflect the rapid advances in our knowledge concerning the tropical circulation at both the global and regional scales a major rewrite of the chapters on the general circulation of the tropics (Chapter 5) and seasonal variations of regional circulation systems (Chapter 7) has been undertaken. Likewise new material has been added concerning tropical disturbances (Chapter 8). Without the results from the many tropical atmosphere measurement and monitoring programmes over the last twenty years, in which satellites have played an extremely important role, rewriting these chapters would have been a difficult task. In fact much of the new material in this book reflects the great advances that have been made over the last two decades in the measurement and modelling of the tropical atmosphere and tropical climates.

In many cases material on the general climatology topics of insolation, temperature, water in the atmosphere and precipitation has remained little changed from the first edition. However, these topics have been expanded with the addition of material on radiation and energy balances, urban climates, evaporation measurement and estimation, measures of humidity, tropical clouds and climate, satellite cloud climatology, and the interannual variability of rainfall and drought. A regional treatment of the last topic is also given in the chapter describing tropical climates which includes new climate maps for many parts of the tropics. The original chapter on applied tropical climatology has been modified with an explicit focus on tropical climates and agriculture (Chapter 11).

The readership of the book remains essentially the same as that for which the first edition was intended; second- and third-year non-technical students in geography and environmental sciences who have some background in climatology. The updated reference list does however provide a possible entry point for non-specialist postgraduates into the field of tropical climatology.

We hope that this second edition is as successful as the first in terms of facilitating a non-technical appreciation of the workings of the tropical atmosphere, the explanation of tropical climates and how the tropical atmosphere and its related processes play a role in the lives and activities of low latitude inhabitants.

Birmingham, UK, June 1997   Glenn McGregor
Guelph, Canada, June 1997   Simon Nieuwolt

Acknowledgements

First and foremost I would like to acknowledge my co-author Simon Nieuwolt, who, despite suffering from a prolonged period of illness, worked with great energy and care throughout the early preparation stages of this book. Unfortunately, Simon passed away on July 14, 1997 before this book could be published. Accordingly, and as an acknowledgement of his contribution to the field of tropical climatology in the geographical tradition, I have dedicated the second edition of *Tropical Climatology* to Simon. I would also like to thank the Nieuwolt family for allowing me to put my name as the first author as an acknowledgement of the time I have spent rewriting major parts of what follows and ensuring that the second edition of *Tropical Climatology* reached the publication stage.

Hearty thanks are extended to the editorial staff at John Wiley and Sons who have been very supportive throughout this project. Many thanks also to the people in NASA, NOAA, the Hong Kong Observatory and the Australian Bureau of Meteorology who supplied or granted permission for the use of the satellite and radar images that appear in this book. I am also grateful to the various copyright holders for granting permission for the use of many of the figures reproduced herein. The graphic and photographic expertise of Kevin, Anne and Geoff of the School of Geography, The University of Birmingham is also acknowledged.

I would also like to thank past and present mentors and colleagues who have stimulated and supported my interest in climatology, especially Ian Owens, Phillip and Marjorie Sullivan, Bernie Owen, Ian Jackson and Brian Giles.

Lastly this book would not have been completed without the encouragement and support of my wife Miyoko and the rest of my family.

Glenn McGregor
December, 1997

CHAPTER 1

Introduction

DEFINITION OF THE TERM "TROPICS"

This book deals with the climatic conditions in those parts of the world that are commonly referred to as the "tropics". This term has no exact meaning, so we will define it here and indicate which regions are covered in this book. The word "tropics" is derived from the Tropics of Cancer and Capricorn, the parallels at 23.5°, which indicate the outer limits of the area where the sun can be in a zenith position (Figure 1.1). It is generally understood that the tropical areas are mainly found between these two lines. They are therefore the regions of the "low latitudes".

These areas have some common characteristics which distinguish them from the rest of the world. As latitude is a major factor controlling climatic conditions, the most important of these common features are those of climate. Other typical characteristics of the tropics, such as those of natural vegetation, soils, agriculture and economic development, are all, directly or indirectly, related to their common climatic features. We can therefore define the tropics as those areas where typical tropical climates prevail.

CLIMATIC BOUNDARIES OF THE TROPICS

The Tropics of Cancer and of Capricorn are unsuitable as boundaries of tropical climates. They are too rigid; latitude is not the only factor in climate and major climatic boundaries frequently deviate from parallels. Some regions with clear tropical features are at latitudes of more than 23.5°, while some obviously non-tropical areas are found much closer to the equator.

A better approach to determine climatic boundaries is to define a major common feature. Undoubtedly, the absence of a cold season is the most important common climatic trait of the low latitudes, as illustrated by the old adage "where winter never comes". To define and delimit the tropics by the lack of a winter season cannot be done with a simple temperature limit, as for instance "the mean temperature of 18°C for the

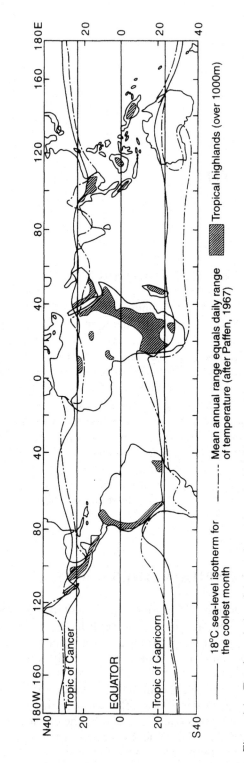

Figure 1.1 Two boundaries of the tropics and tropical highlands

coldest month of the year" (Koeppen, 1936). This method, often used in older climatology texts, has the drawback that it excludes the tropical highlands, where temperatures remain frequently below this limit; yet these areas are truly tropical because they experience no cold season. They can be included in the tropics by using not the actual temperatures, but those reduced to sea level (Figure 1.1).

It must be admitted that temperatures reduced to sea level are fictitious figures in many continental areas, and that they are prone to errors because they are based on the standard rate of temperature decrease with elevation, whereas the actual decline varies considerably both with season and location (Chapter 4). However, on a world scale these errors are relatively minor and the boundary has the advantage that it can be interpolated easily from many reliable temperature data. Its general form is simple and it delimits the tropics as a belt around the equator, varying in width from about 40° to 60°, about 80% of which is ocean surface (Figure 1.1).

Another indicator of the absence of a cold season is *the annual range of temperature.* This figure is usually only one or two degrees near the equator and increases generally with latitude, but it also exhibits a strong influence of continentality. While in the mid-latitudes the annual range generally exceeds the mean daily range of temperature, in the tropics the reverse prevails, as illustrated by the old adage: "the nights are the winter of the tropics". This is an important climatic feature of the tropics, and sometimes the line where the annual and daily ranges are about equal has been taken as the outer limit of the tropics (Figure 1.1). However, this comparison is only possible over land surfaces. Over the oceans, where air temperatures are almost entirely controlled by the surface water temperature, diurnal ranges are very small. For most of the oceanic areas, the line shown in Figure 1.1 is therefore based on temperature readings over islands, but these are far from representative for the ocean surfaces around them (Chapter 4).

Meteorologists often use another boundary of the tropics: *the axis of the subtropical high pressure cells,* which is the dividing line between atmospheric circulations dominated by easterlies (in the tropics) and by mid-latitude westerlies (Hastenrath, 1985, p. 1). This boundary has the disadvantage that it moves latitudinally with the seasons and it includes many areas with subtropical climates which exhibit typical tropical features only during the summer, but have cool winters with mean temperatures below 18 °C for the coldest month. It should be remembered that, except in mountainous terrain, climatic differences with place are generally very gradual because atmospheric conditions vary a great deal from year to year. The actual limit of the tropics fluctuates and may move over large distances around the average position indicated on a map by a line. The boundary line therefore represents a large transition zone rather than an abrupt change of climate.

Some geographers reserve the term "tropics" for regions where sufficient rainfall is received to carry out most forms of crop agriculture without irrigation (Gourou, 1935). Sometimes this condition is clearly stated by the use of the term "humid tropics", but more often the assumption is made tacitly. It is difficult to determine the amount of rainfall necessary to sustain crop agriculture without irrigation, as it depends not only on a number of climatic factors, like temperature, wind speed, sunshine and the seasonal distribution of the rainfall, but also on soil conditions, drainage and agricultural methods. Most estimates vary between 450 and 600 mm per year (Koeppen,

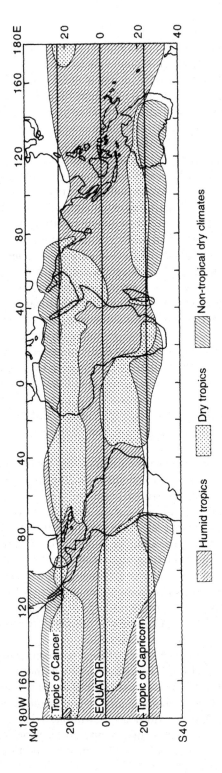

Figure 1.2 Humid and dry tropics and adjacent dry areas

1936). In this book, a corresponding approximate limit is used to subdivide the tropics into humid and dry parts (Figure 1.2). There are many other methods to delimit and subdivide the tropics, either by climatic elements or by the effects of climate on the natural vegetation. The boundaries used in this book have the advantage that they can be determined from easily available temperature and rainfall data and that they are more or less constant over time.

THE IMPORTANCE OF TROPICAL CLIMATES

As will be shown in the following chapters, tropical climates have effects that reach far beyond the limits of the regions where they actually prevail: the tropical radiation balance (Chapter 3) and the resulting temperature conditions (Chapter 4) are the major engine of the general circulation of the atmosphere (Chapters 5–7), and a large part of all atmospheric water content originates in the tropics (Chapter 9). More intermittent tropical phenomena, like ENSO, influence the climates over very large areas, including many parts of the mid-latitudes (Chapter 6).

Tropical climates are also of general importance, because they control the lives and economic activities of the population in their regions to a much greater extent than mid-latitude climates do. The inhabitants of these areas number around 2400 million people, or about 45% of the world population. Almost all these people live in the humid tropics, around 60% of them in southern and eastern Asia (*World Bank Atlas*, 1992, pp. 8–9).

Many tropical countries belong to the group of less developed, or developing nations, characterized by low standards of living and a strong economic concentration on agriculture and the production of raw materials rather than industrial products. The poverty of many tropical countries is shown by the fact that of 82 countries situated entirely or largely in tropical climates, 64 (78%) had a per capita GNP below US$1500 in 1991. On a world scale, only 43% of all countries are in this category (*World Bank Atlas*, 1992, pp. 18–20). Other indicators of poverty are low real purchasing power, low percentage of food needs available, limited access to safe drinking water, poor quality of life and high child mortality. In all these statistics, tropical countries provide the majority in the poorest categories (Kidron and Segal, 1991, pp. 36–55). The economic importance of the tropics is related to the enormous consumer market, which in many areas is only a potential one because of the low purchasing power of a large part of the population. The tropics also produce many raw materials, such as oil (Venezuela, Nigeria, Indonesia), tin (Malaysia, Brazil), bauxite (Guinea, Jamaica, Brazil, India, Surinam), copper (Zambia, Zaire), iron ore (Venezuela, India) and natural gas (Indonesia, Malaysia, Brunei). Industrial production is important in only a few tropical countries (India, Philippines, Indonesia) and is usually strongly concentrated in only a few regions. Some tropical countries provide important commercial and financial services (Singapore, Malaysia).

In most of the poorer tropical countries agriculture is still the main sector of the economy. Its main function is to produce food for the local population. Tropical food crops such as rice, maize, cassava, coconuts, bananas and sorghum are largely produced on a subsistence basis. Climate is usually one of the main factors in this production (Chapter 12). Due to the large increases in population in the last 50 years, in many parts of the tropics a precarious balance between needs and production of food prevails. Whenever normal production is disturbed by exceptional weather conditions, or by civil strife, war or strikes, serious famines may occur. Particularly vulnerable in this respect are Bangladesh, Ethiopia, Somalia, and the countries of the Sahel. Other tropical countries manage to produce food surpluses: Thailand is the largest exporter of rice in the world.

Cash crops for the world market are important in many tropical countries, where they are a major source of badly needed foreign exchange. In countries like Uganda, Somalia, Sudan, Malawi and Mozambique more than 90% of exports are agricultural products (including forestry and fishing), and in Mali, Ivory Coast, Ethiopia and Ghana the percentage exceeds 75 (*World Market Atlas*, 1992, p. 55). For many tropical products there are no substitutes in the mid-latitudes, as for coffee, tea, pepper and many other spices, cocoa, palm oil and pineapples. Other tropical products have a clear price advantage on world markets because of low production costs: natural rubber, cane sugar, cotton, groundnuts and tobacco. Finally, tropical woods are an important export product, because of special qualities and low production costs. However, tropical deforestation, which has climatic implications (Chapter 13), has become a major problem, as in some parts of Indonesia, Malaysia, Thailand and Brazil (*World Market Atlas*, 1992, pp. 59–69).

TROPICAL CLIMATOLOGY

Tropical climatology is the science of tropical climate. Meteorology is the science of weather and provides a physical basis for the study of climate and climatological relationships. Climatology has a number of goals which are of relevance to the tropics, including the following:

1. providing climate inventories which lay the basis for climatic classifications, the nature of which will depend on the purpose of the classification's application, for example for agriculture or human health and comfort;
2. describing the horizontal and vertical fluxes or movements of heat, momentum and moisture. Climatology as a physical science therefore aids in developing an understanding of the causes of climate from the global to local scale, and
3. understanding the extent to which human activities and a range of environmental processes are climate sensitive. Climatological information can therefore be applied to the benefit of society, environmental management and the assessment of the availability and distribution of climate-related natural resources such as water. Such climate sensitivity analyses also provide the basis for assessing the probable impacts of natural or human-induced climate change and variability.

These three main aspects of tropical climatology will be covered in this book.

The study of tropical climatology is as old as human settlements in tropical regions. Where the main source of food is agriculture, the importance of climatic conditions is evident. Experiences over many generations indicated to farmers when and where to plant crops and graze their animals. Experience also taught which crops or animals would do best in each region. These traditions still live on in most rural areas. They are usually accurate, but fail to give predictions over periods longer than one season.

A more scientific and quantified form of tropical climatology became possible with the establishment of meteorological services in most tropical countries during the 19th and early 20th centuries. In many parts of the tropics these were set up by colonial governments and frequently concentrated on collecting data which were important for plantations. These services have now been taken over by national governments, but some colonial traditions have survived, such as the use of the metric system (the current standard measurement system) in former French, Belgian and Dutch colonies, while in the British realm inches and Fahrenheit continued to be used for a long time. Data collected by these organizations are used in many parts of this book, particularly for the description of regional climates in Chapter 11. In addition to the information collected at standard climatological and meteorological stations, other types of data have contributed to the development of tropical climatology as a modern science and therefore the understanding of tropical climates and weather systems. This "new" type of data has come from satellite observation programmes such as the International Satellite Land Climatology (ISLCP) and Cloud Climatology Projects (ISCCP) and special measurement programmes implemented to increase our understanding of the physical and dynamical properties of the tropical atmosphere; the Atlantic Tropical Experiment (GATE) as part of the Global Atmospheric Research Programme (GARP), the Monsoon Experiment (MONEX) and the Tropical Ocean Global Atmosphere (TOGA) programme to name but a few. Although such data are of use for shedding light on the nature and causes of tropical climates, a fundamental explanation of the behaviour of the atmosphere comes from an understanding of the basic laws of weather and climate, the subject of the next chapter.

CHAPTER 2

Understanding the Laws of Weather and Climate

Traditional climatology can be defined as the science involved with describing the mean state of the atmosphere at a location through time and/or over space. This involves characterizing a location's climate, usually using a number of statistical descriptors, in terms of how wet, windy, cool, dry, warm or calm it is. The variables used in these descriptions are normally temperature, atmospheric moisture (humidity), wind speed and direction, and atmospheric pressure. If we want to know something about the vertical characteristics of the atmosphere, often air density and vertical wind velocity are also taken into account. These seven variables do not operate in the atmosphere independently of each other as changes in one of the variables usually produce changes in the others. It is this interactive nature of the atmospheric variables that produces the weather experienced at any point in time or space.

Climate is related to weather such that the climate of a location is the mean weather conditions experienced. Of course the climate will also be characterized by its variability and extremes. However, when we talk about the climate of the low latitudes we are basically talking about the mean weather conditions experienced as described by the behaviour of the seven atmospheric variables listed above. In many cases climate descriptions stop at this point with no explanation of why the climate is like it is. Modern climatology goes beyond straight description and seeks an explanation of why a certain location's climate is like it is. An explanation of the general behaviour of the atmosphere can be sought through the understanding of a number of *physical laws* that describe the relationship between the variables that make up a location's climate. These physical laws have been known for two to three centuries and now form the basis of many weather and climate prediction models. For this reason we feel that it is important to present a brief review of these physical laws and concepts as this will help develop an understanding of the workings of weather and climate of low latitude locations and the general circulation of the tropical atmosphere. Emphasis in this chapter will be placed on laws describing the thermodynamic and dynamic behaviour of the atmosphere while the various radiation laws will be described at the beginning of the next chapter.

THE PHYSICAL LAWS

In most climatology and meteorology texts the physical laws take the form of mathematical equations. In many cases these, unfortunately, turn the less mathematically inclined reader off. We will therefore endeavour to present the physical laws in the form of a number of word equations in the style of Atkinson (1981). In this way we hope to avoid the use of mathematics.

The equations that describe the basic behaviour of a parcel of air in terms of its thermal, moisture and motion characteristics are as follows:

—the equation of state;
—the first law of thermodynamics;
—the conservation of mass;
—the continuity equation;
—the horizontal wind equations;
—the hydrostatic equation; and
—the vorticity theorem (Atkinson, 1981).

Before proceeding, let us define a parcel of air. This is an imaginary slice of the atmosphere with well-defined boundaries and a mass or weight that does not change. In terms of shape, the parcel can be thought of something like a slice of bread with horizontal dimensions, greater than vertical dimensions, as is true for the atmosphere as a whole.

THE EQUATION OF STATE

The equation of state describes the state or condition of a parcel of air in terms of its temperature, pressure and volume (density of the air). The equation of state is a combination of several laws and is of use in understanding the relationship between these three air parcel variables. The laws making up the equation of state, named after the scientists who discovered them, have been known for some time.

Boyle's law states that if the temperature of a parcel of air is held constant (it does not change) then an increase in the volume of the air parcel will result in a decrease of the pressure. In word equation form: *for constant temperature, pressure is inversely proportional (opposite) to volume*, i.e.

$$\text{pressure} = 1/\text{volume}$$

Taking this relationship we can say that for an air parcel with a given weight or mass, the more space occupied by the parcel in the atmosphere, the less pressure it exerts on the earth's surface. This is because the available weight is now spread out over a greater area. Remember that this is for the case where the temperature does not change.

Charles, another scientist working in the period of the scientific revolution, like Boyle, was also interested in the relationship between temperature, pressure and volume. In contrast to Boyle, Charles investigated the relationships between these variables in a slightly different manner. Instead of holding temperature constant for his experiments, as had Boyle, Charles held pressure constant.

Charles' law states that the volume occupied by a gas is directly proportional to its absolute temperature for a constant pressure. In word equation form: *for constant pressure, volume is proportional to temperature.* If pressure does not change, but temperature does, the amount of space occupied by a parcel of air in the atmosphere will increase. Considering this relationship we can think of one further relationship between pressure, temperature and volume. To do this we need to rearrange the above word equation and now make volume constant such that *pressure is proportional to temperature.*

While the above laws were developed in laboratory situations, the place where weather and climate happens is very different to that of the laboratory where it is easy to control the atmospheric conditions. In the atmosphere, temperature and pressure, the constant variables of Boyle and Charles, usually vary together such that neither one is very rarely constant if the other changes.

Using algebra, the above three laws can be combined into one equation to describe the relationship between the temperature, pressure and volume of a parcel of air. This single equation is *the equation of state,* which can be written as follows:

$$\frac{\text{pressure} \times \text{volume}}{\text{temperature}} = \text{constant}$$

This expression tells us that the product of pressure and volume is inversely proportional to temperature and for given combinations of pressure, volume and temperature this equation will always produce a constant. Another way to think about this equation is that if the value of one of the variables is changed then the other variable values must also change in a compensatory fashion.

The work of a group of scientists in the early parts of the 19th century revealed that, for dry air, the value of the constant is 287 J K^{-1} kg^{-1}. This is known as the universal gas constant. If we take the value of the constant to be 287 J K^{-1} kg^{-1} and call this value R, then we can use R and some mathematical manipulation to find an equation that represents the relationship between volume and the other two variables. Before we do this we should replace volume with density because volume is inappropriate when dealing with air which is unconfined (density is mass divided by volume). So the final form of the equation of state can be written as:

$$\text{density} = \frac{\text{pressure}}{R \times \text{temperature}}$$

Meteorologists usually refer to this as the *meteorological form of the equation of state.* Because temperature and pressure are usually observed, the equation of state can be used to determine the density of the air. This equation tells us that increasing the temperature expands the air and decreases the density. Conversely, decreasing the temperature contracts the air and increases the density. Temperature increases or decreases therefore can make the air lighter or heavier. It also tells us that increasing the pressure squeezes the air and increases the density. These relationships are extremely important in understanding how air behaves and thus how some weather events that are features of a location's climate are produced.

THE FIRST LAW OF THERMODYNAMICS

The thermal behaviour of air is important for generating a range of weather and climate phenomena. We have seen from the equation of state that when heat is added to (subtracted from) a volume of air, expansion (compression) occurs. Therefore, to understand the behaviour of air in the situation where heat is added or extracted, a relationship needs to be found between the rate of exchange of heat and changes of an air parcel's temperature and volume. The first law of thermodynamics helps us with this task and can be stated simply as:

heat added = increase in total energy + work done against outside pressure

What does this mean in terms of a parcel of air. First let us look at the components of the thermodynamics equation. The form of this equation tells us that what happens on the left must be met with a response from the two components on the right-hand side. Heat added is straightforward and may refer to that heat gained by heating of an air parcel by sunlight. According to the thermodynamics equation, any heat or energy gain must be balanced by changes in the nature of the components on the right-hand side of the equation. These required changes can be achieved in two related ways: firstly an increase in the internal energy of the air parcel, and secondly an expenditure of energy as the air parcel expands and pushes out against the air around it. The first of these will result in an increase in the kinetic energy or temperature of the parcel of air, while the second will result in the air parcel expanding and thus increasing its volume and thus decreasing its density (it becomes lighter). So what the first law of thermodynamics is describing for us is the behaviour of a parcel of air when heat is added. Simply, if we add energy to a mass of air, the energy is used to raise the temperature of the air parcel and/or to increase its volume.

We can use the first law of thermodynamics to think about the situation where no heat is added to dry air. In this case the laws states:

$$\text{zero heat added} = \frac{\text{zero internal}}{\text{energy increase}} + \frac{\text{zero work done against}}{\text{outside pressure}}$$

In other words, if no energy is added then temperatures will not increase and the density will not decrease.

Continuing with this line of thinking and still adding no energy to the air, if in some way, the air parcel is allowed to expand, then the first law of thermodynamics will state:

$$\text{zero heat added} = \frac{\text{internal energy}}{\text{decrease}} + \frac{\text{work done against}}{\text{outside atmosphere}}$$

In other words, to preserve balance, the energy expended in increasing the volume of the air parcel by pushing out against the surrounding air must be matched by a loss of internal energy. This means, for the situation where no heat is added, an increase in the volume of the air parcel (a decrease of density) will be matched by a decrease in its temperature. This is the process that occurs when air rises, such as in a thunderstorm cloud. Of course, these relationships can work in reverse. For the situation where an air parcel sinks back down to earth, as is typical in the areas of semi-permanent subtropical anticyclones, compression of the air parcel produces warming.

Because the rate at which energy is added to air parcels in the atmosphere by radiation or mixing is extremely low, the relationships as described by the first law of thermodynamics are useful for explaining the thermal behaviour of dry air.

The process by which the thermal behaviour of air changes due to changes in internal energy and volume (density) and in the absence of outside sources of heat, is referred to as an *adiabatic process*. Temperature changes resulting from the adiabatic process are called *adiabatic temperature changes*. The adiabatic process results in temperature changes in the atmosphere of $9.8\,°C\,km^{-1}$ (rounded up to $10\,°C\,km^{-1}$). This is referred to as the *dry adiabatic lapse rate*. Dry air rising from sea level at $25\,°C$ will therefore adiabatically cool to $15\,°C$ by the time it reaches $1000\,m$ above sea level. Conversely, air forced back down from the $1000\,m$ level at $15\,°C$ will warm to $25\,°C$ by the time it reaches sea level. This rule sounds ideal for estimating the temperature of air as it rises or falls through the atmosphere. In fact, it is fine for air that is not saturated.

Because the atmosphere is not always dry and does contain moisture, the actual measured lapse rate is often less. The actual lapse rate is referred to as the *environmental lapse rate* and has a global average of $6.5\,°C\,km^{-1}$. This lower lapse rate highlights the role of atmospheric moisture in controlling atmospheric thermal characteristics. The application of the environmental lapse rate to the problem of estimating vertical temperature gradients is fine in a moist but non-saturated atmospheric layer.

Saturation can be brought about by cooling of the moisture in an air parcel as that air parcel rises. When the moisture cools, its *saturation point* is often reached. This is the temperature at which the air parcel is replete with moisture; it is called the *dew point temperature*. At this point condensation usually occurs. This is important because condensation which is associated with water changing phase from a gas to a liquid, results in the release of heat called *latent heat*. The effect of this released heat is a warming one. In fact, at air temperatures of $0\,°C$ for every kilogram of water that is condensed out of the atmosphere, 2.5 mega-joules (MJ) of energy are released (Figure 2.1). This is equivalent to about one-quarter of the average solar energy received at the earth's surface. Therefore, in a saturated air parcel, when condensation occurs, the released latent heat warms the surrounding air. This has an effect on the atmospheric temperature lapse rate such that in a saturated atmospheric layer the lapse rate is lower than the dry adiabatic lapse rate. This lapse rate is called the *saturated adiabatic lapse rate*. In a warm moist atmosphere, typical of the humid tropics, the saturated adiabatic lapse rate may be as low as $4\,°C\,km^{-1}$. This is because moist air with high temperatures generally has greater moisture contents than cooler air (Chapter 9). As a result, a great amount of moisture is available for condensation; the resulting latent heat released is greater and thus the degree of atmospheric warming greater.

So what has this all to do with the first law of thermodynamics? Recalling from above that this law applies to dry air and states that the temperature is controlled by the change of volume or density (pressure) as a result of an air parcel rising, then for a moist atmosphere, where condensation is occurring, there is an internal heat source. This complicates the simple relationships described in the first law of thermodynamics.

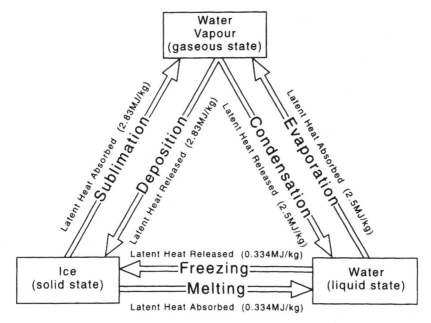

Figure 2.1   The three states of water and associated energy exchanges with change of state

Consequently the first law equation must be modified if it is to describe the thermal behaviour of a parcel of moist air.

Because in an adiabatic process the heat added term is zero, we can rearrange the first law equation. In rearranging the equation let us use temperature for energy, and volume for work done against outside pressure. The first law equation then takes the following form:

$$\text{temperature change} = \frac{1}{\text{change of volume}}$$

The rearranged equation, using temperature and volume as terms, now tells us that as volume increases in an air parcel, temperature decreases and vice versa. Taking this relationship for a rising saturated air parcel in which condensation is occurring and latent heat is being released, the first law becomes:

$$\text{temperature change} = \frac{\text{energy added due}}{\text{to condensation}} - \frac{\text{energy lost due to}}{\text{volume increase}}$$

In other words, energy added by condensation offsets the energy losses due to adiabatic pressure change. Simple reasoning will indicate that if the energy gained from condensation is greater than that expected due to the dry adiabatic cooling process, then the temperature change will be less in the case of a saturated atmosphere. This is why the saturated adiabatic lapse rate is less than the dry adiabatic lapse rate.

Having explained the thermal behaviour of dry and saturated air in terms of the first law of thermodynamics, let us now consider the equation for the conservation of moisture.

THE MOISTURE EQUATION

The moisture equation deals with the concepts of evaporation, condensation and mixing in the atmosphere. It takes the following form:

$$\text{change in moisture} = \text{change due to evaporation} + \text{change due to condensation} - \text{change due to vertical mixing}$$

What this means is that moisture cannot be gained or lost but it may change its form from liquid to gas (Figure 2.1). This idea is important as despite a *change of state*, the mass of moisture is still preserved. So wherever evaporation occurs it must be balanced by condensation somewhere else. This relates to the idea of the *global water balance* which says that all moisture losses from the earth's land surface in the form of runoff and evaporation must be made up by gains due to precipitation that results from condensation. Over the oceans all losses due to evaporation must be balanced by gains due to precipitation, and runoff from the land. So despite moisture continually changing its state there is never any gain of mass. This conservation of mass, despite a change of form, is related to the next law described by the equation of continuity.

THE EQUATION OF CONTINUITY

The equation of continuity describes the principle called the *conservation of mass*. This says that no matter how much air moves about or how much it is compressed or stretched, air will not change its mass. So if we take air to be a fluid, this principle states that nowhere in the climate system can fluid be created or destroyed. Therefore, compressing air in one place must be met by expansion in another place. This is a bit like squeezing a water carrier bag filled with water. If you sit on the bag you are compressing the fluid inside the bag. It responds by changing its shape and as a result the bags expands. However, all the time throughout this squeezing process the water inside the bag still keeps its mass.

In meteorology the equation of continuity is usually stated as:

$$\text{vertical convergence} = \text{horizontal divergence} + \text{a density change}$$

This equation can also be stated in reverse such that:

$$\text{horizontal convergence} = \text{vertical divergence} + \text{a density change}$$

For most cases density in the above two situations can be considered to be insignificant. Therefore the equations now say that convergence must be balanced by divergence and vice versa.

Let us think of some examples relevant to the atmosphere. Firstly, consider the case of air which sinks back down to the ground; the case in subtropical areas where the weather and climate is dominated by large anticyclones or high pressure cells. In this case the sinking air is vertically converging on the ground surface. As the air that is piling up at the surface cannot escape into the ground, it escapes sideways. In other words, it diverges. This is the case of vertical convergence resulting in horizontal divergence (Figure 2.2(a)). The climatic effects of this are that as the air sinks it adiabatically warms, resulting in drying of the air due to the process of evaporation. As

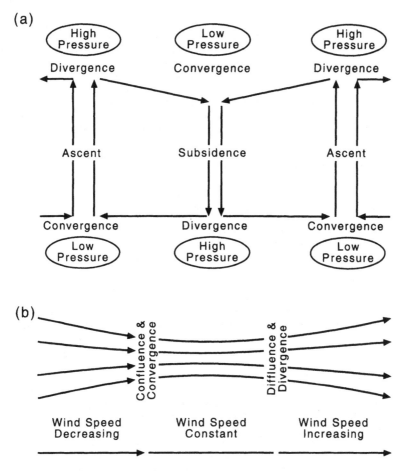

Figure 2.2   (a) Patterns of convergence and divergence associated with surface and upper pressure patterns; (b) convergence and divergence as related to confluence and diffluence of airstreams

a result, the atmosphere is generally cloud-free in areas of subsiding air associated with divergence in the lower parts of the atmosphere. Because divergence in the lower atmosphere is occurring, then according to the equation of continuity, convergence in the upper atmosphere must be occurring. Upper level convergence often occurs downstream of upper level divergence, which brings us to our second example.

   Consider the case where flows of air from opposite directions in the lower atmosphere are running into each other (Figure 2.2(a)). This is the case of lower level convergence. This must be met by divergence as the only place air can go is up; the air ascends. With the ascent of air, adiabatic cooling starts (the first law of thermodynamics). If the air is moist, condensation usually follows, which leads eventually to precipitation (moisture equation). Convergence of streams of moist air in the lower atmosphere therefore usually give rise to cloudiness and precipitation. Ascent will finally give way to divergence in the upper atmosphere. When this happens the air starts to spread out again and move horizontally. Eventually it will run into another stream of upper level air coming from the opposite direction and upper level convergence will occur. This of course must be matched by lower level divergence and is the beginning of the situation described in our first example; upper level convergence is associated with lower level divergence (Figure 2.2(a)).

Convergence can also occur when wind speeds decrease in the direction of flow. It is also enhanced by confluence when airstreams flow into the same area. Conversely, when air speeds up in the direction of flow then divergence and thus subsidence result. Divergence is also enhanced by diffluence when airstreams part (Figure 2.2(b)).

EQUATIONS OF MOTION

The equations of motion help with an understanding of why the winds blow. In any system, motion only occurs when there is some form of force applied to a mass. In the case of the atmosphere, if there were no forces operating, then there would be no motion of air and thus no winds.

The forces that generate air motion are gravity, the pressure gradient force and the Coriolis force. One other force is also important, not because it causes air to move, but because it retards air movement. This is the frictional force.

We are all familiar with the concept of *gravity*. This is the force that holds that thin envelope of gases we call the atmosphere in place. Sir Isaac Newton defined gravitation as the force attracting any two bodies towards each other. The magnitude of the gravitational force is proportional to the mass of the two bodies. Because the mass of the atmosphere is small compared to that of the earth, gravitation operates to pull the atmosphere towards the surface of the earth. Gravity would be important for air motion if there was an unequal distribution of it over the earth's surface. As gravity only varies by an insignificant amount (0.5%) from the equator to the poles (gravity is greater at the poles), the gravitational force is not usually considered in introductory discussions about atmospheric motion. We will therefore move on to the other forces.

Pressure, commonly measured in millibars (mb), is a measure of the weight of the atmosphere above a point in the atmosphere, or above the earth's surface. Given this it is easy to understand why atmospheric pressure decreases with height in the atmosphere: the greater the altitude, the less air (mass) above a given point (Table 2.1). On weather maps, locations with equal pressure are connected by lines called *isobars*. In the atmosphere, surfaces with equal pressures are referred to as *isobaric surfaces*.

Atmospheric pressure does not decrease in a constant or linear fashion as temperature and atmospheric composition, which affects air density, are also involved (see the equation of state).

If we were to stop the earth rotating, bring all atmospheric motion to a halt and watch to see what might re-initiate air motion, we would discover that the unequal distribution of solar energy across the earth's curved surface would play an important role. This is because surface heating causes air to rise. Therefore atmospheric mass would be transferred upwards, resulting in a fall of the surface pressure. At a place further away receiving much less heating, the mass and thus pressure would be greater. These geographical contrasts in pressure would give rise to *pressure gradients* which are a manifestation of excess air at one place and not enough at another. In the

Table 2.1 Altitude pressure and temperature relationships in an average atmosphere in the tropics

Altitude (m)	Pressure (mb)	Temperature (°C)
100	1000	25.8
550	950	22.5
1 020	900	19.6
1 510	850	17.0
2 020	800	14.6
2 570	750	12.0
3 145	700	8.9
3 760	650	5.5
4 410	600	1.9
5 110	550	− 2.1
5 855	500	− 6.4
6 685	450	− 11.4
7 570	400	− 17.1
8 550	350	− 24.0
9 655	300	− 32.5
10 920	250	− 42.7
12 390	200	− 54.8
14 170	150	− 68.2
16 550	100	− 77.9
20 780	50	− 65.9
31 070	10	− 38.0

presence of gravity the excess air would move down the pressure gradient from a point of high pressure to low pressure. This is analogous to water flowing down a hill. The force to which the air responds, with motion resulting, is the *pressure gradient force* (*PGF*). It can be simply represented as follows:

$$PGF = \frac{\text{pressure at A} - \text{pressure at B}}{D \times (\text{air density at A} - \text{air density at B})}$$

where A and B are points on the earth's surface and D is their distance apart. From the PGF equation it can be seen that the greater the pressure at A compared to B, the greater will be the pressure gradient force and therefore the faster the air motion (Figure 2.3).

Tropical cyclones, one of the fiercest weather systems on the earth's surface (Chapter 8), often have horizontal pressure gradients of 1 mb km^{-1}. This contrasts markedly with the subtropical anticyclones (Chapter 5) centred around 30° from the equator with pressure gradients of around $1 \text{mb } 100 \text{ km}^{-1}$. These pressure gradient differences help account for the contrasting wind regimes of these two types of tropical weather system.

If the earth were to start rotating, the air moving as a result of gravity and the PGF would be subject to another force, called the Coriolis force.

The *Coriolis force* is a deflectional force arising from the rotation of the earth around its axis. On any rotating body, objects moving from rest will be deflected in a curved

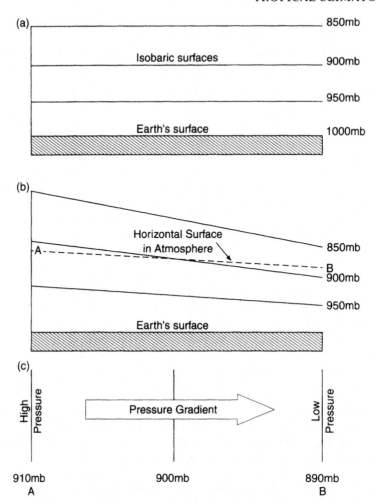

Figure 2.3   Isobaric surfaces in the case of (a) uniform surface pressure; (b) unequal distribution of surface pressure; (c) the situation as in (b) but in plan form demonstrating the pressure gradient between points A and B

path. In the case of air moving under the influence of a PGF, the degree to which the air will be deflected is related to the velocity of the air and the rate of the earth's rotation at the position of movement. This is referred to as the Coriolis force (CF) and is expressed as:

$$CF = \text{velocity} \times 2 \text{ (angular velocity} \times \text{the sin of the latitude)}$$

In the above equation the quantity in the brackets is the component of the angular speed at any latitude. This varies from zero at the equator to a maximum at the poles as angular speed is measured in relation to the local vertical. This is because at the poles a horizontal surface rotates around the earth's axis; at the equator horizontal surfaces are parallel to the earth's axis and themselves do not rotate. As the rate of rotation becomes larger as the poles are approached, so does the CF (Figure 2.4).

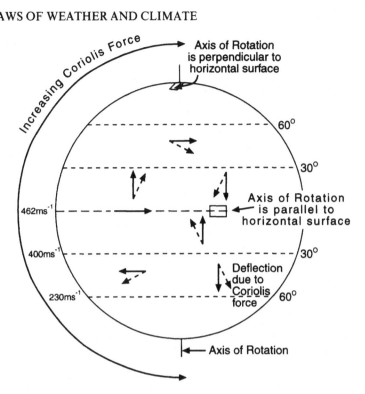

Figure 2.4   The Coriolis effect and axis of rotation relationships

The total expression indicated by CF is often called the Coriolis parameter, which varies from zero at the equator to a maximum at the poles. Air moving from rest at 30° from the equator will be deflected far more than air moving from 5°. In the northern hemisphere the angle of deflection is to the right, whereas in the southern hemisphere the angle of deflection is to the left. At the equator, because the Coriolis parameter is zero, no deflection of air occurs (Figure 2.4). The Coriolis force accounts for the directional flow of the trade winds and ocean currents in the subtropical latitudes (Chapter 5). Because velocity also appears in the CF expression then for a given latitude air moving at a high velocity will be deflected more than air moving at a slow velocity.

The earth's surface imposes *friction* on the lowest layers of a moving atmosphere. This friction is transferred vertically through the atmosphere in the form of turbulent eddies which impede further the movement of air. Frictional effects are greatest over non-uniform terrain because these are eddy-generating surfaces which contrast with the relatively smooth surface of the ocean where frictional effects are minimal. With the interplay of the PGF, the CF and friction (F), the movement of air can be affected both in terms of its speed and direction of movement. The resultant direction of flow will depend on whether the isobaric surfaces are planar or not.

Atmospheric motion on non-curved or planar pressure (isobaric) surfaces is caused by the interplay between the PGF, the CF and F. This interplay produces a resultant force. This is a balance of the forces at play and is the one that determines the direction

of the air's motion. As well as direction, the acceleration and the velocity characteristics of the air are also important. Such wind vector characteristics (speed and direction) are related to the sum of the forces operating on the air and the mass of the air.
These relationships can be summed up in the single equation of motion called
Newton's second law of motion which is stated as:

$$\text{acceleration} = \frac{\text{sum of the forces}}{\text{mass of the air}}$$

This states that the acceleration of air, the rate of change of velocity with time, is
directly proportional to the sum of forces operating on a mass of air. The three forces
as discussed are the PGF + CF + F. As the forces are not all acting in the same
direction the acceleration of the air will be in the direction of the resultant force. The
above equation is a vector equation, i.e. it has properties of both direction and
magnitude. This vector equation can be broken down into its three component
equations which represent the magnitude of acceleration in the two horizontal dimensions and one vertical dimension. For horizontal motion the two horizontal equations
of motion can be written as:

$$\text{acceleration}_x = \text{sum of PGF} + \text{CF} + \text{F in the } x \text{ direction}$$
$$\text{acceleration}_y = \text{sum of PGF} + \text{CF} + \text{F in the } y \text{ direction}$$

where x and y are the west to east and south to north directions respectively (Atkinson,
1981).

The remaining equation for the vertical direction will be discussed under the
heading of the hydrostatic equation.

Above the earth's surface or friction layer, motion is largely determined by the
balance between the PGF and CF. This is because friction is unimportant. Assuming
that the PGF and the CF are the only two forces operating, the resultant of these will
determine the direction of flow. In the low latitudes the CF tends towards zero so as the
equator is approached the direction of the winds becomes very much influenced by the
PGF (Figure 2.5(a)). In addition, as low latitudes are approached, the PGF decreases
so that wind velocities also decrease. As the Coriolis force is also related to wind
velocity, this force is weakened further by the slackening of the pressure gradients. As a
result, winds tend to flow down pressure gradients and on weather maps the winds cut
across the isobars. Such flows are called *ageostrophic flows* (Figure 2.5(a)). This
situation is very different to that of higher latitudes where the PGF and CF tend to
balance to give *geostrophic flow*. In this case winds in the upper atmosphere tend to
flow parallel to the isobars (Figure 2.5(b)).

Near the surface, friction becomes important. In the low latitudes where the Coriolis
force is negligible due to latitudinal and wind-speed effects, the direction of the wind is
determined very much by the balance between the PGF and F. As the oceans are
smooth compared to the land surface, the frictional effects are far greater over the land
compared to the oceans. On average, over the oceans winds are deflected 10–20°, while
over the land surface deflection is between 25° and 35°. This deflection is in the
direction of the PGF.

For the low latitudes there are two large-scale weather situations that are an

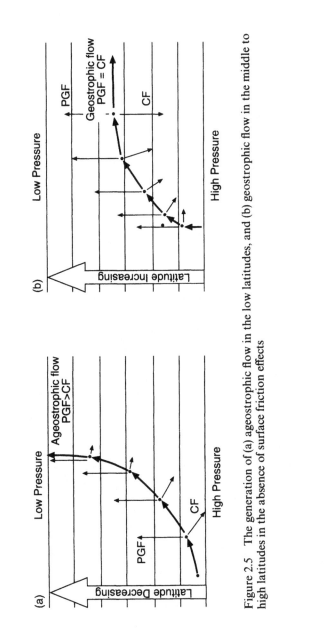

Figure 2.5   The generation of (a) ageostrophic flow in the low latitudes, and (b) geostrophic flow in the middle to high latitudes in the absence of surface friction effects

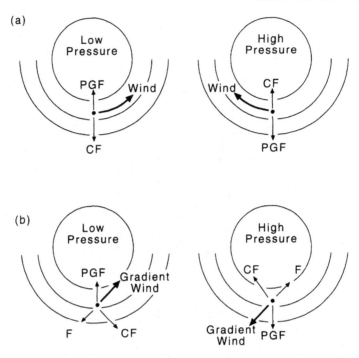

Figure 2.6   Wind flow around low and high pressure systems in the northern hemisphere in the case of (a) no surface friction effects which produces geostrophic flow, and (b) surface friction effects which produce a gradient flow. Note that the gradient wind represents the balance between the four competing forces of the pressure gradient (PGF), the Coriolis force (CF), friction (F) and centrifugal force (cf)

important feature of the climate of the low latitudes for which atmospheric motion is dominated by the balance of forces acting on *non-planar or curved pressure* surfaces. These two weather systems are tropical cyclones and anticyclones. For these, a force called the *centrifugal force* (cf) or, alternatively, *centripetal acceleration* must be taken into account in addition to the PGF, CF and F forces. This is because, for winds moving in curved paths around a centre of low or high pressure, an outward force is generated such that air tends to be hurled out. This is the centrifugal force, which is similar to the force experienced when running round a pole hanging on to a rope. This force is equivalent to the centripetal acceleration towards the centre of rotation, which is the pole in our example. Therefore, while flying around a central pole hanging onto a rope, although experiencing a centrifugal force which flings you out, you are continually being accelerated inwards because of centripetal acceleration.

   In the case of anticyclones the pressure gradient force is directed outwards from the centre, as is the centrifugal force. In this case the extra centrifugal force increases the outwardly directed forces. If balance is to be maintained then the Coriolis force must be increased. The effect of this is a wind which is faster than geostrophic. Such winds are referred to as *supergeostrophic*. However, although it appears that the winds blowing from the centres of anticyclones should be strong, they are not. This is because for anticyclonic situations the pressure gradients are usually very weak so that the centrifugal force is only small. As a result, the Coriolis force does not have to overly compensate for this additional force (Figure 2.6(a) and (b)).

For airflow around a tropical cyclone, the centrifugal force is directed outwards and the centripetal acceleration is towards the centre of the rotation. However, the centrally directed PGF is also extremely large. To balance the large PGF, the centrifugal force and Coriolis force would have to be large, but in low latitudes, the Coriolis force is much reduced which leads to the PGF dominating the nature of the flow around a low pressure. This, with the aid of the centrally directed centripetal acceleration, causes a balanced flow called *cyclostrophic flow* which can achieve extremely high speeds. In both cases of flow around centres of high and low pressure, friction also plays a role in determining the resultant wind (Figure 2.6(b)).

THE HYDROSTATIC EQUATION

In our discussion of factors determining horizontal motion for a planar isobaric surface, the PGF, CF and F force were highlighted as important. When considering vertical motion we usually neglect the CF and F because these are negligible. This leaves two forces that need consideration. They are the vertical PGF and gravity, which act in opposite directions. The PGF operates upwards whereas gravity operates downwards. For air that is moving vertically the rate of acceleration will be determined by the relative strengths of these two opposing forces such that:

vertical acceleration = magnitude of the vertical PGF − gravitational force

Measured and calculated rates of vertical acceleration in large-scale weather systems have revealed that these are small. This shows that gravitational force nearly balances the vertical PGF and thus keeps the atmosphere in *hydrostatic equilibrium*. In fact, in the absence of gravity, an air parcel would accelerate vertically under the influence of the vertical PGF at approximately $10 \, \mathrm{m \, s^{-1}}$, reaching a speed of around $5000 \, \mathrm{km}$ per hour at a height of $100 \, \mathrm{km}$ (Wells, 1986). However, because of hydrostatic equilibrium and the influence of gravity, vertical velocities in large-scale weather systems rarely exceed $10 \, \mathrm{m \, s^{-1}}$. For small-scale weather systems such as individual thunderstorms, squall lines, tropical cyclones and the situation where air is forced over orographic barriers, the hydrostatic equation does not apply. That is, the force exerted vertically upwards far outweighs the gravitational force. In such a case the atmosphere is not in hydrostatic equilibrium and air continues to accelerate away from the earth's surface. As a result, the air parcel does not settle back down to its original height.

An important determinant of the maintenance of hydrostatic equilibrium is the vertical temperature lapse rate. If, in the case of dry air, the adiabatic lapse rate of approximately $10 \, ^{\circ}\mathrm{C \, km^{-1}}$ is exceeded, then the atmosphere may become hydrostatically unstable. For moist air the saturated adiabatic lapse rate represents the critical threshold between hydrostatic stability and instability (Chapter 10). If either of these lapse rates is exceeded, depending on the moisture status of the air, then strong vertical motions will occur in the atmosphere. These often lead to impressive storms.

VORTICITY

In the low latitudes, for most of the time, closed circulations of wind are not common. However, with the onset of the summer or warm season months, in tropical and subtropical maritime areas, closed circulations in the form of tropical cyclones and monsoon depressions are a common occurrence. To understand the evolution of these important low-latitude weather systems, the concept of vorticity and its relationship to convergence and divergence must be introduced.

Vorticity is the name given to turning or spinning motions in the atmosphere. Changes in the amount of turn or spin are related to the degree of convergence and divergence of air.

Absolute vorticity is made up of two components, namely *earth vorticity* and *relative vorticity*. The *earth vorticity* is due to the rotation of the earth itself. This vorticity is passed to the atmosphere in the form of friction. As the degree of turning or spin is greatest at or near the axis of spin then the earth vorticity is greatest near the poles where the axis of the earth's rotation is. As the centre of rotation is approached (i.e. at the poles), earth vorticity increases. This applies to all air moving in a curve away from the equator.

Relative vorticity is the name given to the degree of turning seen in the atmosphere such as indicated by the swirls or vortices of cloud seen on satellite images associated with tropical cyclones. The term "relative" is used because these vortices are turning relative to the earth. Relative vorticity in the atmosphere can be produced when wind speeds vary with height or distance (wind shear) or where air is moving in curved paths across the earth's surface. The total relative vorticity is the sum of vorticity due to wind shear and curvature of motion. Following standard notation, cyclonic curvature and thus vorticity are positive while anticyclonic curvature and vorticity are negative (Atkinson, 1981).

Relative vorticity can change under two circumstances. If a rotating vortex experiences a force operating parallel to its rotation this may increase or decrease vorticity depending on the direction of the force. This is especially important in the situation where there are two parallel streams of air with contrasting velocities. In a cyclonic circulation if wind speeds increase (decrease) towards the outer limits of the vortex then shear and vorticity are increased (decreased) (Atkinson, 1981).

The second situation where vorticity change can occur is if the distribution of mass within the rotating system changes. This mechanism is especially important for closed circulations in the tropics. Generally, if there is a concentration of mass towards the axis of rotation, as would occur with flow of air towards the middle of a rotating tropical cyclone or monsoon depression, then vorticity increases and vice versa. This brings us on to how vorticity is related to convergence and divergence and how the principle of the conservation of momentum can help explain increases or decreases of vorticity.

To understand the principle of the conservation of momentum we need to introduce two new concepts. They are *angular velocity* and the *moment of inertia*. Angular velocity is the rate of rotation. The moment of inertia is a measure of the distribution of mass within a rotating system. The greater the rate of rotation, the greater the angular velocity. The more concentrated mass is near the axis of rotation, the smaller is the moment of inertia. The *principle of the conservation of momentum* states that for a rotating object that is not experiencing any other forces:

$$\text{angular velocity} \times \text{moment of inertia} = \text{constant}$$

Applying this principle, if there is a decrease in the moment of inertia, to preserve the constancy of the product in the above equation there must be an increase in the angular velocity. Therefore, if there is convergence of air towards the axis of rotation (a decrease in the moment of inertia) then the angular velocity must increase (increase in vorticity). This relationship has given rise to the *vorticity theorem*, which states in terms of divergence how absolute vorticity changes such that:

$$\text{rate of change of absolute vorticity} = -(\text{absolute vorticity}) \times \text{the divergence}$$

Because divergence (convergence) can be expressed in terms of vertical velocities and these are represented with positive (negative) values for sinking (rising) air, then according to the above theorem divergence (a positive) multiplied by absolute vorticity (negative) will give a negative (positive) rate of change of vorticity which means a decrease (increase) in the vorticity.

Let us expand on this law and see how it is related to divergence and convergence in the atmosphere. If we think of air that is converging as a vertically stretching column, then as it stretches vertically, its depth increases. This is because of the continuity of mass principle, i.e. matter cannot be created or destroyed. In contrast, in the case of diverging air there is vertical shrinking. This is because air is spreading out and the depth of the air is decreasing. Let us relate this back to the theorem of the conservation of momentum. In a stretching (converging) air mass there is a decrease in the moment of inertia, i.e. there is a concentration of mass towards the centre, as the depth of the column increases. Therefore, the absolute vorticity must increase to maintain balance. In a shrinking (diverging) air mass the opposite happens. As the column of air shrinks due to subsidence (divergence) there is an increase in the moment of inertia. Again, to preserve balance something must change, so there is a decrease in absolute vorticity. This brings us to the general statement that for latitudes where sufficient absolute vorticity exists to initiate rotation, for converging air, there is an increase in vorticity, while for diverging air, there is a decrease in vorticity (Figure 2.7).

As absolute vorticity decreases as the equator is approached, because of a decrease in earth vorticity and the curvature component of relative vorticity (which is related to the Coriolis force), then weather systems typified by rotating air motion are not likely to be found in the very low latitudes. The vorticity theorem thus helps explain why tropical cyclones and monsoon depressions do not occur in equatorial latitudes because, despite the presence of convergence, the absence of the all-important turning forces (earth vorticity and Coriolis force) prevents these vortex-type weather systems developing.

As mentioned above, absolute vorticity is a conservative property such that a change in earth vorticity must be compensated by a change in relative vorticity. This concept, referred to as the conservation of absolute vorticity, along with the concept of the *conservation of angular momentum*, is useful for explaining the behaviour of upper level winds in the tropics.

Angular momentum is related to the rotation of the earth about its axis. It is a

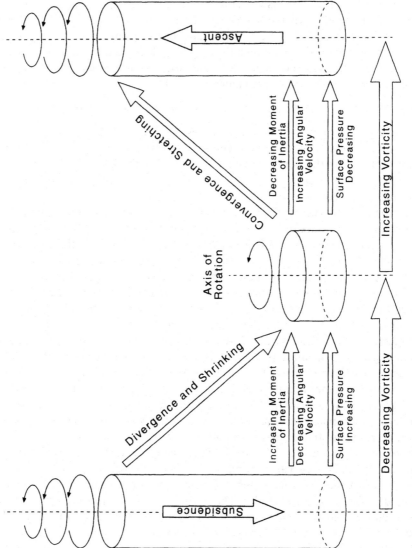

Figure 2.7   Vorticity and convergence/divergence relationships

measure of the effect of momentum in a rotating system. Because the earth is a sphere, the atmosphere's momentum is angular rather than linear. As the earth rotates, so does the atmosphere; it moves from west to east under its own momentum. Greatest momentum is found at the equator while at the poles momentum is zero. This is because the distance from the earth's axis of rotation is at a maximum at the equator (the earth is rotating the fastest here; Figure 2.4), while at the poles it is zero as the poles and the axis of rotation are the same. Angular momentum may be defined as follows:

$$\text{angular momentum} = \text{mass} \times \text{velocity} \times \text{radius of rotation}$$

Like vorticity, angular momentum is a conservative property such that a change in any one of the variables in the above equation must be offset by changes in one of the others. For example, if an air parcel moves equatorward (radius of rotation increases) without changing its mass (mass stays constant), then the velocity must decrease to sustain constancy of the value of angular momentum. The reverse applies to air moving away from the equator as it does in the upper atmosphere in the tropics. This is because increasing distance from the equator means decreasing radius of rotation, which must be compensated for by an increase in velocity. Therefore poleward moving air in the upper atmosphere of the tropics increases in velocity. This air is also acted on by the Coriolis force with the result that in the northern (southern) hemisphere it is turned to the right (left) to become a westerly flow characterized by its high velocity.

SUMMARY

The object of this chapter was to present those physical laws which help to explain different scales and types of atmospheric behaviour. By way of a summary, the laws are listed again with their main relevance to atmospheric behaviour and thus weather and climate noted.

The equation of state explains the relationship between temperature, pressure and density of an air parcel. This is relevant when understanding how a parcel of air will behave when heated or cooled, compressed or dilated. In brief, warmer air is less dense and therefore exerts less pressure and will expand, while the characteristics of cool air are the opposite.

The first law of thermodynamics explains the various dry and saturated adiabatic processes and why the DALR is greater than the SALR. The first law is relevant for understanding the dynamics of a parcel of air as it moves vertically in either direction through the atmospheric column, especially if there is internal heat gain due to condensation. Generally, rising air expands and thus cools which, if the air is moist, will lead to condensation, the precursor of precipitation formation (Chapters 9 and 10).

The moisture equation helps us to understand the idea that moisture is always present in the atmosphere in some form and that it is neither created nor destroyed but only changes form or position. With change of form there is either energy release or absorption, while change of position by transport results in energy transfer.

The continuity equation is similar to the moisture equation in that it represents the principle of the conservation of mass. It is relevant because it helps to explain why convergence at one location in the atmosphere is matched by divergence elsewhere, as occurs in the lower and upper levels of the atmosphere dominated by the intertropical convergence zone and the semi-permanent subtropical anticylones (Chapter 5) and in storm systems such as tropical cyclones (Chapter 8).

The horizontal wind equations describe how the interaction between a number of forces, as described by Newton's second law of motion, are responsible for the generation of atmospheric motion. In the low latitudes, because one of the forces (the Coriolis force) is weak, compared to the others, atmospheric motion at the surface tends to be dominated by the pressure gradient, friction and centripetal forces.

The hydrostatic equation states that the vertical position of any air parcel is a result of the interaction of an upwards-directed PGF and a downwards-directed gravitation force. When either of these forces is out of balance, hydrostatic instability will occur and an air parcel that may have been at rest in the vertical will either rise or fall. Hydrostatic equilibrium is dependent on the magnitude of the environmental lapse rate in comparison with the adiabatic lapse rate for the given atmospheric temperature and moisture conditions.

The vorticity theorem embodies the principle of the conservation of momentum and is relevant in understanding the rotational characteristics of an air parcel in relation to its geographic position and to regimes of convergence or divergence. It is especially helpful in understanding why tropical cyclones do not develop or decay in regions of very low latitudes as the all-important turning forces acting on air (absolute vorticity and Coriolis force) are very weak or absent in these latitudes.

Having hopefully developed an appreciation of the laws that aid us in understanding the behaviour of the variables that make up weather and climate, we are in a position to start considering the general characteristics of climate in the tropics. This will begin with a description of the radiation conditions in the tropics.

CHAPTER 3

Radiation Conditions in the Low Latitudes

RADIATION LAWS

Before discussing radiation characteristics in the tropics, we will introduce a number of laws which aid in the understanding of the behaviour of radiation in the atmosphere and at the earth's surface. The chief radiation or energy source for the earth is the sun. This energy is carried to earth not in the form of a wind or by conduction, but by *electromagnetic radiation*. It travels at the speed of light in the form of waves which, like sea or lake waves, have different *wavelengths* (the distance between the wave tops). Radiation wavelengths are measured in micrometres (μm). Radiation from the sun is referred to as *shortwave* radiation and has wavelengths between 0.15 and 3.0 μm. Radiation with wavelengths between 3.0 and 100 μm is referred to as *longwave or infrared* radiation. In the earth's atmosphere or at the earth's surface, radiation can be absorbed, reflected, transmitted or re-emitted, following absorption.

Several laws exist which describe the characteristics of emitted radiation. These radiation laws are based on the *blackbody* concept. A blackbody is a hypothetical body or a coherent mass of material with uniform temperature and composition; it can be a layer of the atmosphere or the surface layer of a mass of solid material such as the earth's surface. A further characteristic of a blackbody is that all radiation falling on the blackbody is absorbed and that maximum possible emission occurs in all wavelengths and directions (Wallace and Hobbs, 1977).

The amount of radiation emitted by a blackbody at a particular wavelength depends on its absolute temperature (kelvin; $0\,°C = 273\,K$). *Planck's law* describes this relationship, which is displayed graphically in Figure 3.1. For a given temperature there is a single peak of emission at one wavelength. The wavelength of peak emission may be estimated using *Wein's displacement law*. This is expressed simply as:

$$\text{wavelength of peak emission} = \frac{2987}{\text{absolute temperature (K) of emitting body}}$$

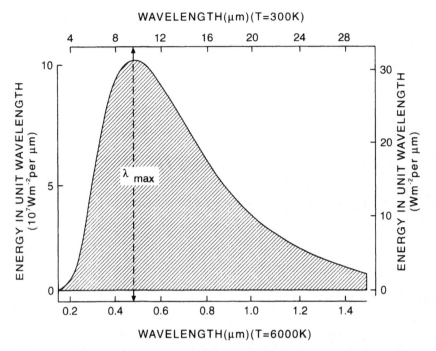

Figure 3.1   Distribution of radiant energy from a blackbody at 6000 K (left-hand side vertical and lower horizontal axes) and 300 K (right-hand side vertical and upper horizontal axes). Note peak emittances are in the shortwave range of 0.4 to 0.6 μm for the sun (6000 K) and the longwave range of 8–12 μm for the earth (300 K). From Oke (1987) *Boundary Layer Climates*. Reproduced by permission of Methuen & Co.

The higher (lower) the absolute temperature, the shorter (longer) the wavelength of peak emission and the further to the left (right) the peak in Figure 3.1 will be *displaced*. This can be seen in Figure 3.1 for the cases of 6000 K and 300 K which are representative absolute temperatures for the sun and the earth's surface (Oke, 1987).

In Figure 3.1 the area under the curve is the total amount of energy, i.e. the total energy of all individual wavelengths. Total energy emitted from a blackbody for all wavelengths is described by the *Stefan–Boltzman law*. This states that the amount of radiation emitted is proportional to the product of emissivity of the emitting body, the Stefan-Boltzman constant (5.67×10^{-8}) and the fourth power of the absolute temperature (T^4), and is expressed as:

$$\text{total energy emitted} = \text{emissivity} \times 5.67 \times 10^{-8} \times T^4$$

Emissivity is included in the above law as not all objects are perfect blackbodies or emitters. *Emissivity* is a measure of the ability of a body to emit as a blackbody, a value of 1.0 representing a perfect blackbody. Typical emissivity values for a range of natural and human-made or processed materials are given in Table 3.1.

Before going on to discuss insolation and its importance, let us define a couple of other terms commonly used in the radiation literature as defined according to Oke (1987). Calculation of the energy emitted using the above laws gives a value which is referred to as the *radiant flux* (rate of flow of radiation measured in joules per second;

Table 3.1   Emissivity values for a variety of surfaces and
materials (Oke, 1987)

Surface/material	Emissivity
Soils (dark wet to light dry)	0.90–0.98
Desert	0.84–0.91
Grass (long to short)	0.90–0.95
Agricultural crops	0.90–0.99
Deciduous forests	0.98
Water	0.92–0.97
Asphalt roads	0.95
Walls	
Concrete	0.71–0/90
Brick	0.90–0.92
Stone	0.85–0.95
Wood	0.90
Roofs	
Tar and gravel	0.92
Tile	0.90
Corrugated iron	0.13–0.28
Whitewash paint	0.85–0.95

J s^{-1} = 1 Watt) from a unit area (m^2) of a plane surface into the atmosphere. The radiant flux per unit area of a quantity is called the *flux density*. This is measured in watts per square metre (W m^{-2}). Lastly, *irradiance* is the radiant flux density incident on a surface and *emittance* is the radiant flux density emitted by a surface (Oke, 1987).

INSOLATION AND ITS IMPORTANCE

Insolation is the solar radiation received by a surface. About 99.9% of all energy in the earth's atmosphere has its origin in radiation from the sun. A minute fraction of atmospheric energy is supplied from the earth itself, either by volcanic activity or by the decay of radioactive substances and the burning of organic materials, but climatologically it has only local importance. All movements and changes in the atmosphere are ultimately caused by variations, over time and place, in the amount of insolation received – the irradiance. These variations are therefore the main cause of climatic differentiation.

The *solar constant* is the average amount of insolation received on a surface facing the sun at the outer limit of the atmosphere with the sun at its mean distance from the earth. This amounts to about 1367 W m^{-2} or 1.96 calories per square centimetre per minute. The wavelengths of this radiation vary from about 0.25 to 25 μm (Plancks' law), but 95% of the total energy is made up from wavelengths under 2.2 μm (Figure 3.1). The maximum intensity (Wein's law) is in the visible light wavelengths of 0.39–0.77 μm (Iqbal, 1983, pp. 52–53). Variations in the solar constant, caused by changes in the sun's activity, rarely exceed 2% of the average value and they can be disregarded in climatology. A further measure of the amount of radiation received is that of *extraterrestrial radiation*. Unlike the solar constant, this is for a surface oblique to the sun (horizontal surface above the atmosphere). It is much less than the solar constant with a global average of approximately 342 W m^{-2} (Oke, 1987).

The total amount of radiation received at any place on earth depends on two factors: the duration and the intensity of insolation. Both factors are controlled by the movements of the earth: its rotation around its axis and its annual orbit around the sun.

DURATION OF INSOLATION

The duration of insolation is, of course, the length of day. It is controlled by the earth's rotation around its own axis, but because this axis makes an angle of 67.5° with the plane of the earth's orbit around the sun, places in the summer hemisphere enjoy longer days than those having winter. The total annual exposure time to the sun is the same for all places on earth, but the difference between summer and winter days increases with latitude (Figure 3.2). The extreme is reached near the poles, where six months of continuous daylight are followed by six months of night. Only during the equinoxes, on 23rd March and 22nd September, are days and nights of equal length everywhere on earth.

At the equator, all days of the year are 12 hours and 7 minutes long. Astronomically, the duration would be 12 hours exactly, but it takes 3.5 minutes for the upper half of the sun to disappear under the horizon at sunset, and similarly at sunrise 3.5 minutes before the centre of the sun's disc is at the horizon, while the upper half of it already provides insolation.

In the low latitudes the difference between the shortest and the longest day of the year increases by about 7 minutes per degree of latitude; it is about 71 minutes at 10°

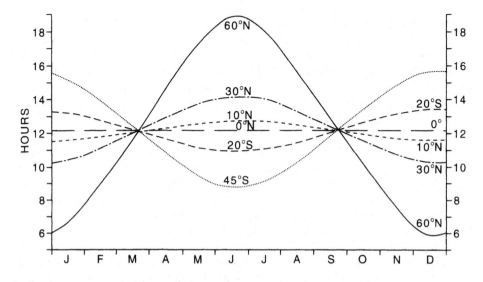

Figure 3.2   Length of day during the course of the year at various latitudes

and 146 minutes at 20° latitude (List, 1958, p. 507). In the low latitudes seasonal variations in the length of day are very small (Figure 3.2). It is this feature which often surprises newcomers to the tropics: in the mid-latitudes high temperatures are associated with long summer days, but in the tropics the days are always short and the sun rarely sets later than 7.00 p.m. local time. Many human activities, like markets, are carried out in the long and relatively cool tropical evenings.

INTENSITY OF INSOLATION

The earth's orbit around the sun is elliptical (not a perfect circle), so the distance between the earth and the sun varies in the course of the year. Around 3 January it reaches its annual minimum (perihelion) of 147 million km, and on 4 July the distance is at its maximum (aphelion) of 152 million km (Iqbal, 1983, pp. 1–3). As a result, solar radiation at the top of the atmosphere in January is about 7% more intensive than in July, and this difference is the same at all latitudes. In theory, this factor would make the summers of the southern hemisphere hotter, and its winters colder than in the northern hemisphere, but these effects are completely obscured by the effects of the stronger continentality of the northern half of the earth. Climatologically, much more important are the differences in intensity of insolation caused by variations in the sun's elevation. This is its position in the sky above the horizon. It is usually indicated for noon local time, when the sun reaches its daily maximum elevation.

There are three reasons why a high position of the sun causes more intense insolation than a low elevation. The first is that rays coming from a high sun are spread over a smaller surface than oblique rays from a low sun. The intensity of insolation varies proportional to the sine of the angle of incidence, which is, for a horizontal part of the earth's surface, equal to the elevation of the sun.

The second reason is that a high position of the sun means a relatively short passage of the solar radiation through the atmosphere, with consequently lower scattering losses caused by atmospheric dust particles. This effect is clearly demonstrated by the fact that to look at a low sun is quite harmless to the unprotected human eye, but highly dangerous when the sun is high in the sky. This is because the proportions of *diffuse* and *direct* shortwave radiation are different. Scattered radiation which reaches the earth's surface is termed diffuse shortwave radiation. Direct shortwave radiation is that which has not been absorbed, scattered or reflected. Scattering affects particularly the shorter wavelengths of the sun's radiation, so the longer reddish rays prevail when scattering is strong. Hence the red colour of the sun at sunset or sunrise. A similar colour effect is often caused by dust particles in the atmosphere, originating from industrial pollution, dust storms, volcanic explosions or large forest fires.

A third effect related to the sun's elevation is the *albedo*, the percentage of incoming solar radiation which is reflected unchanged by the earth's surface. While the albedo is mainly controlled by the physical properties of the surface, in particular its colour, under the same circumstances it decreases with higher elevations of the sun. This effect is especially strong over water and therefore climatologically of great importance in the tropics, where more than three-quarters of the earth's surface is occupied by seas or oceans (Table 3.2).

Table 3.2   Albedo of water surfaces

Sun's elevation (degrees)	Albedo (%)	General value over the oceans	
		Latitude (degrees)	Albedo (%)
60	2.1	0	6
50	2.5		
40	3.4	30 (summer)	6
30	6.0	30 (winter)	9
20	13.4		
10	34.8	60 (summer)	7
5	58.4	60 (winter)	21
0	100.0		

Sources: List (1958, p. 444); Sellers (1965, p. 21).

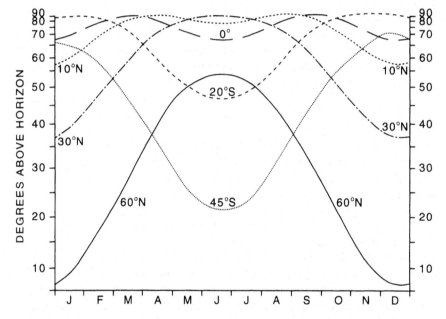

Figure 3.3   Elevation of the sun at noon local time, at the same latitude as in Figure 3.2 during the course of the year

The elevation of the sun at noon during the course of the year, at various latitudes, is illustrated in Figure 3.3. Because the intensity of insolation is approximately proportional to the sine of the sun's elevation, the vertical axis of this diagram is scaled accordingly, so that the curves indicate relative intensities of insolation. Losses in the atmosphere and at the earth's surface are disregarded. The diagram shows that the low latitudes have high intensities throughout the year. In the mid-latitudes the intensity of the solar radiation during the summer is very similar to that in the tropics, but in winter it is much lower. Latitudinal differences are therefore small in summer, but large during the winter.

COMBINED EFFECTS

The total influence of duration and intensity of insolation indicates that the long summer days of the higher latitudes more than compensate for the relatively low elevations of the sun: the higher latitudes receive more insolation during the height of the summer than is ever received in the tropics (Figure 3.4). At the time of the equinoxes, when day length is the same for all latitudes, maximum insolation is received near the equator. Around the solstices, there is an increase of total solar radiation with latitude, from South to North in June–July, and in the opposite direction in December–January. The asymmetry in the diagram is caused by differen-

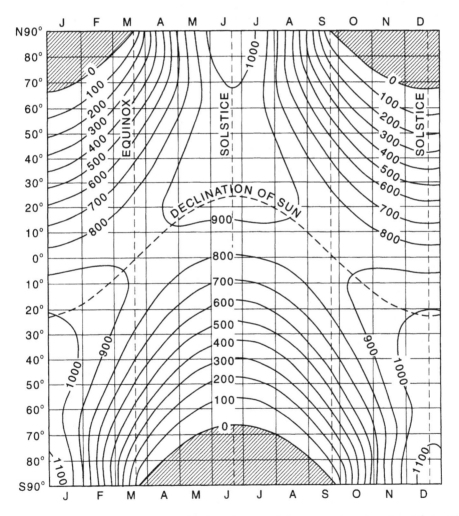

Figure 3.4   Total solar radiation received at the top of the atmosphere in cal cm^{-2} day^{-1}. Shaded areas indicate continuous darkness. Declination of sun indicates the location where the sun is in the zenith. From List (1958). Reproduced by permission of Smithsonian Institute Press, Boston

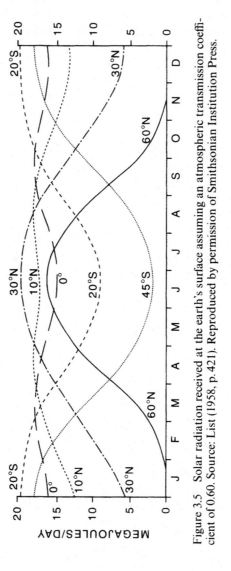

Figure 3.5 Solar radiation received at the earth's surface assuming an atmospheric transmission coefficient of 0.60. Source: List (1958, p. 421). Reproduced by permission of Smithsonian Institution Press.

ces in the distance of the sun: generally higher values occur around the time of the perihelion in early January than during the aphelion in June.

The tropical latitudes, while never receiving the high daily maxima reached near the poles, receive relatively large amounts of insolation throughout the year. When insolation losses in the earth's atmosphere are considered, latitudinal differences become smaller. Figure 3.5 is based on losses, standardized according to length of path through the atmosphere (List, 1958, p. 421). The summer maxima in the high latitudes, caused by long days but a relatively low sun, are much more reduced by these losses than the tropical values, derived from higher elevations of the sun. However, most insolation losses in the atmosphere are by reflection and scattering caused by clouds. Cloudiness is generally high near the equator, especially around the equinoxes, and low in the dry latitudes of about 20–30°, but these are generalizations from which considerable departures are possible. Continentality, relief and ocean surface temperatures all may cause large differences in cloudiness (Chapter 9).

Another source of insolation losses is dust and smoke in the atmosphere. Industrial areas, volcanic eruptions and duststorms may all increase losses temporarily and regionally. In high elevations, above the more humid and polluted parts of the atmosphere, the intensity of insolation is therefore much higher than at sea level.

A useful way of considering the transfer of shortwave radiation through the atmosphere is to assess the magnitudes of the losses due to reflection, and the gains as a result of absorption, both within the atmosphere and at the earth's surface. This type of analysis is similar to that performed in assessing the gains and losses from a bank account and is called the *shortwave radiation balance*. This can be represented diagrammatically (Figure 3.6). Generally about 25% of the extra-terrestrial radiation received at the outer limits of the atmosphere is absorbed by the atmosphere (clouds 5%, atmospheric gases and particulate matter 20%), while 28% is lost due to reflection (clouds 19%, atmospheric gases and particulates 6%, earth's surface 3%). This leaves a *net shortwave balance* of 47% of the original extra-terrestrial radiation available to be absorbed at the earth's surface. These percentages are annual estimates for the whole earth; actual values vary considerably, particularly with the nature of the surface albedo. Water surfaces generally absorb solar radiation much more readily than land areas. Seasonal and regional differences are also quite large, but the general picture remains that a very large proportion of solar radiation passes through the atmosphere to be absorbed by the surface of the earth.

TERRESTRIAL RADIATION

At the earth's surface, the absorbed insolation turns into heat, a process that takes place almost every day. However, the earth's surface is not getting hotter in the long run, because the energy is re-radiated. This is the *terrestrial radiation*, also called longwave or infra-red radiation, because its wavelengths are mainly between 3 and 50 μm, or about 20 times those of insolation. Peak intensities for longwave emission from the earth are in the region of 8–10 μm (Figure 3.1 and Wein's law). Longwave radiation is sometimes called nocturnal radiation, because it dominates the radiation balance at night, while during the day insolation is often stronger. This term is,

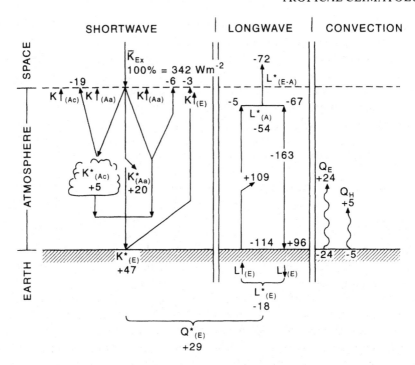

Figure 3.6   Shortwave and longwave transfers. Figures are percentages of extra-terrestrial radiation (342 Wm^{-2}). From Oke (1987) *Boundary Layer Climates.* Reproduced by permission of Methuen & Co.

however, misleading because terrestrial radiation is active on many days, particularly when there is heavy cloud cover. The latitudinal distribution of terrestrial radiation is closely related to surface temperatures (Stefan–Boltzman law) and shows a maximum of its annual average in the tropics. Satellite-based measurements of outgoing long-wave radiation (OLR) are often used as a proxy of tropical convective cloud cover as greater amounts of OLR mean clear skies and vice versa. OLR has also been used to make estimates of tropical rainfall (Chapter 10).

Because of its longer wavelengths, terrestrial radiation is readily absorbed, the consequence of which is atmospheric warming. This absorption depends on the gaseous content of the atmosphere. The main absorbers are water vapour, carbon dioxide and ozone. Each gas is quite specific in the range of wavelengths absorbed, but the aggregate absorption covers almost the entire spectrum of terrestrial radiation (Liou, 1992). The heating of the atmosphere is therefore an indirect process: only 20% of solar radiation (Figure 3.6), but 97% of terrestrial radiation is absorbed in the atmosphere. Because of its similarity to what is perceived to take place in glass- or greenhouses (i.e. the trapping of energy), this process is called the *greenhouse effect,* and the absorbing gases are often called the *greenhouses gases.* The atmosphere is therefore analogous to a thermal blanket as it prevents heat escaping to space. Absorbed heat is subsequently re-emitted back down to the earth's surface in the form of longwave radiation often referred to as *atmospheric or sky radiation.* With no natural greenhouse effect (i.e. with a transparent atmosphere), average global temperatures would be approximately $-18\,°C$, which is in stark contrast to the actual global average at $15\,°C$.

The absorption of terrestrial radiation takes place mainly in the troposphere, the lowest layer of the atmosphere, where the majority of the greenhouse gases are concentrated. Atmospheric warming as a result of longwave absorption is greatest in a cloudy or turbid atmosphere. Cloud height is also important for surface temperatures; lower clouds have a much greater effect on surface temperatures than higher clouds as only a relatively thin layer of air below low clouds is affected by downward directed longwave. In such a case a balance is reached between downward longwave (atmospheric) radiation and upward longwave (terrestrial) radiation; consequently temperatures become stable. When no clouds are present, terrestrial radiation heats up a much thicker layer of the atmosphere. It is maintained by the temperature difference between the earth's surface and the lowest layers of air. On clear nights, temperatures near the earth's surface may decrease rapidly because of large amounts of longwave radiation loss, especially in highland and desert environments.

Globally, only about 5% of terrestrial radiation is lost to outer space. This is lost because the atmosphere is not efficient at absorbing radiation with wavelengths of $8–11\ \mu m$, the range in which peak terrestrial radiation (OLR) intensities are found. This range of wavelengths is often referred to as an *atmospheric window* because such radiation can escape to space as if heat is escaping through an open window. However, the atmospheric window is not a hole in the atmosphere! Current concerns about climate change relate to increases in the concentration of greenhouse gases due to the burning of fossil fuels. As gaseous concentrations rise, less heat can escape to space thus increasing atmospheric warming. This results in an enhancement of the natural greenhouse effect, the cause of predicted global warming (Chapter 13).

The cascade of longwave radiation in the earth–atmosphere system is portrayed diagrammatically in Figure 3.6. In accordance with a mean surface temperature of around 288 K ($15\,°C$), the earth emits 114% of the extra-terrestrial radiation (Stefan-Boltzman law). This may seem odd but is possible as the terrestrial radiation absorbed by the atmosphere (109%) causes the surface temperature to rise above that for an atmosphere with no atmosphere (Oke, 1987). Some 5% of the terrestrial radiation escapes to space (114% − 109%). Absorbed longwave heats the atmosphere, and consequently the atmosphere emits longwave (163%) both upwards (67%) and downwards (96%). Taken together the total longwave losses to space are 72%; 5% from the earth's surface and 67% from the atmosphere. At the earth's surface the *net longwave balance* is negative (− 18%); 144% out and 96% in (Figure 3.6).

THE RADIATION BALANCE

The total radiation inputs and outputs from the earth–atmosphere system are referred to as the *radiation balance*. For the condition that radiation inputs to the earth's surface or atmosphere are greater (less) than the outputs, then the *net radiation balance* is positive (negative). This can be summarized by a simple word equation which applies to both the atmosphere and the earth's surface:

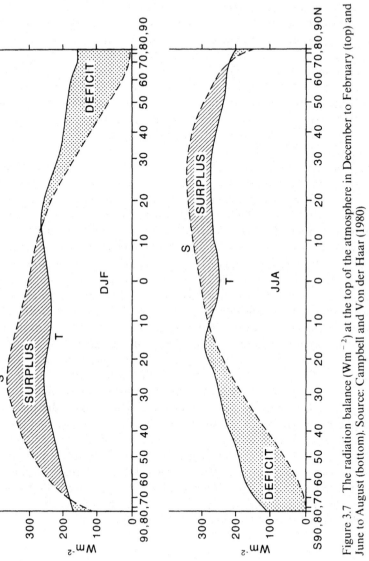

Figure 3.7   The radiation balance (Wm^{-2}) at the top of the atmosphere in December to February (top) and June to August (bottom). Source: Campbell and Von der Haar (1980)

net radiation = (net shortwave) + (net longwave)

$$= (\text{inputs}_{\text{shortwave}} - \text{outputs}_{\text{shortwave}}) + (\text{inputs}_{\text{longwave}} - \text{outputs}_{\text{longwave}})$$

Taking the values in Figure 3.6, the net radiation balance for the atmosphere and earth's surface can be calculated in terms of the percentage of the extra-terrestrial radiation.

For the *earth's surface*:

$$\text{net radiation}_{\text{earth}} = (50 - 3) + (96 - 114) = 29$$

For the *atmosphere*:

$$\text{net radiation}_{\text{atmosphere}} = (50 - 25) + (109 - 163) = -29$$

From the above it can be seen that the net radiation balances for both the earth's surface and atmosphere are the same value but of opposite sign. This may lead to the conclusion that the atmosphere and earth are in perfect balance. However, this is not the case, as some means must exist for the surplus energy at the earth's surface to be transported to the atmosphere to offset the negative balance there (see next section).

The situation described so far is for the globe as a whole and the figures used are global averages which mask geographical variations in the net radiation balance (Figure 3.7). Because conditions near the upper limit of the troposphere are relatively uniform, latitudinal gradients in outgoing radiation are small. During the extreme seasons outgoing radiation is highest over the tropics, and lowest over the winter pole. Minor variations are caused by differences in cloudiness and surface albedo. Absorbed solar radiation possesses a strong gradient in the winter hemisphere, but a much weaker one in summer. The very low values near the poles, even during summer, are caused by the high albedo in these latitudes, related to the low incidence of the incoming solar radiation and also to the prevalence of snow coverage (Piexoto and Oort, 1992).

Deficits and surpluses in the radiation balance change place during the course of the year, but near the equator is a zone where a continuous surplus of solar radiation over emitted outgoing radiation prevails. Yet the tropics do not heat up uncontrollably nor the poles cool uncontrollably. A constant flow of energy, from the earth to the atmosphere and from the tropics to the higher latitudes, compensates for both vertical and latitudinal differences in the *energy balance*.

THE ENERGY BALANCE

As noted above, a mechanism is required to transport the surplus energy at the earth's surface into the atmosphere where a deficit exists. This is achieved by the process of *convection* which transports *sensible* heat and *latent* heat away from the earth's surface; 24% and 5% respectively (Figure 3.6). The balance of net radiation surplus at the earth's surface and the upward fluxes of sensible and latent heat are referred to as the *energy balance*. The main components of the energy balance are the fluxes of net radiation, sensible heat, latent heat and subsurface heat. The symbols Q, H, E and G respectively are often used to represent these. Sensible and latent heat are *convective*

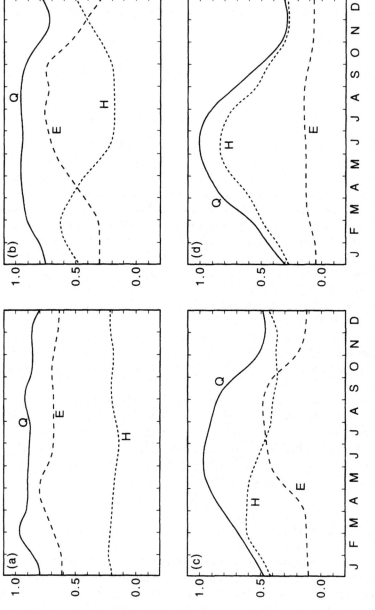

Figure 3.8 Representative energy balances for various low latitude climate types: (a) humid equatorial; (b) wet and dry monsoon climates; (c) dry grassland and steppe climates; and (d) tropical deserts. From Kraus and Alkhalaf (1995). Reproduced by permission of the *International Journal of Climatology*, Royal Meteorological Society

fluxes, subsurface heat is a *conductive flux*. The energy balance may be written using the figures in Figure 3.6 as:

net radiation$_{earth}$ = (sensible heat + latent heat + subsurface heat)

$$Q = (H + E + G)$$
$$29 = (25 + 4 + 0)$$

The subsurface heat flux is zero on an annual basis as flows into and out of the ground are approximately zero.

The energy balance is an important concept for interpreting climates at any temporal or spatial scale as it basically says that any net radiation surplus must be consumed by either sensible heating of the atmosphere, by evaporation (latent heat) or heating the subsurface. Generally, positive (negative) values of Q mean a flow of energy to (away from) the surface, while positive (negative) values of H, E and G indicate energy flows away from (to) the surface. At the regional and local scale the apportionment of the net radiative flux Q between the convective fluxes H and E is very much dependent on the nature of the surface. Generally dry surfaces have high sensible heat fluxes as there is little moisture available for evaporation; a low latent heat flux. The reverse is true for moist surfaces. The ratio between sensible and latent heat is a useful tool for climate analysis. This is called the *Bowen ratio* and is expressed as:

$$\text{Bowen ratio} = \frac{\text{sensible heat flux } (H)}{\text{latent heat flux } (E)}$$

Bowen ratio values greater than one indicate that sensible heating dominates (dry environments) while values less than one indicate the dominance of evaporation (moist to wet environments). Values therefore vary in accordance with the type of surface: tropical oceans, 0.1; wet tropical forest, 0.1–0.3; semi-arid desert, 2–6; desert > 10 (Oke, 1987); and tropical urban surfaces, 1.5–5.

The seasonal march and the magnitude of the various energy balance components varies in nature according to the climate type (Figure 3.8). For humid equatorial climates where there is plentiful rainfall in all months, all year round high and quasi-constant levels of Q, H and E in descending order of importance are a feature (Figure 3.8(a)). Minor fluctuations may occur due to changes in atmospheric circulation related to the migration of the intertropical convergence zone. For wet and dry monsoon climates, characterized by marked seasonal changes in rainfall, Q is fairly constant throughout the year but reaches a broad maximum in the dry season. However, H and E display inverse behaviour on a monthly basis; their relative importance changing around the time of the transition between the wet and dry seasons (Figure 3.8(b)). Dry climates of the tropics occur over extensive tropical grassland or steppes and deserts. For tropical grassland and steppe environments, Q displays a marked peak due to low cloud cover. For the majority of months, H exceeds E, apart from the time of a short rainy season (Figure 3.8(c)). Tropical deserts (Figure 3.8(d)) display a strong seasonal variation of Q with negligible fluxes of E. As a result, H is of similar magnitude to Q throughout the entire year (Kraus and Alkhalaf, 1995). Diurnally, the dry grassland and desert environments are the only ones which are likely to display negative Q. This happens at night as clear skies allow large longwave (terrestrial) radiation losses. This helps explain why deserts often have very cold nocturnal temperatures. For the other climates, on a daily basis, H and E may change their relative importance due to rainfall; following rainfall, E will increase but gradually fall relative to H as drying progresses.

At the equator there is an excess of energy which must be exported. The form in which this energy is transported depends on the nature of the earth's surface in the tropics (Sellers, 1965, pp. 103–115). On a global scale, about half of this energy is carried in the form of sensible heat by the advection of warm air masses to colder areas by the winds of the general circulation of the atmosphere (Chapter 5). About 35% is transported as latent heat. This is moved by the same winds. About 15% is carried by warm ocean currents which bring warm tropical water to the high latitudes. These ocean currents are also largely driven by the major wind systems (Chapter 5). Because there is an excess of energy in the low latitudes and this is moved to higher latitudes, the tropics are often called the general circulation's heat engine. This movement of energy helps maintain the global energy balance and also helps explain global temperature patterns.

SUMMARY

Radiation from the sun is the chief energy source for processes that operate in the climate system. The total amount of radiation received at any place depends on the duration and intensity of the insolation which are both controlled by the earth's rotation around its axis and orbit around the sun, while that which remains to be absorbed by the surface, depends on the nature of that surface, especially its albedo characteristics. Because the low latitudes receive more radiation than they lose, the surface net radiation balance is positive. This energy is then available for either sensible heating of the atmosphere, evaporation or heating of the subsurface. The apportionment of the available net radiation between these three processes is referred to as the energy balance. The main control on this apportionment is the availability of water such that tropical climates covering the wet to dry spectrum have characteristic energy balances with the relative importance of sensible and latent heat fluxes changing as expressed by the Bowen ratio. To prevent the low latitudes from heating up uncontrollably, excess energy is exported to higher latitudes where the net radiation balance is negative. This energy transport is achieved by the atmospheric and ocean general circulations.

CHAPTER 4

Tropical Temperatures

Unless mentioned otherwise, temperatures discussed in this chapter are air temperatures, observed in a screen at the standard exposure of about 1.65 m above the ground. Although these temperatures are largely controlled by the radiation and energy balance conditions described in the previous chapter, they are also affected by a number of other factors. Their distribution patterns are therefore much more complicated than those of radiation. A strong correlation with latitude makes temperatures in the tropics definitely different from those of other latitudes.

SEASONAL UNIFORMITY

Tropical climates are defined by the absence of a cold season, therefore temperature differences between seasons are generally small in the tropics. This can be indicated by the mean annual range, i.e. the difference between the mean of the coolest and the hottest month of the year. Figure 4.1 shows that for most of the tropics this difference is less than 4 °C, and over the equatorial oceans less than 2 °C, which means that there are practically no temperature seasons. For the construction of this map sea surface water temperatures were used, because their observations are more reliable, uniform and numerous than those of the air temperatures over sea, which are based on ship observations, quite numerous over the sea lanes of international shipping, but very rare outside these relatively narrow areas. Normally, the difference between the two is only a fraction of a degree, as the sea surface is the main controlling factor of air temperatures at about 1.65 m above the sea. For mean values, as used here, the two are practically the same (*US Navy Atlas*, Vol. IX, 1981).

The actual march of temperature during the course of the year normally follows more or less the pattern of the insolation curve, with the highest temperatures during the period around the overhead position of the sun (Figure 4.2). The stations shown in Figure 4.2 are all close to sea level and at marine locations. They show that the seasonal differences increase clearly with latitude, and also that there are important deviations from a general scheme due to local factors. Mozambique and Beira, for

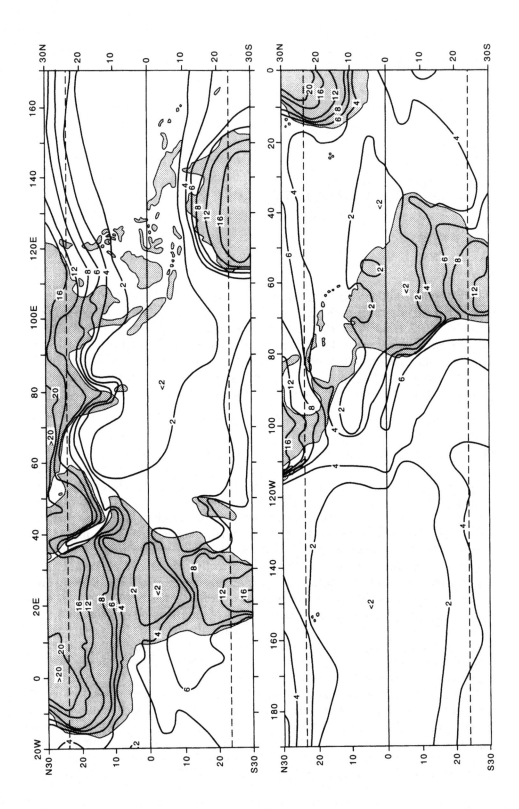

instance, though about 5° latitude apart, show very similar temperatures throughout the year, due to their common proximity to the warm Agulhas current; Carnarvon, in Western Australia, is much colder because it is close to a relatively cool ocean current. Its winter is colder than in Durban, 5° further south, but near a warm ocean current. Minor variations in the seasonal temperature curves are caused by cloudiness: in Pontianak, at the equator, the highest temperatures are from May to July, when there are few clouds as the south monsoon comes mainly from the land side.

Many tropical stations at latitudes over about 12° show a peculiar temperature regime, in which the spring and sometimes the autumn are hotter than the summer (Figure 4.3). This "Ganges-type" of curve derives its name from its association with monsoon climates, as it is particularly well developed in monsoon Asia (Koeppen, 1931). However, the graphs show that it also occurs in non-monsoonal continental stations in Africa, like Timbuktu, Khartoum and Kabwe. This feature is mainly caused by cloudiness: the main rainy season is in the summer, but spring and autumn are less cloudy and temperatures reach high values despite the somewhat lower position of the sun compared to mid-summer. This type of temperature regime is rare near the equator and at most marine locations, where clouds form frequently even during the drier parts of the year.

SPATIAL UNIFORMITY

Isotherm maps show that temperature differences with place are very small in the tropics: almost everywhere the means for the extreme months are between 25 and 30°C (Figure 4.4). There are two main reasons for this feature. First, there are only very small differences in the amount of net radiation received: between the equator and 10° latitude almost the same amount is received at all places (Figure 3.5). Secondly, most of the tropics consist of ocean surfaces, great heat storage reservoirs. Over these enormous areas of water, cold air masses only form over cool currents. Advection of cold air is therefore very rare in the tropics.

The maps also indicate where the highest mean temperatures occur in each season. These are indicated by the "thermal equator", which is not an isotherm, but the line connecting the maximum temperature at each longitude. Over the Atlantic and Pacific oceans, the thermal equator remains close to the geographic equator throughout the year, but over the continents it makes large seasonal movements of as much as 40° latitude. The largest movements are over highlands, as in Central Asia, Mexico, Ethiopia and Iran, where the thermal equator is far from the tropics during the summer. Actual temperatures at the higher elevations are high during the day, but nights are cool; causing means which, reduced to sea level, produce somewhat unrealistic values. Near cold ocean currents, like the Humboldt and Benguela currents to the west of South America and Africa, the thermal equator remains in the northern hemisphere throughout the year. At most other longitudes, it will be near the geographic equator around April and October.

Figure 4.1   Mean annual range of temperature. Sources: *US Navy Atlas* (1957–81), Hastenrath and Lamb (1977, 1979)

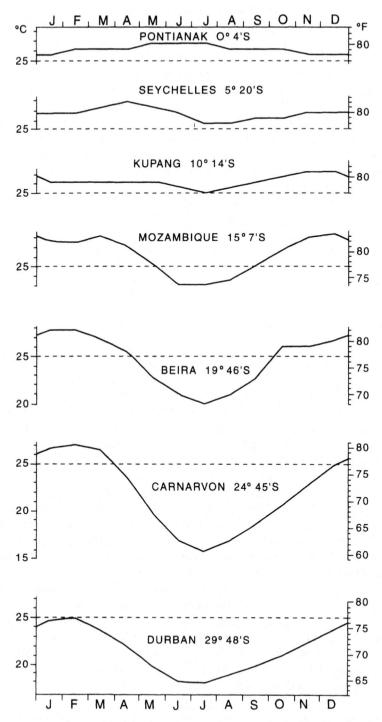

Figure 4.2   Monthly mean temperatures at selected stations

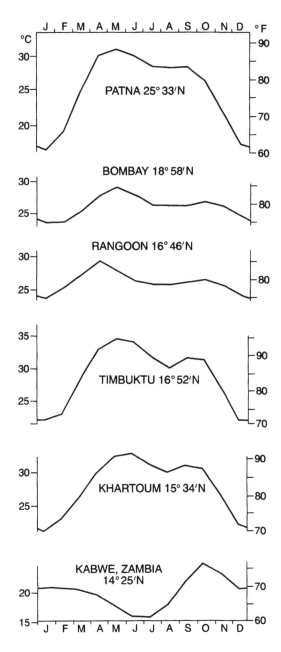

Figure 4.3 Monthly mean temperatures at some monsoonal and continental stations in the tropics

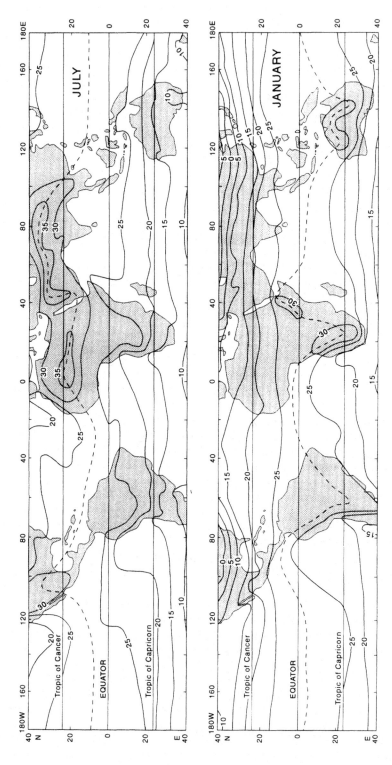

Figure 4.4   Sea level temperatures in July and January (°C) and the position of the thermal equator (temperatures reduced to sea level)

DIURNAL TEMPERATURE VARIATIONS

In most of the tropics, the seasonal temperature differences are so small that they are of little consequence. Temperature conditions are almost entirely dominated by the diurnal cycle, which is indicated by the mean diurnal range, i.e. the difference between the daily maximum and minimum temperatures (Figure 4.5). Over oceans this feature is largely absent; it amounts generally to less than 1°, as during the day the heated water at the surface is frequently replaced by cooler water from below by wave action, and during the night cooled water sinks quickly and is replaced by relatively warm water. Therefore the main influence on the diurnal range is continentality – it rapidly increases with distance from the sea or even from large lakes, and reaches its highest values in continental interiors. A second effect is that of elevation: over extensive highlands, where radiation is strong both at night and during the day, the highest values are reached. A third factor is again cloudiness: drier regions have higher diurnal ranges, as indicated by tropical Africa north of about 15°N, and central Australia (Figure 4.5).

In general, the diurnal range of temperature is quite regular, as day to day differences in temperature are much smaller than in the mid-latitudes, due to the absence of cold fronts and cold air invasions. It is this regularity of the diurnal march of temperature which often leads to expressions such as "monotony" and "every day the same weather" in descriptions of tropical climates by observers from the mid-latitudes. However, this regularity is sometimes disturbed. Short-term temperature variations are almost always related to thunderstorms (Figure 4.6, top). A drop of up to 7° within half an hour, as illustrated in the diagrams, is of great significance in a climate where the mean daily range of temperature is around 5°. Longer lasting differences are between sunny and cloudy days, which especially affect the daily maximum temperature (Figure 4.6, bottom).

To illustrate the combined effects of both the seasonal and the diurnal variations of temperature, thermoisopleths can be used (Figure 4.7). The Singapore diagram shows a typical pattern for an equatorial station: conditions are completely dominated by the diurnal cycle. At this marine location, the diurnal range is rather small, yet it is the main influence on temperatures. Similar conditions prevail at most equatorial stations. At Lusaka the situation is more balanced: the diurnal and the seasonal cycles are of about equal significance. Such conditions occur mainly at stations near the outer limit of the tropics, but many local deviations from this general scheme occur. Lusaka shows a clear spring maximum in October and a slight secondary maximum in March–April. This is typical for a continental station at this latitude; marine stations would show only one maximum in the summer. Neither Lusaka nor Singapore represent extreme situations, yet the diagrams illustrate the differences that exist between the various tropical climates in respect of temperature regimes.

THE EFFECTS OF ELEVATION

In all climates, elevation reduces temperature by about 0.65° per 100 m, but its influence is particularly significant in the tropics, where it interrupts the spatial

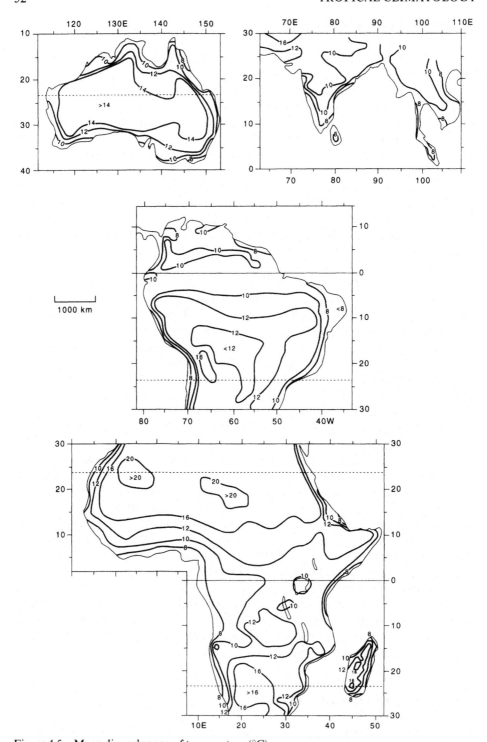

Figure 4.5   Mean diurnal range of temperature (°C)

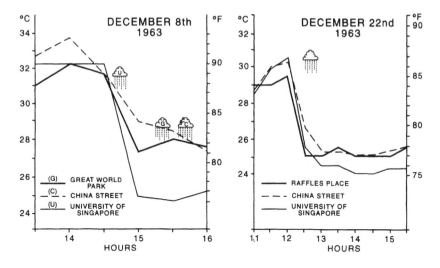

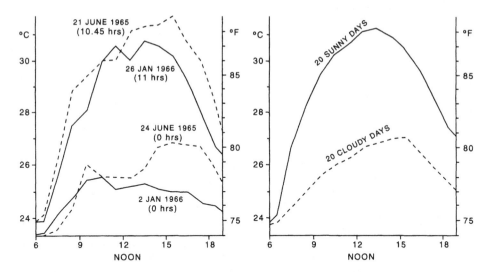

Figure 4.6   Temperature differences at Singapore. Top: rapid changes during the early stages of a thunderstorm. Bottom left: differences between sunny and cloudy days (hours of sunshine are given in brackets). Bottom right: temperature differences for 20 sunny days (more than 9 hours of sunshine) and cloudy days (less than 1 hour of sunshine). Boths groups of days were distributed equally over the different months of the year. From Nieuwolt (1966b, 1968). Reproduced by permission of *Singapore Journal of Tropical Geography*

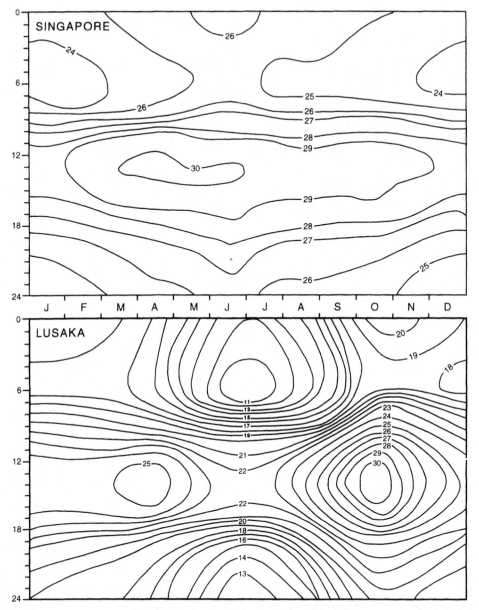

Figure 4.7   Thermoisopleths (°C) for Singapore (1°21'N, 8 m a.s.l.) and Lusaka (15°25'S, 1154 m s.a.l.). Based on records for 1951–60 (Singapore) and 1941–51 (Lusaka)

uniformity of temperature. It often causes large temperature differences over short distances.

Generally, the seasonal march of temperature is unaffected by elevation; on a world scale the annual range shows no general influence (Figure 4.1). The diurnal cycle can be strongly modified by elevation, but the effects depend on landforms and cloudiness.

Over extensive highlands, where radiation controls the temperature conditions and cloudiness is generally low, daily ranges reach maximum values, as has already been pointed out, but over mountain ranges and especially isolated tops, exchange with the surrounding air of the free atmosphere prevents temperatures from rising much during the day and equally from falling at night. Moreover, cloudiness often reduces the effects of radiation, so the diurnal range is little different from lower elevations (Figure 4.5).

The importance of elevation lies in two fields: in agriculture it creates unique possibilities (Chapter 12) and it also offers relief from the oppressive heat of the tropical lowlands. Some highland stations are primarily based on this feature: Bandoeng (717 m) near Djakarta, Simla (2200 m) near New Delhi, and Fraser's Hills (1300 m) near Kuala Lumpur are all examples of settlements which provide a cool environment for people from lowland capitals. And when the British established Nairobi (1660 m) as the capital of Kenya, the pleasant climate made it the obvious choice over Mombasa.

OCEAN CURRENTS

The large tropical oceans are generally quite warm, and they form an enormous reservoir of heat, which reduces short-term temperature changes on islands and coastal areas in the tropics, particularly the mean diurnal range (Figure 4.5).

There are, however, a number of cold currents in the tropical oceans (Chapter 5) which have quite a different effect on temperatures. The main ones are the Humbolt–Peru current west of South America, the Benguela current off the south-west coast of Africa, the West Australian current and the Somali current off the north-east coast of Africa (Figures 5.14–5.16). These bring cool and in some cases subpolar waters to tropical latitudes. The strongest effects are those of the Humbolt–Peru current which may bring water from as far as 50 °S. The effects on temperatures are clearly illustrated by Figure 4.4 – the cooling reaches well north of the equator. It also influences the mean annual range, which reaches 6 °C (Figure 4.1). It seems that the current has transported temperature conditions from about 25 °S to the equator.

The Benguela current is weaker, and its effects may not reach the equator. Similarly, the West Australian current does not affect equatorial latitudes. Both currents affect the coastal regions on which they border, not only by bringing lower temperatures, but also by increasing the annual range of temperature (Figures 4.1, and 4.4).

The Somali current is of a different origin: upwelling of cool waters from depth. During the south-west monsoon the winds along the Somali coast are off-land; this reduces the effects of the cool current to a very narrow coastal strip. During the north-east monsoon the winds are mainly parallel to the coast; this still causes upwelling, and the effects are felt over a larger area.

PHYSIOLOGICAL TEMPERATURES

The temperature, as experienced by living organisms in general, and by human beings in particular, depends mainly on the rate of heat loss from the body. The human body is kept at a constant temperature of 36.7°, and defence mechanisms prevent excessive loss of heat or too much heat absorption. Under warm conditions heat disposal takes place mainly from the skin and the lungs. If this is not sufficient, additional loss of heat is caused by the evaporation of body fluids in the form of perspiration.

Physiological temperatures therefore depend not only on the air temperature, but also on humidity and the circulation of air around the body. Humid air restricts the evaporation of perspiration and causes oppressive conditions. Dry air, on the other hand, allows a rapid cooling of the skin by evaporation, and high air temperatures are endured easily.

The circulation of air around the body also influences the physiological temperature. When air is stagnant, the layer of air around the body heats up to approximately the body temperature, but soon becomes saturated with water vapour, thus restricting cooling of the skin by evaporation. With sufficient circulation, the constant replacement of air around the body prevents this situation arising. Direct exposure to the sun may also increase the physiological temperature by direct absorption of radiation.

There have been various attempts to design empirical (based on measurement) indices for the assessment of human thermal comfort. Such indices attempt to integrate the thermal and humidity effects of the environment and represent in one value, the level of thermal stress. These have been applied at a variety of scales from the global to the local (Gregorczuk and Cena, 1967; Olaniran, 1982; Jauregui, 1991; McGregor, 1995a). Recently in the field of bioclimate assessment, there has been a move away from the use of empirical indices to the use of more complex biophysical models. These attempt to model the man–environment exchange of heat based on an evaluation of the human energy balance equation. Because these models are demanding in terms of their data requirements, many tropical locations do not record some of the required model input variables, their application in tropical bioclimate assessments has been minimal. However, when applied their power as tools for bioclimatic analysis is very clear (de Dear, 1989).

Two of the most widely used empirical thermal comfort indices are the temperature–humidity index (THI), sometimes called the discomfort index (Thom, 1958; US Weather Bureau, 1959) and the effective temperature (ET) (Houghten and Yaglou, 1923). These indices express thermal stress by indicating the temperature which, combined with a relative humidity of 100%, would create the same thermal comfort reactions (too warm, comfortable, too cool, etc.) by test persons as the existing combination of temperature and humidity (the ambient conditions). For example, in the case of the ET, if the existing air temperature is 22°C and the relative humidity 50%, this will produce the same level of thermal comfort as would be experienced at 100% humidity and 20°C which is the ET. Other combinations of temperature and humidity can also produce an ET of 20°C; for example, an air temperature of 25°C and a relative humidity of 10%. From these two examples it can be seen that as temperature rises (falls) there must be a fall (rise) in the humidity for the same comfort level (ET) to be maintained. Despite some concerns about the faithfulness of these empirical indices in terms of representing thermal stress (de Dear, 1989), their attractiveness lies in their ease of calculation and application at a variety of scales using standard climatological data (Jauregui, 1991).

The formulae for the THI (°C) and the ET (°C) are:

$$\text{THI} = 0.8t + \frac{rh \times t}{500} \quad \text{or} \quad \text{THI} = 0.55t + 0.2d + 5.3$$
$$\text{ET} = t - 0.4(t - 10) \times (1 - rh/100)$$

where t and d are the air and dew point temperatures (in °C) and rh is the relative humidity (in %).

Generally, a THI of around 21 °C is associated with most people feeling comfortable; at values around 24 °C about half of a population experience some form of thermal stress; and when the THI reaches 26 °C almost all feel uncomfortable. With further increases of the index the efficiency of workers rapidly deteriorates (US Weather Bureau, 1959). In the case of the ET, the boundary of sultriness and discomfort is at 24 °C, while 35 °C is considered the upper level of tolerance. These figures are based on experiments in the mid-latitudes. People who live constantly in tropical lowlands probably can tolerate higher values of the THI or ET somewhat better, as nutrition, clothing and general speed of physical activity are all adjusted to a hot climate. For example in Delhi, Indian studies have shown that people have acclimatized to an ET of 27 °C with the availability of fan ventilation (Lahiri, 1984). Similarly, results of studies in the Australian tropics suggest acclimatization of European populations to north Australian tropical climates (Auliciems and de Dear, 1986; Williamson, Coldicutt and Penny, 1991). For equatorial Singapore it also appears that native Singaporeans have acclimatized to a temperature well above that expected to produce thermal comfort for mid-latitude dwellers (de Dear, 1989; de Dear, Leow and Foo, 1991).

As temperature and moisture conditions vary diurnally and seasonally, so do comfort conditions. The diurnal and seasonal variations of the THI at two stations are illustrated in the form of isopleth diagrams (Figure 4.8). A comparison with the temperature diagrams for the same stations shows that the general patterns are similar, but that day–night differences are strongly reduced (Figure 4.8). This is because the relative humidity is generally much higher at night, often reaching close to 100%, and the THI therefore is close to the air temperature. During the day a lower humidity reduces the THI up to 3 °C at Singapore. At Lusaka, seasonal differences are also less than for temperatures alone because the hot seasons there are also the driest. As the THI plots show, oppressive conditions (THI above 26 °C) occur at Singapore during the afternoon throughout the year, but never at Lusaka. However, cool nights might be a problem there for the homeless or people living in squatter settlements. The plots show again the importance of elevation (Lusaka is at 1154 m a.s.l.) in reducing the oppressive heat of the tropical lowlands; however in some cases, very large cities located in upland areas may have greater night-time discomfort than smaller lowland cities because of the urban heat island effect.

URBAN HEAT ISLANDS

Urbanization has a marked impact on climate as it results in the replacement of natural surfaces with artificial surfaces, a consequence of which is an alteration of the radiative, thermal, moisture and aerodynamic properties of the atmosphere. Such unintentional modifications to the land surface and the atmosphere disrupt the

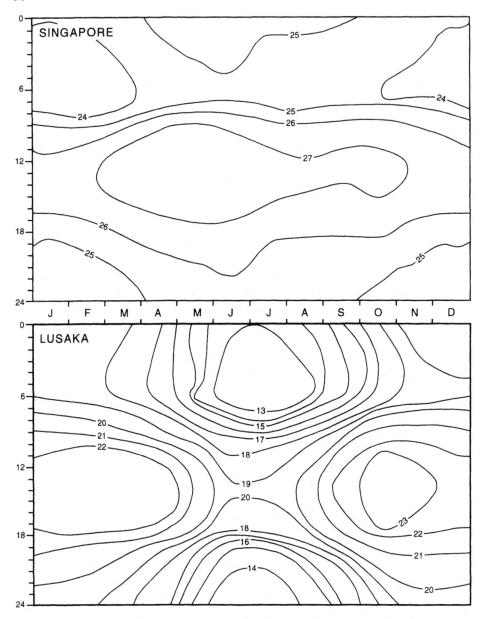

Figure 4.8   Isopleths of the temperature humidity index for Singapore and Lusaka

radiation, energy and water balances of an area and lead to the development of *urban heat islands*. These are defined as a concentration of heat in urban areas; an island of heat in a sea of cool rural air. The urban heat island occurs at the bottom of the urban boundary layer in the *urban canopy* layer, that layer of the urban atmosphere below roof level (Oke, 1987).

Perhaps there is nowhere in the world where urbanization has been as rapid as in

Table 4.1 Generalized differences of surface thermal, radiative, moisture and aerodynamic properties likely to exist between typical temperate climate cities and those in tropical dry or humid climates (Oke, 1986)

Climate zone comparison	Thermal admittance	Albedo	Emissivity	Moisture availability	Roughness
Temperate vs dry tropical	⩽ *	<	⩾	>	⩾
Temperate vs humid tropical	⩾	⩽	⩾	<	⩾

* Greater/less than is relative to the tropical value.

tropical countries. For example, in 1940 there were six tropical cities with populations of more than one million inhabitants, but by 1970 this had increased by over eight times to 52, with associated increases in the physical size of cities. Currently, of the 34 cities with populations in excess of five million, 21 are tropical cities. Furthermore, it has been estimated that by the year 2025 approximately 60% of the populations in Africa, Asia, Latin America and Oceania will be urban. By this time tropical cities such as Mexico City, São Paulo, Lagos, Cairo, Karachi, Delhi, Bombay, Calcutta, Dacca, Shanghai and Jakarta are expected to have populations in the range of 20–30 million (Peterson, 1984). This holds serious implications for the future climates of tropical cities. Unfortunately, the current state of knowledge about the climates of tropical cities is poor.

Because temperate and tropical city urban morphologies and building materials differ, the surface thermal, radiative, aerodynamic and moisture properties are also likely to differ. This has implications for how tropical cities will respond climatically to rapid urbanization (Oke, 1986). A comparison of these properties is given in Table 4.1. The basic causes of urban heat islands in tropical cities are likely to be the same as for temperate cities; however, their relative importance may differ because of the contrasts outlined in Table 4.1, especially albedo and moisture.

Several causes of the heat island have been identified which relate to the surface energy balance (Oke, 1987). Polluted urban atmospheres are likely to act in a way similar to the greenhouse effect. Outgoing longwave radiation from urban surfaces will be absorbed in the polluted atmosphere and be re-emitted back down to the urban surface. Outgoing longwave is also likely to be reduced due to a reduction in the sky view factor (the amount of open sky as seen from ground level), the importance of this being very much dependent on the urban form; tall buildings spaced close together will have the maximum effect. This "canyon effect" will also reduce albedo and thus increase shortwave radiation absorption in urban areas. For some tropical cities this may not be important as the vertical extent of the urban canopy is generally shallow; exceptions are major cities such as Singapore, Kuala Lumpur, São Paulo and Mexico City. Due to the thermal properties of building materials there is likely to be daytime heat storage and night-time heat release; night-time cooling rates will be depressed in cities. In the tropics the importance of this factor may be dependent on the nature of the building materials used in squatter settlements which are frequently found on the outskirts of tropical suburban areas or occasionally close to the city centre. In

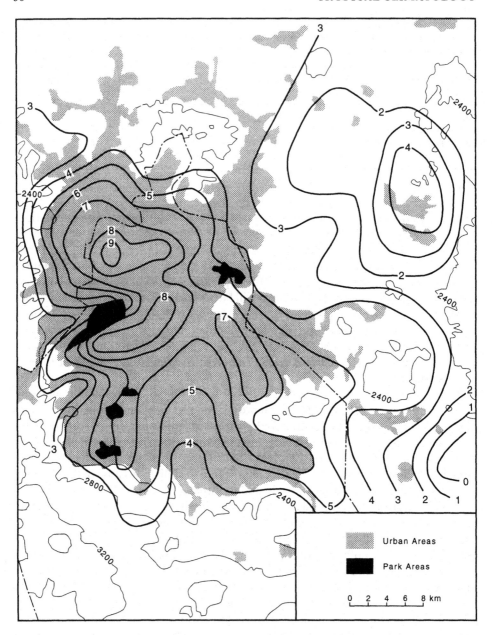

Figure 4.9   Mexico City's heat island. Isotherms are in °C. Note the cool area around central parkland. Reproduced by permission of the World Meteorological Organisation

high-altitude cooler cities such as La Paz, Bogota and Mexico City, squatter housing is made predominantly from more solid material than that in warm to hot lowland or coastal cities such as Rio de Janeiro, Bombay and Singapore where light materials like cardboard, bamboo and metal sheets are more prevalent (Jauregui, 1986). Heat may

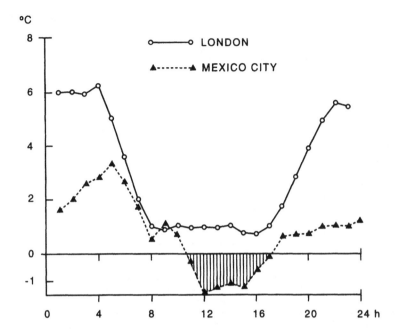

Figure 4.10   Diurnal variation of heat island intensity for Mexico City and London. Note the different onset times for rises and falls in heat island intensity. Reproduced by permission of the World Meteorological Organisation

also be released from buildings and contribute to heat island development. In cities such as Singapore and Hong Kong where air conditioning is widely used, this is likely to be important. As vegetation is removed in the urbanization process, especially in areas of squatter settlement, decreased evaporation from the urban surface will increase the importance of sensible heat flux over that of latent heat; the Bowen ratio will increase. However, the importance of this will depend very much on the area of wet surfaces in tropical cities. In some cities where there are open waterways such as in Bangkok, evaporation is likely to be large; the thermal admittance of water (i.e. the ability to accept or release heat) is also higher than soil surfaces. Because wind speeds are reduced in cities (friction increases roughness) there will be decreased loss of sensible heat. Low ventilation rates also have implications for atmospheric turbidity. Calm conditions are conducive to pollution buildup which will affect the radiative transfer of short and longwave radiation thus enhancing the mini-greenhouse effect.

The main heat island characteristics are morphology and intensity. Heat island morphology is very much dependent on a city's character and physical structure. Generally, the highest temperatures (lowest humidities) occur in areas where building densities are greatest while the steepest temperature gradients (the heat island "cliff") are found where there is a rapid change from rural to urban conditions. This is clearly seen in the case of Mexico City where under conditions of clear skies and little wind two heat island cores occur (Figure 4.9). These are separated by a cool area of parkland highlighting the importance of such surfaces as cool oases. Steep temperature gradi-

Table 4.2   Seasonal variations in the mean heat island intensity for Ibadan (Oguntoyinbo, 1986)

Season	Month	Synoptic situation	Heat island intensity ($^{\circ}$C)
Wet	June	Cloud bands, intermittent sunshine	3.3
	July–August	Uniformly cloudy, daytime stratiform clouds	0.6–0.8
Dry	December	Dust polluted sky (Harmattan)	1.0 –2.5
	February	Clear cloudless sky	5.0–7.0

ents occur to the north and west of the metropolitan area where katabatic flows of cool air from the surrounding mountains occur (Jauregui, 1986).

Heat island intensity is the difference between the urban core and ambient rural temperature. This measure is used as an indication of the magnitude of the climatic modification brought about by urbanization. Generally, for cities of equal size, tropical city heat island intensities are smaller than temperate city intensities because of the warm conditions in the low latitudes. However for very large cities, intensities may approach those of temperate cities, e.g. intensities for Delhi, Bombay and Pune in India are 6.0°C, 9.5°C and 10°C respectively (Padmanabhamurty, 1986), while for Shanghai and Mexico City intensities are on average around 6.5°C and 8°C respectively (Chow and Chang, 1984; Jauregui, 1986). Under constant weather conditions the heat island intensity demonstrates a diurnal variation. Its style of diurnal evolution in tropical cities may differ from that for mid-latitude cities, with maximum growth rates and intensities occurring later than those in mid-latitude cities (Figure 4.10). Heat island intensities also vary seasonally and with the general synoptic scale weather conditions. Greatest intensities occur in the cool or dry season and for calm clear weather conditions (Table 4.2). In coastal locations sea breezes may reduce the urban heat island effect.

As for the case of temperate cities, it appears that tropical heat island intensities are dependent on population size. The form of this relationship may be non-linear, unlike the temperate case, with intensities increasing at a much more rapid rate for cities with populations in excess of 10 million (Figure 4.11). This may be related to marked changes in urban morphology, because as cities grow so does the size of their squatter settlements, along with reduced areas of green and taller central city buildings (Jauregui, 1986).

SUMMARY

At the scale of the global tropics, especially in the very low latitudes, temperatures demonstrate a high degree of both seasonal and spatial uniformity, a product of small seasonal and geographical differences in insolation. However, the spatial uniformity of temperatures in the low latitudes is not universal as a location's thermal regime may also be controlled by factors such as elevation, proximity to large water bodies, which determines the apportionment of available energy, and ocean currents. These factors can produce clear geographical contrasts in temperatures within the tropics which has

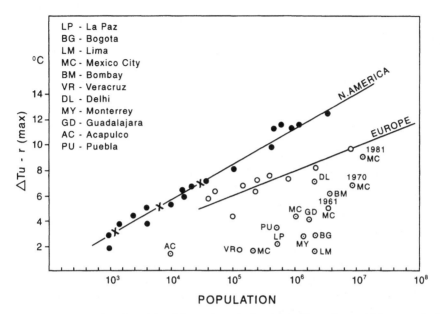

Figure 4.11 Population–heat island intensity relationships. Note the rapid increase in heat island intensity for tropical cities with populations in excess of 10 million. Reproduced by permission of the World Meteorological Organisation

implications for the geographical distribution of human thermal climates. Through the process of urbanization and the alteration of the surface energy balance, temperature contrasts can be found between cities and their cooler surrounding rural areas. This phenomenon, generally known as the urban heat island, will be accentuated as tropical urban populations grow, potentially leading to physiologically stressful climate conditions in large tropical cities. Such problems may be exacerbated by predicted climate change.

CHAPTER 5

General Circulation of the Tropics

In describing the nature of the general circulation of the atmosphere one usually considers air movements, changeable both in direction and velocity, over largely different periods of time in a three-dimensional medium. This complicated structure can be illustrated only by a combination of maps and cross-sections, drawn for different seasons; but even so the general circulation and its main changes with the seasons are difficult to visualize in their entirety.

Meteorologists, when illustrating conditions in the atmosphere, usually exaggerate the vertical dimensions and movements, without actually stating so. In doing this, an impression that the atmosphere possesses a great depth is created. However, the troposphere, in which the majority of the general circulation takes place, is extremely shallow: its maximum depth around the equator is around 20 km, while its horizontal dimensions, in any direction, are measured in thousands of kilometres. Because of its shallow depth, horizontal air movements prevail in the troposphere and masses of air transported parallel to the earth's surface are many times larger than those carried by vertical currents. Nevertheless, the latter are of great importance, because they transport heat and moisture away from the lowest levels and also because they profoundly change the stability conditions of the air masses involved. This is the main reason why vertical movements are emphasized in most meteorological literature.

An understanding of the nature of the atmosphere's general circulation has traditionally been built on the analysis of upper air observations. Compared to the mid-latitudes there are few of these in the tropics. However, with the aid of satellite technology and the implementation of several major research projects designed to measure the heat, moisture and momentum exchanges between the ocean and the atmosphere, there has been rapid progress in the understanding of the workings and general patterns of the tropical general circulation.

An outcome of the increased measurement and monitoring of the tropical atmosphere is the discovery that there are many deviations from the general pattern of tropical circulation that was thought to exist prior to the 1970s. These variations are effective over very different areas and times: some affect whole continents or ocean basins, others are purely local, limited to coastlines, islands or mountain ranges; some are seasonal in character, others are diurnal in occurrence and still others come and go at irregular intervals. Despite increased observations, usually presented in the form of meridional (north–south) or zonal (east–west) cross-sections, the statement by Riehl (1954, p. 24) that "No single meridional cross-section can be considered as representative of the tropical circulation" still remains true. This is because the tropics are characterized by a number of interrelated circulation systems and features which vary in terms of their spatial and temporal characteristics. Of these, the meridional Hadley cell circulations of the two hemispheres and the product of their interaction, the intertropical convergence zone; the subtropical anticyclones; the trade winds; and the upper subtropical westerly and tropical easterly flows have the greatest geographical coverage and occurrence. The purpose of this chapter is to present a general description of these circulation systems and their associated climatological features.

The deviations from the general circulation pattern will be discussed in Chapters 6 and 7 while the smaller-scale disturbances, superimposed on the large-scale movements of air, will be dealt with in Chapter 8.

THE TROPICAL CIRCULATION IN A GLOBAL CONTEXT

The tropics, as delimited in Chapter 1, cover more than 40% of the earth's surface, but the tropical circulation often extends beyond these boundaries. Moreover, the thickness of the troposphere is higher over the low latitudes than further away from the equator. The tropical circulation therefore involves the main bulk of the troposphere and plays a major role in determining the nature of the global circulation at any one time. It is therefore important to consider briefly the relation of the general tropical circulation to that of mid- and higher latitude circulation systems. This is perhaps best done by considering a model of the general global circulation (Figure 5.1).

The low latitudes are dominated by the meridional circulation of the Hadley cells with ascending branches on their equatorward sides and descending branches on their poleward sides. Return flow to the equator at the surface is in the form of the trade winds. On the poleward side of the upper atmosphere return branch of the Hadley cell is the subtropical jet, composed of high-velocity westerly winds which on average encircle the earth. Centred on approximately 30° from the equator are the axes of the subtropical highs which have been used as a natural boundary in the definition of the geographical extent of the tropical atmosphere (Figure 5.1).

Poleward of the tropical atmosphere, vertical circulation is dominated by slantwise transport of warm air from the tropical latitudes into the middle troposphere of the mid-latitudes. This transport of warm and moist air is achieved by rising currents on the western sides of the subtropical highs, tropical storms venturing into the mid-latitudes and mid-latitude storms importing warmer low latitude air into their circulations. Most of the low to mid-latitude heat transport occurs in the latent form associated with condensation and rainfall production. Matching the poleward transport of warm air in the area of slantwise circulation is equatorward transport of cold air. This mostly occurs on the eastern sides of the major ocean basins in the equatorward returning circulation of the subtropical highs.

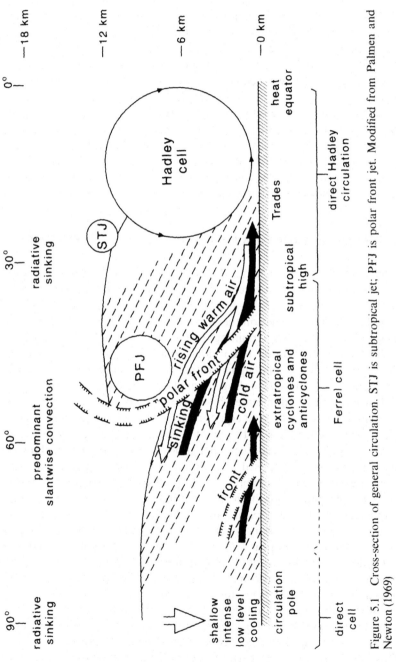

Figure 5.1 Cross-section of general circulation. STJ is subtropical jet; PFJ is polar front jet. Modified from Palmen and Newton (1969)

The boundary between tropical and mid-latitude atmospheres is characterized by a major area of baroclinicity or rapid horizontal temperature change. Such a baroclinic atmosphere helps maintain the vertical slantwise transport of warm tropical air to the mid-latitudes. Beyond the baroclinic zone, characterized by the polar front, cold polar air dominates. On the equatorward side of the polar front in the upper atmosphere is found the polar front jet. This at times flows in close proximity to the subtropical jet and is responsible for injecting cold air into the upper subtropical atmosphere, creating instability.

From this model it is quite clear that the tropical atmosphere does not operate independently of its mid-latitude counterpart. The continual irreversible vertical and horizontal exchanges of cold and warm air between the mid-latitude and low latitude atmospheres represent major energy transfers. These are kinetic energy or motion producing. As a result, the importance of the tropical circulation is not only related to its control of the main characteristics of tropical climates, but also to its influence on conditions outside the tropics. These long-distance correlations between the behaviour of the tropical atmosphere on the regional to global scale and climatic effects in tropical or extratropical regions are referred to as *teleconnections.*

The main features of the tropical atmospheric circulation appearing in the global circulation model (Figure 5.1) are presented in plan form in Figure 5.2 and are discussed below.

THE HADLEY CELLS

The Hadley cell circulation is composed of a thermally driven rising limb of air in the equatorial zone, a poleward moving flow in the upper atmosphere, a sinking limb in the region of the subtropics and a returning trade wind flow at the surface which converges with its counterpart from the northern or southern hemisphere. The Hadley cells show seasonal variation in their intensity, geographical extent and latitudinal position. Furthermore, they are not restricted to their respective hemispheres but display regular cross-equatorial incursions into the opposite hemisphere.

Perhaps the best way to understand the structure, dynamics and geography of the Hadley cells is by examining a number of longitudinal cross-sections of the atmosphere in the region of the tropics. In examining these it must be remembered that the vertical exaggeration is large, which sometimes creates the false impression that the vertical components of the atmospheric circulation are more important than the horizontal ones.

Figure 5.3 shows meridional cross-sections of vertical velocities, the vertical equivalent of horizontal wind speed, in the atmosphere for annual average conditions, as well as December to February (DJF) and June to August (JJA). These figures convey a clear picture concerning the strength of vertical motions in the atmosphere. Negative (positive) values indicate ascending (descending) motion. On an annual basis the atmosphere between 10°N and 10°S is characterized by strong vertical motions which are a product of the convergence in and the confluence of the trade flows (this chapter). Strong ascending motions occur centred on about 5°N. The position of the annual vertical motion maximum coincides with the mean annual position of the intertropical

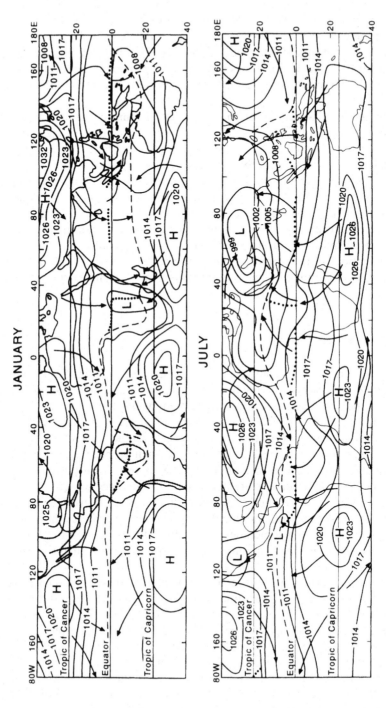

Figure 5.2   Mean sea level pressure and predominant surface winds in January and July. Dashed line is the mean position of the ITCZ. Dotted lines are secondary convergence zones

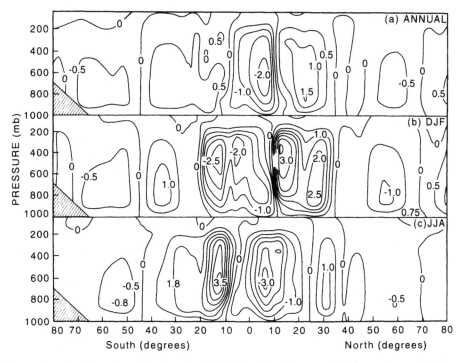

Figure 5.3 Meridional profiles of vertical wind velocities: (a) annual; (b) December to February (DJF); (c) June to August (JJA). From Piexoto and Oort (1992). Reproduced by permission of American Institute of Physics, New York

convergence zone (see next section). The regions between 15° and 40° either side of the equator are dominated by sinking air (positive values) associated with the descending branches of the Hadley cells.

The vertical velocity profiles for DJF and JJA reflect the seasonal behaviour of the Hadley cells. In the southern hemisphere summer (DJF) the strongest ascending motions occur between 10 and 20°S with maximum vertical velocities centred on 15°S. The vertical velocity maximum represents well the mean position of the ITCZ (inter-tropical convergence zone) at this time of the year. At the same time of the year in the northern hemisphere strong subsidence dominates the latitudes between 10 and 30°N (Figure 5.3(b)).

During the northern hemisphere summer (JJA) the zone of maximum vertical velocities is displaced north of the equator with strong ascending motion centred on 5°N while the low latitudes of the southern hemisphere become dominated by sinking motions. These are especially intense around 10 to 15°S (Figure 5.3(c)). Poleward of here the intense southern hemisphere winter anticyclonic systems that almost straddle continuously the entire southern hemisphere can be found. From these centres the trade flows emanate (Figure 5.2(b)).

Figure 5.4 portrays the mean speed of meridional wind components for a range of latitudinal bands. On an annual basis poleward of 5°N and south of the equator (Figure 5.4(a)), surface winds are opposite in direction. Between approximately 22°N and 5°N the winds are northerly (negative values). These are the north-east trades of

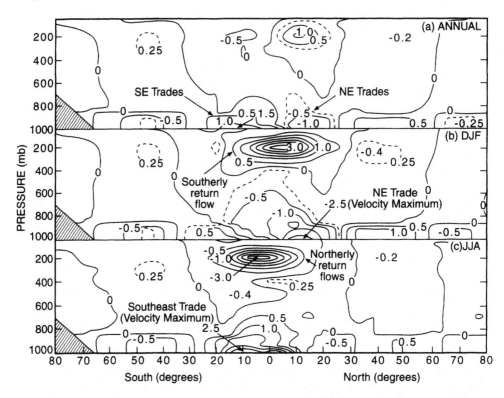

Figure 5.4   Meridional profile of horizontal wind velocities. Negative numbers are northerly and positive numbers are southerly winds. (a) Annual; (b) December to February (DJF); (c) June to August (JJA). From Piexoto and Oort (1992). Reproduced by permission of American Institute of Physics, New York

the northern hemisphere. South of 5°N to approximately 25°S the winds are southerly (positive values) and flowing towards the north. These are the south-east trades of the southern hemisphere. Both the equatorward flowing streams of air are the lower branches of the Hadley cells (Figure 5.1). Geographically, the south-east trades dominate and manage on an annual basis to push across the equator into the northern hemisphere, which explains the asymmetry in the lower branches of the Hadley cells. Velocities in the lower branches of the Hadley cells decrease in the direction of the equator. This deceleration is a major factor in causing convergence in the trade flows. In the middle to upper atmosphere the zone of surface deceleration is matched with the zone of maximum vertical velocities (Figure 5.3(a)). On an annual basis the depth of flow in the lower branches of the Hadley cells is shallow, only reaching up to approximately 700 mb (Figure 5.4(a)). In the upper atmosphere the poleward returning winds of the upper branches of the Hadley cells in either hemisphere are found.

On a seasonal basis the geography of the meridional wind flows shows strong inter-seasonal contrasts. In the northern hemisphere winter (DJF) the surface flows are dominated by the north-east trades which push into the southern hemisphere (Figure 5.4(b)). Surface wind speeds decrease rapidly equatorward of 10°N. In the upper atmosphere a strong return flow of air to the north (Figure 5.4(b)) supplies the air that subsides in the poleward branch of the Hadley cell (Figure 5.3(b)). At the surface, south of 10°S, the meridional circulation in the lower branch of the southern hemisphere Hadley cell is weak and of limited geographical importance compared to its northern hemisphere counterpart. The geographical patterns of the meridional flow for DJF are reversed for the southern hemisphere winter (JJA) (Figure 5.4(c)).

Carried with the air that moves in the cellular Hadley circulations are massive amounts of heat and moisture. Such heat and moisture movements are important for maintaining the global energy balance. In the rising limbs of the Hadley cells heat and moisture are carried into the upper parts of the atmosphere. At about 9–12 km above the earth's surface these are turned polewards in the upper branches of the Hadley cells. The transport of moisture represents an energy transfer as when moisture condenses latent heat is produced (Chapter 2). This represents a large heat source for the atmosphere which is redistributed to higher latitudes by the upper branches of the Hadley cells.

Other forms of energy transported by the Hadley circulations are kinetic energy and angular momentum. Kinetic energy is that energy due to the motion of air. The total kinetic energy in the atmosphere can be broken down into three energy components. They are the zonal mean, transient eddy and stationary eddy components. As the Hadley circulation is essentially a meridional circulation it is not an important generator of zonal kinetic energy. The other two terms are of possible interest in the low latitudes. The transient energy component represents energy produced by air motion in travelling disturbances such as tropical cyclones. The lower branches of the Hadley cell are important in this respect as they often steer tropical cyclones in a westward direction and towards higher latitudes, especially at the height of the summer. In this case energy in its transient eddy form is exported from the low to high latitudes. The stationary eddy component is due to motion in semi-permanent features of the atmosphere such as the Hadley cells. During the winter, because the Hadley circulations become intense, a large amount of kinetic energy in the stationary component form is generated. This is exported to higher latitudes by the upper branches of the Hadley circulation. The northern Hadley cell is the more important for the transport of kinetic energy in its stationary eddy form (Piexoto and Oort, 1992).

Angular momentum is the other property exported from the low latitudes by the Hadley circulations. As we have seen, the winds in the lower branches of the Hadley cell blow in an east to west direction. This is in the opposite direction to the earth's rotation which is from west to east. As a result, a frictional force is created between the atmosphere and the earth's surface. This is called a torque in the negative direction because the frictional force is in the opposite direction of the earth's rotation. As well as the atmosphere exerting a torque on the earth's surface, the earth also exerts a torque on the atmosphere. This is opposite to that of the atmosphere; the earth's surface exerts a torque in the positive direction. For a rotating body, if the torque is in the positive direction, the angular momentum increases. Given this, the earth's surface in the region of the low latitudes is a source of positive angular momentum, meaning that the low latitude regions should speed up. As this does not happen, then the low latitudes must have some way of getting rid of this excess angular momentum. This is achieved by angular momentum transfer to the lower atmosphere which is subsequently transported by the upper branches of the Hadley cell to higher latitudes where there is negative angular momentum (Piexoto and Oort, 1992).

THE CENTRAL ZONE OF LOW PRESSURE AND CONVERGENCE

At the centre of the tropical circulation between the two Hadley cells is a wide belt, characterized by relatively low surface pressure, rising air movements and convergence of air masses. This zone is known by various names, depending on which of its main characteristics is emphasized. When low pressure is emphasized, it is usually called the "equatorial trough", but when the convergence of air masses is considered more relevant, it is called the "intertropical convergence zone" (ITCZ). At times the term "intertropical front" was used for this zone, by analogy with the mid-latitude fronts between convergent air masses. This name is now redundant since it was realized that the convergence zone near the equator rarely displays the characteristics of a mid-latitude front, because temperature differences between the converging air masses are usually very small. Finally, a more neutral name, "meteorological equator", considers a number of important features of the zone of maximum development (Flohn, 1969). Throughout this book we will use the term *intertropical convergence zone* (ITCZ) as this has become a commonly accepted term with which to describe the zone of equatorial cloudiness visible on satellite photographs of the earth (Plate 2).

There has been much debate about the origin of the ITCZ. Many studies have emphasized the fact that the climatic features of this zone – low pressure, maximum surface temperatures, high cloudiness and rainfall, trade wind convergence and confluence – all coincide. However, satellite and surface based observations of the ITCZ have revealed that the zone of low pressure, maximum surface temperatures and surface wind confluence is separated from the zone of maximum cloudiness, rainfall and convergence. In the Atlantic and eastern equatorial Pacific oceans, the separation of these two zones may be 300–400 km, while over the African continent this distance may reach around 1000 km. In both cases the low pressure minimum, which is maintained under clear skies by insolation, is located further to the north than the zone of maximum cloudiness (Hastenrath and Lamb, 1977). Furthermore, the zones of low pressure and maximum surface temperatures (the meteorological equator) appear to follow the path of the annual solar cycle. This characteristic, in addition to the fact that low pressure only exists up to around 3000 m (700 mb), suggests a thermal origin for the equatorial trough. Over the oceans the mixing characteristics of near-surface ocean layers may also aid the development of the low pressure minimum. Within the area of low pressure little mixing of surface oceanic waters occurs because wind speeds are low here; the *doldrums*. As a result, heat input from insolation is kept near the surface which will maintain warm sea surface temperatures. These will in turn encourage the development of low atmospheric pressures through convective activity (Hastenrath, 1985).

Important climatic features of the ITCZ are cloudiness and rainfall. These are mainly due to convergence. Dynamic theories are used to account for these features. Convergence occurs when wind flows slow down or change direction. In either case, air tends to pile up and as a result rises. This process, aided by warm surfaces, produces large-scale convection which transports water vapour up into the atmosphere where it condenses and ultimately forms cloud and rain. In the eastern Atlantic and the eastern Pacific, at the height of the northern summer, the zone of maximum convergence and thus cloudiness lies south of the zone of confluence where the trade winds from the northern and southern hemispheres meet. Here deceleration and convergence is a result of the equatorward-flowing south-easterly trades changing direction to south-westerly winds due to the Coriolis force. In the western Atlantic, deceleration and convergence in the south-easterly flows is due to an equatorward slackening of the pressure gradient where the zone separating regions of confluence and convergence is much narrower.

In the Indian Ocean area, at the height of the northern summer, the equatorial trough is obliterated by the strong south-easterly flow over the Indian Ocean. However, with the onset of the northern winter the equatorial trough re-establishes itself over the Indian Ocean where a zone of low pressure and high sea surface temperatures coincide (Hastenrath and Lamb, 1979). In the western Pacific region, the equatorial trough is partly fragmented at the height of the southern summer because of the effect of the Australian continent. Here the equatorial trough axis dips to a southern maximum of around 18°S over north-western regions where its position is very much determined by the development of the Pilbara heat low. Over north-eastern Australia and the adjacent ocean, the trough axis is oriented in a more west to east direction lying at about 12°S. Over the land and ocean the trough axis coincides with areas of low surface pressure and maximum sea surface temperatures. To the north of the Australian equatorial trough an area of cloudiness is found. This a product of convergence in the north-easterlies as they recurve after crossing the equator from the northern hemisphere.

In addition to the mechanisms for generating the ITCZ, of interest are the mean, variability and extreme positions of the ITCZ. This is because associated with the ITCZ is a band of high rainfall on which many low latitude subsistence and cash-based agricultural economies depend.

Although it is often assumed that the ITCZ stretches uniformly around the globe, the global ITCZ is made up of a number of distinctly different zones (Table 5.1) which show different characteristics in terms of their ITCZ structure and behaviour; a product of distinct atmospheric circulation, surface type and ocean current patterns (Waliser and Gautier, 1993).

Over the oceans contrasts in the spatial structure of the ITCZ zones exist. The Atlantic and eastern Pacific zones are typified by a narrow band of cloudiness whereas the western Pacific and Indian ocean zones possess broad latitudinal bands of cloudiness (Figure 5.5). Between the eastern and western Pacific there is also a zone of cloudiness extending in a south-easterly direction into higher latitudes. This is called the South Pacific convergence zone (Vincent, 1994).

The seasonal movement of the ITCZ, which is generally greater over land, is an important determinant of the timing of the onset, duration and cessation of the rainy season in many tropical areas. Figure 5.6 presents information on the seasonal march of the mean position of the ITCZ for the various regions in Table 5.1. For the African region, the ITCZ displays an almost symmetrical pattern of movement north and south of the equator which mimics the annual solar cycle and zone of maximum

Table 5.1   Seven ITCZ zones (Waliser and
Gautier, 1993)

Zone	Longitude limits
Africa	10–40°E
Indian	60–100°E
West Pacific	10–150°E
Central Pacific	160–160°W
East Pacific	100–140°W
South America	45–75°W
Atlantic	10–40°W

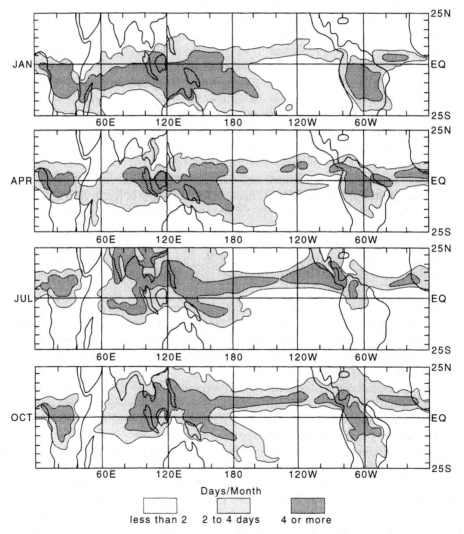

Figure 5.5   Seasonal movement of the ITCZ. ('Days' are the number of days of highly reflective cloud.) Simplified from Waliser and Gautier (1993). Reproduced by permission of the American Meteorological Society

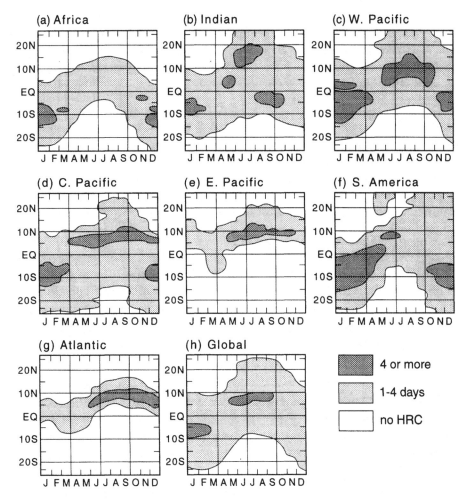

Figure 5.6   Monthly movement of ITCZ for various ITCZ regions (see Table 5.1). ('Days' are the number of days of highly reflective cloud (HRC).) Simplified from Waliser and Gautier (1993). Reproduced by permission of the American Meteorological Society

surface heating. This behaviour contrasts with that of the Indian and western Pacific oceans, where the summer monsoons play an important role in explaining the seasonal behaviour of the ITCZ and cloudiness. For the eastern Pacific and Atlantic ocean areas, the annual cycles are similar, with the ITCZ remaining mainly within the northern hemisphere because of the suppressive effects of oceanic upwelling on convective cloud development. For these two regions the timing of the maximum latitudinal extent of the ITCZ lags behind the solar cycle by about two months due to the slow reaction time of the oceans to surface heating. Because of this, the maximum northward position is reached in the northern hemisphere autumn when ocean temperatures have reached a maximum and the Hadley circulation is at its strongest (Waliser and Gautier, 1993). For the South American area the ITCZ predominates south of the equator. This asymmetric behaviour may be explained by the nature of the underlying surface. When the ITCZ moves from south to north over land, following the height of the southern hemisphere summer, it follows the annual march of the solar cycle closely. However, in the northern summer, the maximum northward position lags behind the solar cycle as the sun moves across mainly oceanic areas in the northern parts of the South American sector. This reflects the role of ocean heat storage as has been seen to be important for other oceanic areas (Waliser and Gautier, 1993).

If all the regional ITCZ behaviours are averaged, this produces a near-symmetrical pattern of global ITCZ migration (Figure 5.6(h)). In terms of intensity and thus convective activity, the ITCZ is strongest during the northern summer and autumn and weakest during the equinoxes. Globally, convective activity is suppressed around and south of the equator. This is because of cool sea surface temperatures (SST) in the equatorial regions due to oceanic upwelling and the influence of the cool northward-flowing Humbolt and Benguela currents in the southern hemisphere on the eastern sides of the Pacific and Atlantic oceans respectively. Further to its mean characteristics, the ITCZ displays considerable year to year variability in its geographical behaviour, with latitudinal departures of up to 6° for some regions and up to 2° for the global average. Such departures, lasting anywhere from 3 to 18 months (Waliser and Gautier, 1993), may produce periods of drought or flood in some tropical regions.

THE SUBTROPICAL HIGHS

Between the latitudes of about 20° to 40° the mean surface pressure patterns are dominated by a number of high pressure areas usually referred to as the *subtropical highs*. Characteristically elliptical in plan form and oriented in an east–west direction, they dominate the surface pressure pattern over the large ocean basins throughout the year (Figure 5.2). In the southern hemisphere they display higher pressure during the winter seasons and extend their influence over the adjacent continental areas, so that an almost closed belt of high pressure is formed. During the summer this belt is interrupted by heat lows over the continents. In the northern hemisphere the seasonal variations are different: over the oceans the subtropical highs display higher pressures and cover larger areas during the summer. In winter they are often connected with the continental highs at higher latitudes by high pressure ridges, but no continuous belt results. The subtropical highs are an important feature of the tropical circulation of the low latitudes because they dominate the climate of large areas of the subtropics and are also the source of the trade winds (Figure 5.2).

Although persistent throughout the year over the large ocean basins, the size, intensity and geographical position of the subtropical highs do fluctuate (Figure 5.7). During the winter months the subtropical high centres reach their most equatorial positions while in the summer months movement is in a poleward direction. These latitudinal shifts are in sympathy with the seasonal cycles in the respective hemispheres. East–west movements are also a characteristic. During the northern hemisphere winter months, all subtropical highs, apart from the South Pacific high, are located in the eastern regions of their respective ocean basins, whereas during the northern hemisphere summer months, they are located to the west. The reasons why

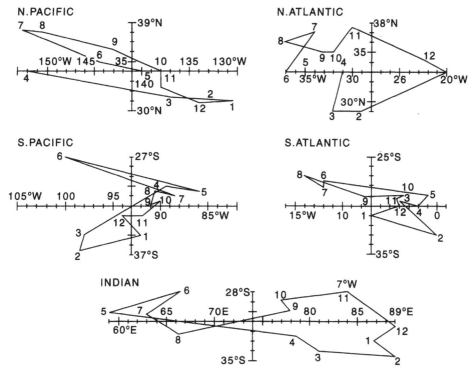

Figure 5.7 Movement of centres of subtropical anticyclonic systems. Numbers represent months of the year. From Hastenrath (1985). Reproduced by permission of Kluwer Academic Publishers

the longitudinal movement of the subtropical highs in either hemisphere mirror each other remains unclear (Hastenrath, 1985).

In addition to their geographical characteristics, the subtropical highs also display characteristics of variable intensity as measured by mean sea level pressure. For the southern hemisphere this is quite clear. For example, the South Pacific anticyclone, in the region of its mean annual centre, tends to be strongest during the southern hemisphere spring and weakest in the autumn. It also appears that the centre of this anticyclone has shown a decline in its strength over the period 1951–85 (Jones, 1991). However, for the northern flanks of the South Pacific anticyclone, Inoue and Bigg (1995) have shown a strengthening of this system between 1950 and 1979 but note that the geographical extent of this strengthening is quite variable. Such variations in the strength of the South Pacific anticyclone have been suggested as a possible reason for explaining the variable strength of the South Pacific trade winds (next section). For the other southern hemisphere anticyclones, Jones (1991) has also shown that the South Atlantic anticyclone tends to be strongest in the winter and weakest in the summer and autumn, with an increase in intensity for some seasons between 1951 and 1985. Interestingly, the South Pacific and South Atlantic anticyclones demonstrate inverse trends in their intensities. For the Indian Ocean anticyclone, intensities are greatest in

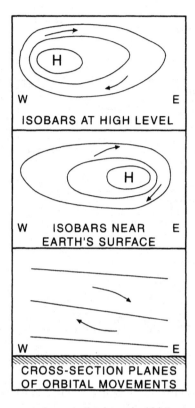

ISOBARS AT HIGH LEVEL

ISOBARS NEAR
EARTH'S SURFACE

CROSS-SECTION PLANES
OF ORBITAL MOVEMENTS

Figure 5.8   The vertical structure of a typical subtropical high pressure cell

the winter and weakest in the summer but, in contrast to the other two anticyclonic centres, there is no evidence of an annual trend in its intensity.

Variations in anticyclone intensity also occur in the northern hemisphere. For example, sea level pressure at Hawaii, which lies in the region of the North Pacific anticyclone, decreased from 1910 to the late 1960s then increased into the early 1970s (Trenberth and Paolino, 1980). For the North Atlantic anticyclone there was a significant increase in surface pressure between 1946 and 1987. These pressure increases, relative to the equatorial regions, are consistent with increases in the strength of the North Atlantic trades (Inoue and Bigg, 1995).

An important characteristic of the subtropical highs is the asymmetry in their internal structure. Near the earth's surface the centre of highest pressure is often to the east while at higher levels the core of maximum pressure is further to the west (Figure 5.8). The plane of the orbital circulation from the centre is therefore not strictly parallel to the earth's surface, but sloping gently towards the west. Consequently, subsiding movements prevail mostly in the eastern half of the cell, while rising air currents occur frequently in the western sections. The consequences of these structural and thermodynamic characteristics are distinct patterns of weather.

Typically the subtropical highs are associated with fine weather with little precipitation which, if present, occurs in the western half of the cell. Here air masses are more unstable and humid than in the eastern half, where very stable air masses prevail almost continuously. The eastern stable air masses produce characteristically cloud-free conditions, or clouds whose vertical development is inhibited, due to a general regime of subsidence. Typical clouds are stratocumulus which can form extensive cloud decks on the eastern sides of anticyclones. Above the stratocumulus decks, the atmosphere is usually clear due to its dryness as a result of adiabatic warming of subsiding air. This dryness results in the evaporation of any condensed atmospheric water.

Air at lower altitudes within an anticyclonic system, in the vicinity of the inversion (next section), has lower potential temperatures than higher altitude air. This suggests that air immediately above the inversion is cooling diabatically, which contrasts with the air at higher altitudes which possesses higher potential temperatures, as a result of dry adiabatic warming. This apparent dilemma can be explained by considering the net radiation balance of air at lower altitudes in anticyclones (McIlveen, 1992). Cooling of air in the vicinity of the inversion is possible because diabatic warming from solar radiation is exceeded by net longwave radiation loss.

The origin of the subtropical highs is rather complex and a number of dynamic and thermal theories have been put forward to explain their development and maintenance. The classical theories explained the high pressure cells simply as the result of the piling up or convergence of air at about 20° in the poleward-flowing "antitrades", the upper atmosphere equivalent of the surface trades. Convergence at high levels would result in downward movement of air and high pressure near the earth's surface. The weak point in this dynamic explanation is the frequent absence of the anti-trades which makes it impossible to attribute the origin of the persistent subtropical highs to this one feature. This theory also violates the hydrostatic approximation (McIlveen, 1992). Another dynamic theory assumes that polar air masses are the main cause of the subtropical highs. Because of the changes of the Coriolis force with latitude, anti-cylonic cells near the polar front show a tendency to move equatorwards, while cyclones generally move polewards (Rossby, 1947). Moving cold anticyclones, by providing outbreaks of cold air, therefore regenerate the subtropical highs frequently. The polar outbreaks would generally show a preference for the eastern parts of the ocean basins, where cold ocean currents prevail, and where friction along the continental coasts, especially in the southern hemisphere, give a strong meridional impetus to these movements. This idea has been suggested as applicable to the explanation of the pulses in the intensity of the subtropical highs on a daily time scale. The interaction between the outbreaks of cold polar air and surface ocean currents is also believed to maintain the higher pressure over the eastern oceans at low levels and the generally stronger development of these high pressure systems over the southern oceans.

Thermal theories consider two different areas of origin: cooling in upper air and cooling at the surface. Upper air cooling takes place as air in the upper poleward-flowing branches of the Hadley cells lose heat by longwave radiation loss to space. These air masses therefore become denser with growing distance from the equator, resulting in higher pressure at all levels. A second area of cooling is found near the earth's surface: the cold ocean currents and the cool continents during the winter. This second source explains the cellular pattern of the subtropical highs and their extension over the continents.

As both dynamic and thermal theories can account for the high central pressures of the subtropical highs, it seems most likely that the subtropical highs have a multiple origin. As noted by McIlveen (1992), the most likely dynamic mechanism that explains the existence of excess atmospheric mass is the Coriolis effect as this imposes dynamic constraints on the flow of air in anticyclonic systems. Therefore, once upper level convergence has been established, which in itself provides the opportunity for mass convergence, the Coriolis effect retards the outflow of air at the surface, leading to atmospheric mass buildup in the anticyclonic centres. Further to this mechanism are also the thermal effects of subsiding air (McIlveen, 1992). Subsidence produces warm air, which in turn suppresses the vertical pressure gradient so that air pressure falls more slowly with increasing altitude. Isobaric surfaces are therefore higher than in surrounding cooler air. This doming of the isobaric surfaces increases with height and can extend into the upper troposphere which produces deep and warm anticyclonic systems (McIlveen, 1992).

The climatological effects of the meteorological processes operative in anticyclone systems are low cloud amounts and thus low rainfall. Many subtropical areas are therefore characterized by dry climates. The aridity of subtropical coastal areas such as western Australia, western North America and the deserts of south-western Africa (Kalahari) and South America (Atacama) is a product of the dominance of anticyclones with their characteristically stable air masses. The Sahara Desert is also a product of anticyclonic dominance as during the winter, as the desert area cools, the North Atlantic anticyclone extends over the northern parts of the African continent maintaining subsidence and thus cloud-free conditions. During the summer, because of intense surface heating a heat low forms over the Sahara. However, this is still capped by anticyclonically subsiding air which prevents the northward penetration, into the central Saharan region, of air flowing from the south off the Gulf of Guinea during the African summer monsoon (Chapter 7).

THE TRADE WINDS

Between the subtropical highs and the ITCZ the low level circulation over the oceans in both hemispheres is dominated by strong and persistent easterly winds. These winds are known as the *trade winds*. The name is derived from Old English in which "trade" means "path" because of the regular course the winds take. Because of their importance for sailing, other languages also have specific words for the trades. The trades are known as *Alisios, Alizees, Alisei* and *Passat* in Spanish, French, Italian and German respectively. In the western Pacific the Melanesian word is *Lahara* while the Australian aboriginal word is *Dir'mala*. Over the continents the trades tend to be dry and in the Saharan area are referred to as the Harmattan, while in eastern Asia, they are referred to as the winter monsoon.

The trade winds are globally important as they cover almost half of the globe's surface, extending over 20° of latitude in the summer hemisphere and approximately 30° in the winter hemisphere. In the northern hemisphere their general direction is east-north-east while in the southern hemisphere it is east-south-east. Their mean velocities are in the range of 3.6–7.2 ms^{-1}, with maximum velocities attained in the winter. A characteristic of the trades is steadiness. Over the oceans the winds have their greatest regularity, this being 60–70% on average. Although characterized by steadiness, the trade winds demonstrate considerable inter-annual variability in their strength, especially in El Nino years when they weaken considerably (Chapter 6). However, analyses of trends in the trade winds for the Atlantic and Pacific basins (Bigg, 1993; Inoue and Bigg, 1995) have shown that there has been a strengthening of these. This appears to be consistent with an increase in the intensity of the ocean anticyclonic systems from which the trade winds emanate.

The trade winds originate from the subtropical anticyclonic systems on the eastern sides of the Atlantic, Indian and the Pacific ocean basins. From the eastern sides of the anticyclonic systems, the trade winds flow in the direction of the equatorial trough where they meet their counterparts from the opposite hemisphere. The trade flows have their greatest regularity towards the middle of their course where steadiness values can rise to 90–95%. During their journey to the equatorial trough region the trades undergo thermodynamic changes. At their source the trades are divergent but, as they approach the equator, they become convergent as the pressure gradient down which they flow slackens and the flow decelerates. Convergence in the trade flows leads to the trade wind air piling up. Convergence, in conjunction with confluence, as a result of the trades from the two hemispheres meeting, produces the strong ascending motion characteristic of the whole ITCZ area. Because the trades flow over vast expanses of ocean they are also great transporters of heat and moisture.

The trades have a distinct three-layer structure, the heights of which increase in the direction of the equator (Figure 5.9(b)). From the surface upwards the three layers are a subcloud layer, a cloud layer and an inversion layer. The lower two layers are often referred to as the lower trades.

The *inversion layer* is a layer of the atmosphere characterized by its increase in temperature with height which is opposite to what would normally be expected. This is perhaps the most distinctive feature of the trade wind system because of its spatial extent and variation in height and intensity.

Four possible mechanisms exist for the generation of inversions. These are turbulence, radiation/heat budget, advection and subsidence (Hastenrath, 1985). Of these, subsidence is considered the most likely mechanism responsible for the origin and maintenance of the trade wind inversion. This is because the spatial and temporal behaviour of the inversion matches well the geographical characteristics of divergence and subsidence associated with the large oceanic subtropical anticyclonic systems. As air subsides it is subject to adiabatic warming. Warm air therefore builds up in the lower parts of the atmospheric column. This can be seen easily in the two atmospheric profiles in Figure 5.9 where the normally expected slope of the temperature profile to the left in the direction of decreasing temperatures is disrupted by a bulge to the right in the direction of warm temperatures. Above the surface, the subsiding air meets air which is being carried away from the earth's surface in convective currents. This air cools adiabatically as it ascends so that air temperature decreases away from the earth's surface. The point at which the ascending (cooling) air meets the subsiding (warming) air is the *inversion base*. This is the point where the temperature profile and lapse rates reverse (Figure 5.9). The height of the inversion base is very much dependent on the relative strengths of subsidence and convection.

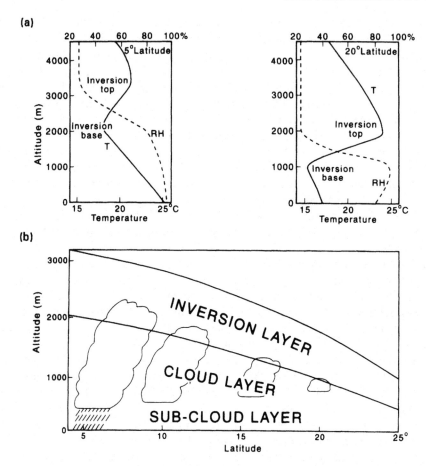

Figure 5.9   (a) Vertical temperature and humidity profiles at 5° and 20° latitude. Note the latitudinal dependence of inversion base, intensity and top. (b) A generalized meridional cross-section showing a three-layer trade wind structure

Inversion intensity is the rate at which temperature increases from the bottom to the top of the inversion. The *inversion top* is that point where temperature begins to decrease again. The height and intensity of the trade wind inversion are closely tied to latitude (Figure 5.9) and distance from the semi-permanent anticyclonic centres. This dependence is related to the changing interplay between the thermodynamic properties of the atmosphere and thermal properties of the ocean as the trade flows move from east to west across the ocean basins.

As the trade flows move equatorward over the ocean, evaporation into their lower layers occurs with the result that atmospheric moisture levels build up. However, because the ocean waters are cooler than the atmosphere in the trade's source region, there is a loss of heat from the atmosphere to the cool ocean surface (Figure 5.10). This process, along with longwave radiation loss to space, results in air temperatures approaching those of the ocean. Consequently, stratus clouds develop. This style of cloud development occurs because moisture convection is suppressed by cool SST,

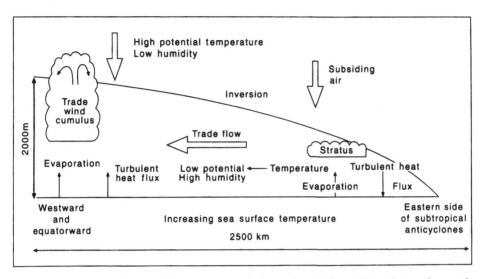

Figure 5.10   Transformation of air mass properties in trade wind flows. Note that as the west is approached the boundary layer increases in thickness, as does the height of cumulus cloud formation; a result of increasing sea surface temperatures. From Wells (1986). Reproduced by permission of Taylor and Francis, London

subsidence in the eastern regions of the subtropical anticyclones and strong stability as manifest by a well-developed trade inversion. Because of these ocean–atmosphere interactions the eastern sides of the large ocean basins, especially the Atlantic and Pacific, are typified by extensive layers of low stratus. As the trade wind air moves towards the equator it flows over progressively warmer ocean waters. This results in a reversal of the ocean atmosphere temperature gradient, with the oceans now being warmer than the overlying cooler air. As a consequence there is a flow of heat from the ocean surface into the lower trades which produces weak convection, the breakup of stratiform cloud, and the development of convective-type cloud called trade wind cumulus (Figure 5.10). Because of the development of convective currents, moist air at the surface is mixed with the overlying dry air in the inversion layer which results in the development of a deeper boundary layer and a higher inversion base (Wells, 1986).

Convective activity and thus convective rainfall amounts increase westward across the ocean basins due to the growing energy supply from the westwardly increasing SST. Consequently, in the western parts of the ocean basins, the boundary layers grow to depths greater than 2000–3000 m which is approximately four to six times the depth of the boundary layer and the inversion base in the eastern parts of the ocean basins (Figure 5.9). For example, on the eastern side of the Atlantic basin, the inversion base can have heights less than 500 m but may rise to more than 1500 m over the western parts of the Atlantic and up to 2000 m in the equatorial zone (Hastenrath, 1985). Matched with these are trends in the inversion intensity. These are greatest on the eastern sides of the ocean basins where in the North Atlantic in the region of the Canary Islands and the South Atlantic between St Helena and the West African coast, the inversion intensities can be 5 °C and 8 °C respectively. In these same regions the decrease of relative humidity from the bottom to the top of the inversions is also the

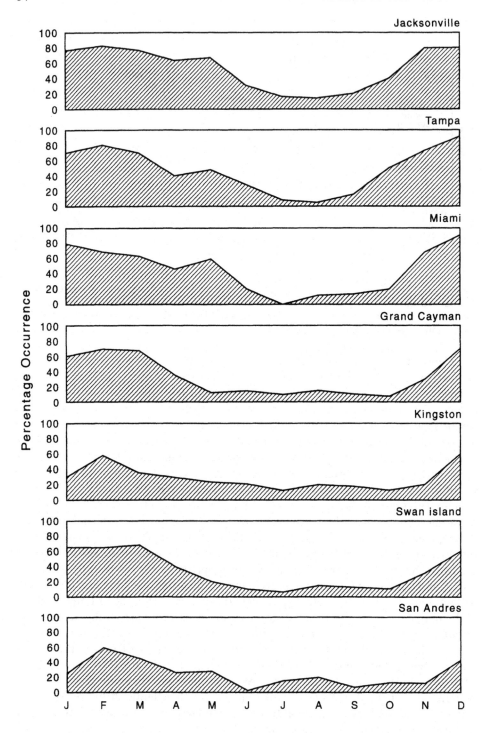

largest for the Atlantic Ocean area. For the Pacific Basin, similar trends to those of the Atlantic can be found. Off the west coast of the USA, in the region of California, the inversion base is around 400 m. This rises to around 2000 m in the vicinity of the Hawaiian Islands. The temperature jump across the inversion varies from more than 10 °C at the Californian coast to about 2 °C in the region of the Hawaiian Islands. For the South Pacific, similar values are found between the west coast of South America and Tahiti. In the Indian Ocean region in winter, the inversion has been noted to be at a height of 1000 m at Durban on the east coast of South Africa rising to over 2500 m over Malagasy (Hastenrath, 1985).

Inversions are most frequent and intense during the winter months; minimum inversion occurrence is in the summer when surface heating and convective activity is strong (Figure 5.11). The intensity and shallow height of the trade inversion in some parts of the tropics exacerbates air quality problems as inversions effectively cap the lower layers of the atmosphere and inhibit the vertical diffusion of pollutants. Persistent inversions can lead to the build-up of pollutants which can have implications for human health especially in large subtropical cities such as Los Angeles, Johannesburg, Mexico City and São Paulo.

In the layers below the inversion, maximum wind speeds are found at approximately the 900 mb level. Minimum windspeeds occur at the bottom of the subcloud layer due to surface frictional effects. The subcloud layer represents that part of the atmosphere where cloud formation processes do not occur due to, on average, the dew point temperature not being reached. In the subcloud layer directional steadiness and potential temperature show little variation with height but these increase in the direction of the equator. Specific humidity decreases with height in the subcloud layer but increases in the direction of the equator. In the east of the ocean basins the cloud layer is dominated by stratus type clouds while in the west convective clouds predominate. Associated with equatorward and westward increasing convective activity are an increase in the thickness and height of the trade wind cumuli in the cloud layer (Figure 5.10). An interesting feature of the trade wind cumuli is their vertical structure. Viewed in profile their orientation is not strictly vertical as they lean backwards slightly. This is because the lower trade wind-speed maximum at the base of the cloud layer tends to push the bottom of the cumuli forward in the direction of the equator, leaving the cloud tops trailing behind.

UPPER LEVEL CIRCULATION IN THE LOW LATITUDE ATMOSPHERE

At the equator the rising limb of the Hadley circulation delivers air to the upper atmosphere which turns to flow poleward. In classical theories of tropical circulation the term anti-trades was used to describe this upper tropospheric outflow of air from

Figure 5.11 Frequency of lower tropospheric inversions in 1980 along 80 °W from southern Florida to Panama. Percentages are the frequencies of upper air atmospheric profiles by radiosonde. From Hastenrath (1985). Reproduced with kind permission from Kluwer Academic Publishers

the equatorial regions. The name anti-trades is an unfortunate one as although upper poleward flows of air do occur, they are not equal in strength to the surface trade flows. Instead, meridional components are weak in comparison to zonal components. This is because as upper level air moves away from the equator it increasingly comes under the influence of the Coriolis force (Chapter 2). As a result, the upper circulation in the low latitudes becomes increasingly westerly at distance from the equator. These westerly flows of air are referred to as the subtropical jet. This straddles the subtropical latitudes throughout the year in the southern hemisphere and for the majority of the year in the northern hemisphere, apart from the summer, when it is replaced by the easterly tropical jet over the southern Asian Indian Ocean area. Both of these winds systems are called jets because they are regions in which winds of high velocity are concentrated. This property may be explained by reference to the law describing the conservation of momentum (Chapter 2).

THE SUBTROPICAL JET

Annually the subtropical jet is found near the poleward boundary of the Hadley cell at around 30° from the equator in both hemispheres (Figure 5.1). This is a region of persistent westerly winds which encircle the globe, occasionally reaching velocities of around $100 \, \text{m s}^{-1}$. Their path describes a triangular wave-like pattern when viewed from either of the poles. The equatorward corners of this triangle, occurring at about 20°W, 150°W and 90°E, are wave troughs. Poleward wave crests occur on average around 70°W, 40°E and 150°E. It is in the wave crests that maximum wind speeds are found (Palmen and Newton, 1969).

The upper-level subtropical westerlies are best developed in the winter season when the Hadley circulation reaches its maximum intensity and equator–pole temperature gradients are at a maximum. At this time of year there is maximum export of absolute angular momentum from the equatorial region which is important for maintaining the westerly flows; velocities may reach $135 \, \text{m s}^{-1}$ (Chapter 2). In association with seasonal changes of strength, the subtropical jet also displays changes in its mean position. This is related to the seasonal variations in the intensity and geographical extent of the Hadley circulation in either hemisphere. In the summer months, when the flows in both the Hadley circulation and the subtropical westerlies have weakened, the mean position is further equatorward than its winter position. Of the two hemispheres, the southern hemisphere subtropical jet shows the most stable position. This is because the Hadley circulation over the predominantly ocean-covered southern hemisphere is more conservative in its geographical behaviour than its northern counterpart.

The mean position of the jet and the year to year variability of the westerly wave crests and troughs are important for determining precipitation patterns in the low latitudes. Generally the three poleward wave crests are located above regions of enhanced equatorial convective activity and rainfall in the northern winter. Riehl (1954, p. 275) has reported on the importance of jet axis position for determining the temporal patterns of rainfall in Hawaii. Kodama (1992, 1993) notes that the SPCZ and the South Atlantic convergence zone and their associated precipitation zones form along the eastern side of troughs in the subtropical jet. In both northern and southern hemisphere cases it seems that maximum convective activity and precipitation occur in a downstream region on the poleward side of the jet axes often referred to as the delta region. Here deceleration and diffluence of air occurs. Removal of air in the delta region pulls air from the surface, which in turn creates uplift as upper level divergence is matched with lower level convergence (Chapter 2). If the rising air is moist, condensation and consequently rainfall production ensues.

THE TROPICAL EASTERLY JET

During the Asian monsoon (Chapter 7) a strong easterly flow of air develops in the upper atmosphere. This is centred on approximately 15°N, 50–80°E and extends from South-East Asia to Africa (Figure 5.12). This upper level easterly wind, which lasts from late June to early September, is referred to as the tropical easterly jet. It is best developed at around 15 km above the earth's surface, with speeds in the core of the jet reaching up to 40 m s^{-1} over the Indian Ocean. The fact that the tropical easterly Jet only occurs in the summer suggests that its development is related to the seasonal cycle of surface heating in the area over which the jet lies.

During the northern summer, surface heating reaches a maximum in the subtropical latitudes of the African and southern Asian continents. Surface heating of these extensive land masses produces extremely well-developed heat lows, especially in the Indian subcontinental area. These encourage ascent of air to great heights in the atmosphere. This low level supply of air to the middle and upper atmosphere is matched by upper level outflow from an area of high pressure. Outflow occurs to preserve vertical mass balance (Hastenrath, 1985). The mass convergence in the area of the upper level anticyclone is so intense that pressure surfaces in this area bulge upwards creating atmospheric thickness differences between the subtropics and the equatorial and mid-latitude regions to the south and north respectively. Related to this, and important in terms of the development of the tropical easterly jet, is the reversal of the normal equator to subtropical temperature gradient; throughout the troposphere warmer upper tropospheric temperatures occur in the subtropics. This is because of the large volumes of warm air rising from the surface in the area of the Indian heat low. Just as air moves parallel to pressure gradients in the upper atmosphere where there are no friction effects, flow in response to atmospheric temperature and thickness differences will also be parallel to thickness and temperature gradients. In such a case, flow will occur with low (high) temperatures and decreasing (increasing) thickness on the left (right) hand side in the northern (southern) hemisphere. Wind generated due to thermal and thickness differences is called a *thermal wind*. This is exactly what the tropical easterly jet is, as it is formed because of the geographical contrasts in heating between the warm subtropical Asian land masses and the relatively cool oceanic equatorial latitudes.

The tropical easterly jet also occurs in the upper atmosphere, south of an area of upper anticyclonic outflow, above the mid-summer Saharan heat low. Velocities in this part of the jet are lower (25 m s^{-1}) than over the Indian Ocean. Convergence within the jet over Africa creates subsidence which possibly plays a role in moderating the advance of the south-west monsoon over West Africa. At the end of the northern hemisphere summer, when the area of maximum surface heating moves southward, the tropical easterly jet over both India and Africa disappears.

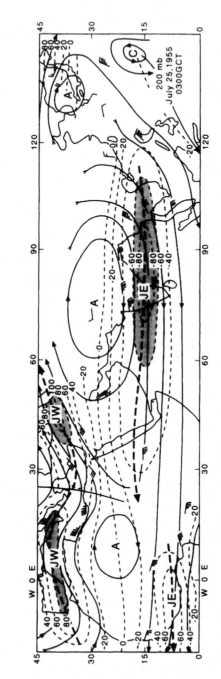

Figure 5.12 The tropical easterly jet. Solid lines are air flows (streamlines). Dashed lines are wind speeds (isotachs) in knots (1 knot = 0.51 m s⁻¹). Jet core is shaded. JE is jet axis; A is anticyclone; C is cyclone; JW is subtropical westerly jet. From Koteswaram (1958)

As for the case of the subtropical westerly jet, the tropical easterly jet is important for determining rainfall patterns in southern Asia and Africa. At the surface, rainfall has been found to be greatest north of the jet entrance in the southern Asian region and south of the jet exit in the West African region (Hastenrath, 1985).

WEST AFRICAN MID-TROPOSPHERIC JET

Due to mid-tropospheric temperature gradients between the warm Sahara Desert area and the cool Gulf of Guinea in southern West Africa, a thermal wind referred to as the West African mid-tropospheric jet or African easterly jet exists over West Africa. This is centred on 15°N at about the 600 mb level (approximately 4500 m). This easterly wind reaches its maximum development at the height of the northern hemisphere summer when mid-tropospheric temperature gradients are greatest; velocities in the jet core may exceed $10 \, \mathrm{m \, s}^{-1}$ in the period April to November. On the equatorward side of the jet maximum rainfall is found. This region coincides with a region of maximum cyclonic shear which may also be the birthplace of easterly waves (Chapter 8).

EAST AFRICAN LOW LEVEL JET

In the mid-1960s an area of low-level high-velocity winds in the vicinity of the East African coast was "discovered". These winds subsequently became known as the East African low level jet, although other names such as the Findlater jet, the Somali jet and the low level jet over the western Indian Ocean, have been used at times.

The core of the jet is best developed at a height of around 1500 m (850 mb) where velocities reach 25–$50 \, \mathrm{m \, s}^{-1}$ on occasions but average at 12–$15 \, \mathrm{m \, s}^{-1}$. Its mean position extends from east of the north-eastern tip of Madagascar to the African mainland in the vicinity of Mombasa and Garissa, at which point it recurves in a north-easterly direction to extend across the Arabian Sea in the direction of western India (Findlater, 1977).

Although the East African jet exists in all months, its maximum development and geographical extent is related to the onset of the African–Asian monsoon. The jet is therefore an integral part of the northern summer monsoon circulation in the African–Indian area. During the winter the jet is confined to areas south of the equator, mainly to the east of Madagascar. By May, at the beginning of the African and Indian monsoons, a narrow zone of high speed winds and rain starts to push across the equator in the region of the Kenyan coast. From May to September, as the African and Indian monsoons develop, so does the geographical extent of the jet (Figure 5.13). In addition to covering large tracts of the western Indian Ocean and the Arabian Sea, branches of the jet penetrate inland over eastern Africa through topographic breaks in the East African Plateau. This barrier also appears to be important in the development and steering of the jet with the jet core lying approximately 150 km east of this major plateau (Findlater, 1974).

Associated with the jet are important meteorological features. Maximum coolness, moisture and cloudiness coincide with the jet core and its eastern regions over the coast of eastern Africa where maximum ascent of air occurs. Minimum cloudiness occurs above the jet core and to the west in the direction of the footslopes of the East

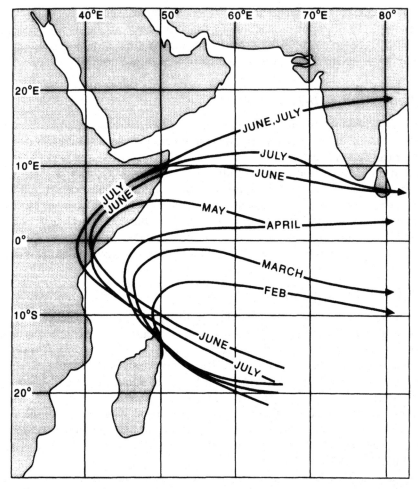

Figure 5.13   Monthly progression of East African Low Level Jet core. Note the bifurcation of jet core over India in June and July. Source; Findlater (1972)

African Plateau where the air is descending. Here the warmest, driest and most stable air is found (Findlater, 1972).

In the northern hemisphere summer, winds in the jet core reach their greatest constancy of around 95% with minor diurnal fluctuations in speed. Maximum daily speeds have been found in the early morning with these waning in the afternoon. As well as diurnal fluctuations in speed, the jet also displays pulses on a monthly time scale which have been found to be positively correlated with rainfall on the coast of western India (Findlater, 1977). Such a relationship exists as when the cross-equatorial flow intensifies, so does the cross-equatorial transport of water vapour.

OCEAN GENERAL CIRCULATION

To this point the general circulation of the atmosphere has been considered as a control on climate. However, an important additional factor in determining tropical climates is ocean circulation, especially its impact on ocean surface temperature patterns and coastal climates (Chapter 4).

The oceanic equivalent of the major streams of air in the atmosphere are ocean currents. The basic surface pattern of these matches that of the wind systems in the tropics. This is because the kinetic energy of motion in the atmosphere is transferred, via frictional drag, to the ocean surface. Such *momentum transfer* represents the basic driving force of the ocean currents. Secondary factors that control surface ocean current patterns are temperature and salinity contrasts.

As winds are the basic driving force of ocean currents, in the absence of land masses, the ocean currents would simply circle the globe. In the equatorial regions the current direction would be from east to west in response to the trade winds, while in the mid-latitudes, current flow would be from west to east under the influence of the westerlies on the northern and southern sides of the northern and southern hemisphere subtropical anticyclones respectively. However, because of the presence of the continental land masses, ocean currents are deflected with the result that global oceanic circulation is characterized by a series of circular ocean currents called *gyres*. The largest of these are found in the subtropics and are characterized by their anticyclonic circulations; clockwise (anticlockwise) in the northern (southern hemisphere). They are formed by water being pushed from the eastern sides of the great ocean basins to the west by the trade winds.

The westward travelling currents are deflected in a poleward direction by continental land masses on the western sides of the ocean basins. As they travel poleward, the Coriolis force begins to deflect the currents to the east. This transport of water away from its basic path of flow due to the Coriolis force is referred to as Ekman drift. At the same time the influence of the strong westerly flows in the mid-latitudes aids in the return of the currents to the eastern side of the ocean basins thus completing the circular motion. Given these factors, the basic model for ocean circulation in the region of the subtropical gyres is a westward current in the equatorial latitudes, eastward currents in the outer subtropical latitudes, and a cool equatorward (warm poleward) current along the eastern (western boundary) of the ocean basins. In between the large subtropical gyres and the westward-flowing equatorial currents, small equatorial gyres exist. These are formed by the eastward return of water following its transport to the west by the trade winds.

This model applies to both the Pacific and Atlantic basins throughout the year as there are no major seasonal wind direction changes, in contrast to the Indian Ocean basin. For the Pacific and Atlantic the only ocean current changes that occur are that of position and strength, and three major low latitude currents can be recognized (Figures 5.14 and 5.15). These are the "north equatorial current" (westward) between approximately 10°N and 20°N and its southern hemisphere counterpart, the "south equatorial current" (westward) spanning 5° either side of the equator; these are the zonal equatorial branches of the subtropical anticyclonic gyres. Between these is the eastward-flowing "north equatorial countercurrent" between approximately 5°N and 10°N.

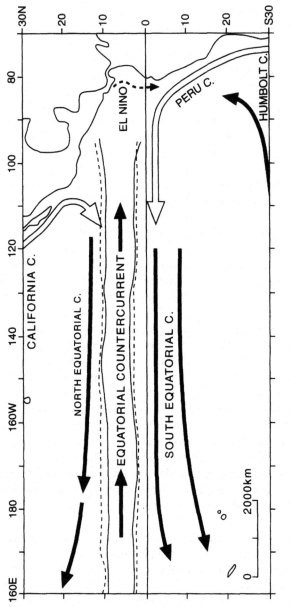

Figure 5.14   Ocean general circulation in the Pacific Basin

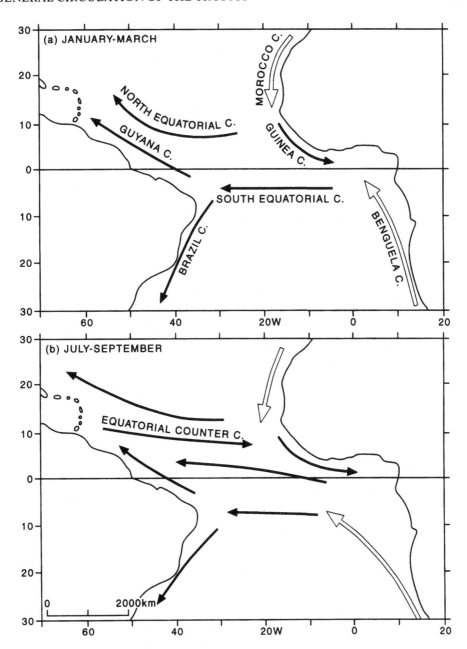

Figure 5.15   Ocean general circulation in the Atlantic Basin

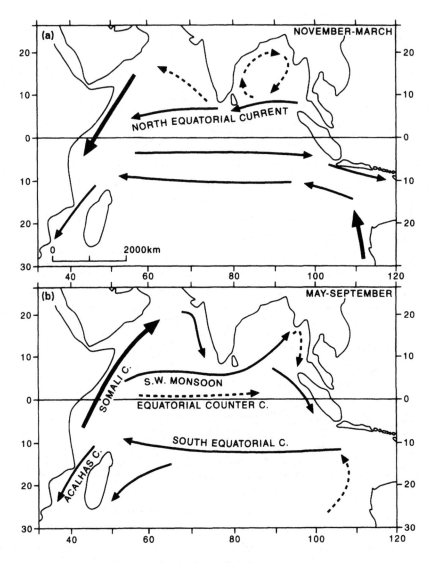

Figure 5.16   Ocean general circulation in the Indian Ocean

The situation over the Indian Ocean changes with the change of wind direction over the north Indian Ocean from the north-east in the northern winter to the south-west in the northern summer; a product of the Indian monsoon (Chapter 7). Changes in ocean current patterns are therefore most pronounced in the equatorial and north Indian Ocean regions.

For the southern Indian Ocean the current pattern is analogous to that of the southern Atlantic and Pacific oceans; an anticyclonic gyre prevails (Hastenrath, 1985). In the east of the southern Indian Ocean the cool equatorward-flowing west Australian current recurves near the equator to become the south equatorial current. This flows westward until it reaches the African coast whereupon it turns south and flows as the warm Agulhas current between Madagascar and the African continent. In the equatorial Indian Ocean regions, the northern winter current pattern is similar to that of the Pacific and Atlantic basins; an equatorial countercurrent is enclosed by southern and northern equatorial currents (Figure 5.16(a)). The only difference is that this current lies further south than its Atlantic and Pacific Ocean counterparts; it is found between 3°N and 8°S (Hastenrath, 1985). Off the East African coast a strong south-ward-flowing current exists.

With the onset of the summer monsoon over the Indian subcontinent the pattern, as described above for the northern winter, changes. As the strong trade flows from the southern hemisphere push across the equator into the northern hemisphere, the equatorial countercurrent between 3°N and 8°S is destroyed. The south equatorial current continues to flow westward but, while maintaining a weakened Agulhas current, develops a strong northward-flowing branch along the East African coast, thus reversing the winter current pattern in the west Indian Ocean (Figure 5.16(b)). This is the Somali current which pushes warm equatorial water into the Arabian Sea thus raising sea surface temperatures along the East African coast. The Somali current recurves in the Arabian Sea to flow eastward. This northern branch of a very much compressed north Indian Ocean anticyclonic gyre, diverges in the region of Sri Lanka, with branches flowing northward into the Bay of Bengal and southward along the Indonesian archipelago (Figure 5.16(b)).

The main climatic impacts of the ocean currents described are as follows: cool equatorward currents induce atmospheric stability and therefore reduce the potential for rainfall production, stratus type cloud cover is usually extensive and coastal fogs are common, onshore climates are cool and dry; warm currents or sea surface tempera-tures increase atmospheric instability and therefore rainfall potential, cloud cover is mostly of the convective type and onshore climates are warm and moist. Because cool equatorward flowing currents tend to dominate the eastern hemispheres of the ocean basins, especially in the case of the Pacific and Atlantic basins, the low latitude climates of eastern oceans tend to be cooler and drier than their western counterparts. In the Pacific Basin, however, this "normal" climate pattern can be reversed during El Nino events when the climatological pattern of wind and ocean currents is reversed. This phenomena will be discussed in the next chapter.

SUMMARY

The low latitudes are dominated by the meridional circulation of the Hadley cells in both the northern and southern hemispheres. These are characterized by ascending branches of air on their equatorward sides and descending branches on their poleward sides. Return flow of air to the equator at the surface is in the form of the trade winds which emanate from the subtropical highs centred on approximately 30°N and 30°S. The trade winds from both the northern and southern hemisphere meet to form a zonal band of cloudiness which varies regionally in its structure and dynamics depend-ing on regional atmospheric circulation, surface type and ocean current patterns. This zone is called the intertropical convergence zone (ITCZ). Its seasonal movements are largely controlled by the seasonal solar cycle. Although often confused with the thermal equator, the ITCZ is usually displaced equatorward of the zone of surface low pressure, maximum temperatures and clear skies, which is sometimes referred to as the equatorial trough.

As well as the main surface circulation features, a number of upper level features are of note in the tropics, namely the various jet streams. On the poleward side of the upper atmosphere return branch of the Hadley cell in both hemispheres is the subtropical jet. This is a perennial feature composed of high-velocity westerly winds which on average encircle the earth. A feature of the upper circulation in the northern hemisphere summer months is the appearance of easterly jets over the Indian Ocean and the northern half of the African continent. These owe their origin to the reversal of upper atmosphere temperature gradients with the subtropical latitudes becoming warmer than lower latitudes as a result of land and sea heating contrasts. As these jets are produced by thermal contrasts in the atmosphere they are referred to as thermal winds.

Closely associated with the pattern of atmospheric general circulation at the surface is the ocean circulation pattern. This in general is dominated by large anticyclonic gyres which have east to west flows along the equator and equatorward flows in the east of the Atlantic and Pacific basins. In the Indian Ocean, however, because there is a seasonal reversal in the atmospheric circulation associated with the onset of the monsoons, ocean circulation patterns vary between the seasons.

CHAPTER 6

Non-seasonal Variations of the Tropical Circulation

The description of the tropical circulation presented in the last chapter is for mean annual conditions. Imposed on this are a number of important non-seasonal and seasonal variations which also vary in terms of their spatial scales. These variations constitute an important part of the tropical circulation and will be discussed in this and the following chapter. Non-seasonal variations discussed in this chapter include the quasi-biennial oscillation, the 40–50 day tropical oscillation, the El Nino southern oscillation phenomenon and diurnal variations.

THE QUASI-BIENNIAL OSCILLATION

In the early 1960s meteorological research revealed that stratospheric winds over the equatorial regions showed marked variability in their direction from year to year. In one year stratospheric winds would be dominated by strong easterlies, while in the following year, strong westerlies would dominate. This oscillation of stratospheric winds between easterly and westerly modes became known as the *quasi-biennial oscillation* (QBO). The word "quasi" was used as the time between the extremes of the easterly and westerly modes was not a perfect 24 months (27 months). Nevertheless, the return of stratospheric winds to a westerly mode from an easterly mode and vice versa on a regular basis has been established as a distinct feature of equatorial stratospheric winds.

Regular observations of the equatorial atmosphere have allowed climatologists to build a clear picture of the characteristics of the QBO. These observations have also facilitated an insight into the mechanisms which are thought to explain the QBO phenomenon (Piexoto and Oort, 1992). Perhaps the best way to understand some of the QBO's characteristics is by studying a time series of zonal winds (east–west) at 30 mb (Figure 6.1).

From Figure 6.1 the following QBO characteristics are clear. The easterlies are

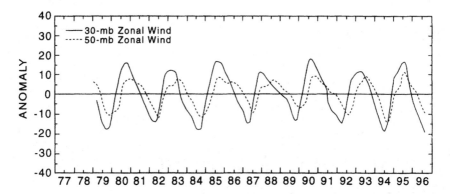

Figure 6.1   Smoothed zonal wind anomalies for equatorial winds at 30 mb and 50 mb. Positive values are westerly, negative values are easterly. Notice how the higher level (30 mb) winds become easterly or westerly before the lower level (50 mb) winds. This indicates that the winds propagate down through the atmosphere. Modified from *Climate Diagnostics Bulletin*, CPC (1996)

much stronger than the westerlies; the time between the easterly and westerly maximum is much shorter than the reverse; the difference between the peaks and troughs (amplitude) of the QBO lies between 40 and 50 m s^{-1}; the periodicity on average is just over two years, and there is considerable variability in both the amplitude and the periodicity of the QBO. A further characteristic is that the easterlies and westerlies propagate downwards through the atmosphere with time. The average rate at which the winds move down is approximately 1 km per month but, of the two winds, it is the westerlies that move down faster. For most levels, the transition from westerlies to easterlies is smooth with time. However, for the region between 30 and 50 mb, the westerlies may persist for several months, so that the transition to easterlies is often delayed. Above the 50 mb level the amplitude of the QBO does not change a great deal, but below this level, it decreases rapidly. The maximum variation between easterly and westerly modes is at 20 mb and the velocity of both easterly and westerly winds decreases with height. A further height-dependent characteristic is the duration of the zonal winds, such that at lower levels the westerlies last longer than the easterlies, while at higher levels the opposite is true.

For some low latitude locations, climatic variables such as rainfall and temperature demonstrate temporal variations that resemble those of the QBO (Ogallo, 1979; Shapiro, 1982; Hastenrath and Rosen, 1983; Chu, 1984). For this reason climatologists have been interested in the utility of the QBO combined with other large-scale oscillations such as the southern oscillation and the 40–50 day tropical oscillation for long-range climate forecasting (Jury, McQueen and Levey, 1994; Jury et al., 1995). Despite considerable research, the role of the QBO in tropical climate is not yet fully understood and considerable ground remains to be made, especially in assessing the role of the QBO in modulating other components of the tropical circulation such as the southern oscillation (Gray, Schaeffer and Knaff 1992), the monsoons (Knaff and Gray, 1994) and tropical cyclone frequency (Gray and Schaeffer, 1991).

THE 40–50 DAY TROPICAL OSCILLATION

In 1971, Madden and Julian (1971) presented the results from an analysis of surface and upper atmospheric data from over the equatorial Pacific basin. This analysis revealed that there was a low-frequency variation in the strength of upper atmospheric winds, in the temperature at a variety of levels and surface pressure. The periodicity of these variations was found to be at a maximum between 41 and 53 days, with the most frequent occurrence near 45 days. This variation became known as the *40–50 day tropical oscillation*, although the oscillation does cover a wide range of periods between 30 and 60 days.

The 40–50 day oscillation possesses a number of spatial and temporal characteristics which help explain some aspects of the low-frequency variability of the tropical circulation and thus climatic variability. Of these characteristics, the eastward movement of the 40–50 day oscillation is of greatest interest. This movement is in the form of an atmospheric wave, most likely associated with large convection cells, which move at speeds between 10 and 39 m s^{-1} from the Indian Ocean to the western Pacific and across the Pacific to South America. The surface effects of such eastwardly propagating convection cells are clear at equatorial locations such as Canton Island (2.8°S, 171.7°W), where over a 40–50 day period, the zonal winds rise and fall in strength along with temperature and surface pressure (Figure 6.2).

Recent studies based on analyses of the patterns of tropical outgoing longwave radiation (OLR), a proxy measure of convection, have shown clearly that not only is the Indian Ocean an important source area from which anomalies in tropical convection emanate, as originally noted by Madden and Julian (1971, 1972), but so are the North Pacific and the North and South Atlantic Ocean regions. Furthermore, the Amazon Basin, the Congo Basin and the central Pacific Ocean appear to be important sink areas for tropical convection anomalies (Anyamba and Weare, 1995).

As well as its notable spatial characteristics, the 40–50 day oscillation also possesses some important temporal characteristics, as distinct intra-seasonal and inter-annual variations in the nature of the oscillation exist. Intra-seasonal variations include those in the 30–50 day, 10–20 day and biweekly time scales (Ding, 1994). Of these, it is the 30–50 day oscillation that is most important because this has significant impacts on the break and active phases of the monsoons (Madden and Julian, 1994). The inter-annual oscillations include the seasonal–annual, the QBO and the southern oscillation.

At the seasonal time scale the oscillation shows no large change in its average period; however, it does show some changes in strength. The oscillation is the largest during December to February and the smallest during June to August. It is the smallest over western Pacific stations and largest at stations in the Indian Ocean. These seasonal variations are considered to be a product of the seasonal migration of convective activity associated with movement of the ITCZ (Madden and Julian, 1994). Although seasonal changes of the average period of the oscillation are small, distinct inter-annual variations in the period do occur (Gray, 1988). For example, the most frequently occurring period in the years 1980–85 was 26 days. This contrasts with the normal 30–50 day average. Analysis of the oscillation's periodicity for warm El Nino southern oscillation (ENSO) and cold La Nina years (see next section) has also

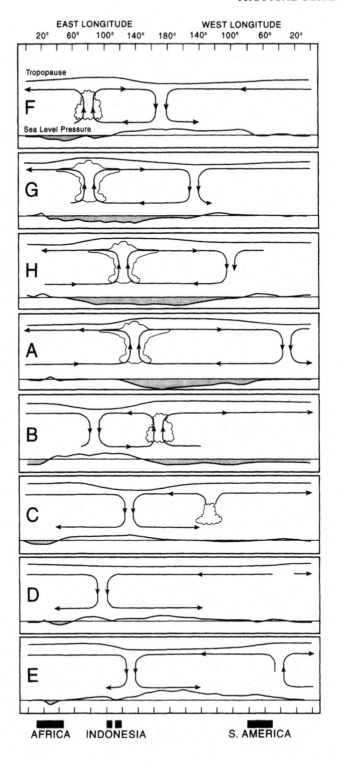

revealed the tendency for the oscillation to have a higher frequency for ENSO years (Gray, 1988).

THE WALKER CIRCULATION, SOUTHERN OSCILLATION AND EL NINO

The Walker circulation is a zonal east to west circulation along the equator. It is characterized by the ascent of air in the western Pacific in the region of Indonesia and descent in the eastern Pacific off the coast of South America (Figure 6.3). This circulation was named the Walker circulation in honour of Sir Gilbert Walker who, in the 1920s, had recognized an east to west variation of atmospheric pressure across the Pacific. This pressure see-saw Walker called the southern oscillation (SO) to distinguish it from other similar pressure oscillations such as the North Atlantic oscillation and North Pacific oscillation. Unbeknown to Walker, the southern oscillation was in fact a manifestation of the variation in the intensity of the circulation that later took his name.

The intensity of the Walker circulation appears to be controlled by sea surface temperature variations in the eastern and western Pacific. Changes in sea surface temperatures and thus the heat content of the ocean, are transferred into the atmosphere in the form of atmospheric pressure changes. As a consequence, alterations of the distribution of pressure across the Pacific Basin occur. Based on these observations, Bjerknes (1969) suggested that both the oceans and atmosphere are strongly *coupled*, such that changes in one of these climate system components would result in changes in the other. This thinking revolutionized the way in which climatologists began to explain variations in climate and weather across the Pacific and locations beyond. An immediate product of this new thinking was the realization that rapid Pacific Ocean warming events off the South American coast were related to periods when the Walker circulation entered one of two possible extreme modes. Such tied ocean and atmospheric events have become known as *El Nino southern oscillation* (ENSO) events.

Before we look further at the nature of ENSO events, let us go back and consider in a little more detail the southern oscillation (SO) as this is the atmospheric component of the ENSO phenomenon. The SO is characterized by east to west pressure variations across equatorial Pacific. This can be best seen in Figure 6.4 which is a plot of the standardized pressure differences between Tahiti (17.5°S, 149.6°W) and Darwin, Australia (12.4°S, 130.9°E). The standardized pressure difference ((Tahiti − Darwin) − (average Tahiti − Darwin) / (standard deviation of Tahiti − Darwin)) between these two locations is called the *southern oscillation index* (SOI). Darwin and Tahiti are used in the construction of the SOI as these locations are close to those geographical areas

---

Figure 6.2   Model of the spatio-temporal disturbances associated with the 40–50 day tropical oscillation. Panel A represents time when pressure is lowest at Canton Island (2.8°S, 171.7°W) while E is the same but for the highest pressure. Cumulus clouds represent areas of enhanced convection. Note the eastward propagation of the disturbances and the associated perturbations in the pressure and temperature fields. From Madden and Julian (1972). Reproduced by permission of American Meteorological Society

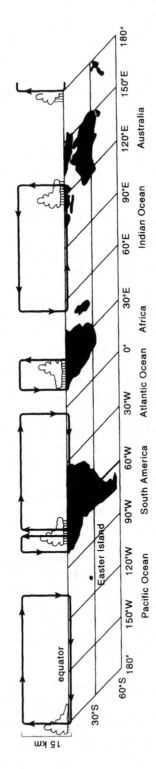

Figure 6.3 Zonal equatorial circulations. Walker circulation is that spanning the Pacific Ocean. Interactions between the zonal circulations over West Africa and the Atlantic/South American sector may also determine the sensitivity of the West African south-west monsoon to ENSO events. From Newell (1979)

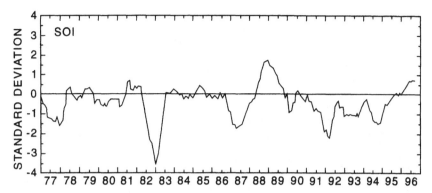

Figure 6.4   The southern oscillation index. Negative values represent times when Tahiti pressure is less than that at Darwin and coincides with ENSO events. Modified from *Climate Diagnostics Bulletin*, CPC (1996)

on either side of the Pacific Basin which experience the greatest variations of pressure on inter-annual time scales. Such areas are often referred to as centres of action.

When the SO is large (small) and positive (negative), atmospheric pressure in the eastern (western) Pacific is well above (below) normal. As the normal situation is for high (low) pressure to exist in the eastern (western) Pacific, a positive SOI means the south-eastern Pacific anticyclonic system has gained in intensity. The same can be said for the low pressure system that dominates the western Pacific. From Figure 6.4 it is clear that there are considerable inter-annual variations in the SOI. The extreme modes have attracted greatest attention because a strong negative (positive) SOI has been associated with the occurrence of *El Nino* (La Nina) events.

El Nino, meaning "Christ Child", was originally used by fishermen to describe the warm current that, around Christmas, ran southward along the coasts of Peru and Ecuador. More recently it has become synonymous with the large sea surface warming or El Nino events that occur aperiodically in the eastern Pacific (Trenberth, 1991). Extreme positive SOI or cold phases have been referred to as La Nina (the girl) events (Philander, 1983).

Analysis of historical climate records (Diaz and Markgraf, 1992) has revealed that ENSO events have tended to occur in groups throughout history in cycles of around 90, 50, 24 and 22 years. As historical analyses depend very much on palaeoclimate data, which are scarce, there has been an emphasis on ENSO events in the current century, especially those which occurred after 1970 when satellite data became available. Of the post-1970 events (1972–73, 1976, 1982–83, 1987, 1991–94) the 1982–83 event has been the most intense this century (Rasmusson and Wallace, 1983; Glantz et al., 1991) while the 1991–94 event has been the most prolonged (Bigg, 1995).

Since the 1970s ENSO events have been frequent, with an average time between events of around four years, and a range of two to ten years. Often referred to as warm events, ENSO events are characterized by a number of distinct changes in the atmospheric and oceanic circulation in the Pacific Basin. During ENSO (La Nina), the normally cool SST of the eastern Pacific (Chapter 4) are replaced by warm (cooler) ones as a result of a complex interplay between oceanic upwelling, the trade wind

(a) ENSO

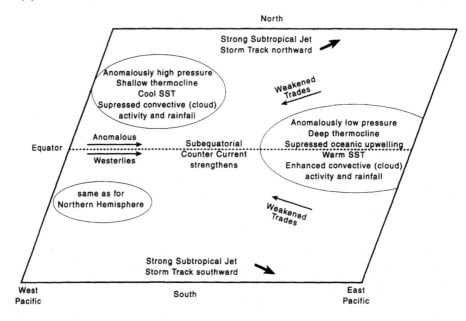

(b) La Nina

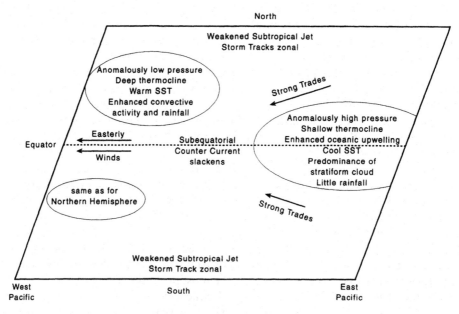

Figure 6.5   Summary of the main Pacific Ocean and atmosphere responses for (a) ENSO and (b) La Nina events. Adapted from Trenberth (1991)

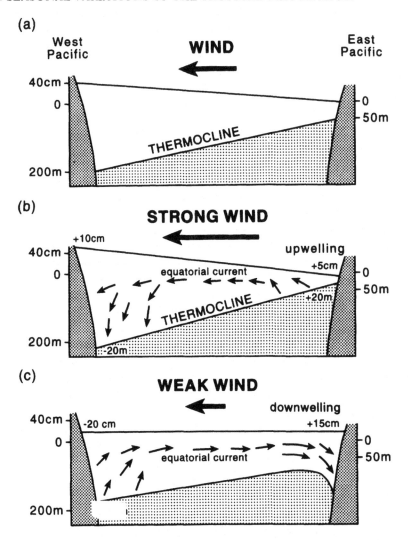

Figure 6.6   Ocean structure for (a) average, (b) La Nina and (c) ENSO events. Note the changes in thermocline depth, sea level and direction of the equatorial current. From Piexoto and Oort (1992). Reproduced by permission of the American Institute of Physics, Boston

system, the oceanic equatorial countercurrent and the pressure distribution across the Pacific Basin (Figure 6.5).

In the eastern Pacific, the south-east trades which emanate from the semi-permanent subtropical anticyclone in this area, drive water to the west. As a result, the westwardly displaced water is replaced by cold water coming up from great depths of the ocean (Figure 6.6(a)). This is the process of *upwelling*, which brings to the surface cold water and along with it nutrients which sustain high oceanic productivity on which fishing nations depend. In areas of upwelling, SST are lower than other areas. Upwelling also has an effect on the oceanic *thermocline*. The thermocline is the depth at which warm well-mixed surface water gives way to cooler thermally stratified deeper water. Upwelling (downwelling) results in a shallower (deeper) thermocline (Figure 6.6(a)). In the eastern Pacific, the thermocline is much shallower at around 40 m on average, than the western Pacific, where it may be at 100–200 m. Furthermore, as water is driven to the west by the trade winds, it tends to warm and also "pile up" in the western Pacific leading to a west to east downward slope of the sea level (Figure 6.6(a)). This difference, which is around 40 cm on average, is very dependent on the strength of the trade winds, such that in La Nina or cold events when the eastern Pacific subtropical anticyclone and thus the trade winds gain in strength, the trans-Pacific sea level differences are accentuated (Figure 6.6(b)). If the anticyclonic system and trades weaken, as they do in El Nino or warm events, the trans-Pacific water slope collapses. Consequently, the warmer water in the western Pacific "slops" back towards the east in a series of internal oceanic waves called *Kelvin waves* (Figure 6.6(c)). These take on average about 60 days to cross the Pacific and cause SST and sea level to rise in the eastern Pacific. Time series of some of the diagnostic Pacific Basin atmospheric and oceanic parameters such as wind strength, outgoing longwave radiation and SST are displayed in Figure 6.7(a)–(c). The departures and trends from the long-term mean of these parameters in ENSO years is extremely clear. Such departures and trends have been used as the basis for developing long-term climate forecasting models for the tropics (Palmer and Anderson, 1994).

One of the great ENSO mysteries is what actually sparks off the sequence of events leading to ENSO occurrence. Although no single mechanism has been identified as the precursor event for ENSO onset, what is clear, both from climate modelling and empirical studies, is that the ocean–atmosphere system in the Pacific is essentially unstable and at certain times of the year, may be primed for ENSO onset. Given this instability, ENSO events may occur as long as a triggering mechanism is present. As pointed out by Trenberth (1991), the triggering mechanism could take one of a variety of forms such as a delayed oscillator in the form of oceanic Rossby waves generated by a previous ENSO event acting as a precursor to an ensuing ENSO event (Graham and White, 1988); a delayed response of the oceans to changes in the tropical wind field especially in the western Pacific (Chao and Philander, 1993) possibly due to changes in Indian monsoon activity (Meehl, 1987; Yasanari, 1990; Webster and Yang, 1992); variations in the activity of the South Pacific convergence zone (Van Loon and Shea, 1985); the tropical 40–50 day oscillation (Madden and Julian, 1994) and Eurasian snow cover (Barnett et al., 1989).

Although the exact nature of the ENSO triggering mechanism remains unknown and all ENSO have their own identity in terms of their style of evolution, time of onset, cessation, duration and intensity (Wang, 1995), the climatic impacts of ENSO are clear. These are in the form of a set of consistent precipitation and temperature anomaly patterns which tend to recur with each ENSO event.

One of the most noticeable changes to weather patterns of the low latitudes during ENSO events is the shift of thunderstorm activity from the Indonesian area eastward into the central Pacific. This results in anomalously wet conditions for the central Pacific islands, while for tropical Australia, New Guinea, Indonesia and the Philippines, abnormally dry conditions prevail in both summer and winter seasons (Figure 6.8). During the northern winter season, dry conditions also prevail in the Nordeste

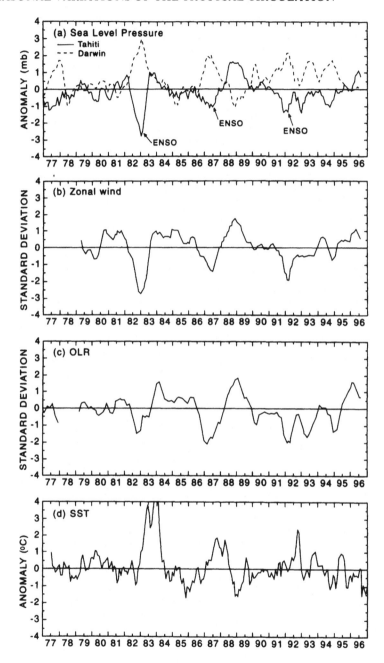

Figure 6.7 Statistically smoothed time series of various diagnostic ocean and atmosphere variables for the equatorial Pacific. (a) Darwin and Tahiti pressure; (b) zonal wind averaged over 5°N–5°S, 175°W and 140°W; (c) outgoing longwave radiation averaged over 5°N–5°S, 160°E–160°W, and (d) sea surface temperature averaged over 0–10°S, 90°W–80°W. Anomalies are relative to a 1951–80 base period for (a) and 1979–95 for (b), (c) and (d). Modified from *Climate Diagnostics Bulletin*, CPC (1996)

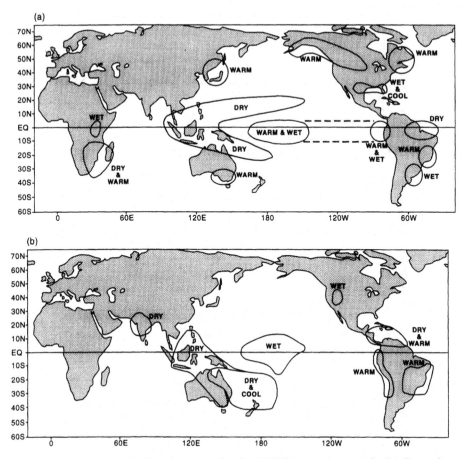

Figure 6.8   Major global climatic anomalies for ENSO (warm) events in (a) December to February; (b) June to August. Note that these patterns are generally the reverse of those for La Nina events (Figure 6.9). Reproduced with the permission of Climate Prediction Center (CPC), USA

region of northern Brazil and in south-eastern Africa (Figure 6.8(a)). Dryness is also a feature in ENSO years during the summer monsoon months in India, especially in the agriculturally important north-western region (Figure 6.8(b)). Of interest is West Africa, as this region appears to be less sensitive to ENSO events, compared to other low latitude regions. However, at times of exceptionally intense warm phases, West Africa can experience ENSO-related climatic impacts in the form of reduced south-west monsoon precipitation amounts; the 1983 ENSO event is a good example. Above-average precipitation amounts in the northern hemisphere months of ENSO years occur along the west coast of tropical South America, southern Brazil and central Argentina, as well as in the subtropical latitudes of North America (Figure 6.8(a)). These anomalously wet conditions lead to flooding, enhanced erosion and landslides, all of which can have disastrous effects for agricultural production, transport systems and human settlements.

Extratropical climatic impacts are also a feature of ENSO events. In the northern

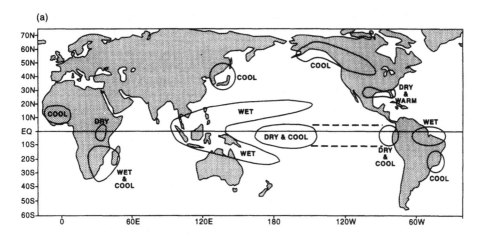

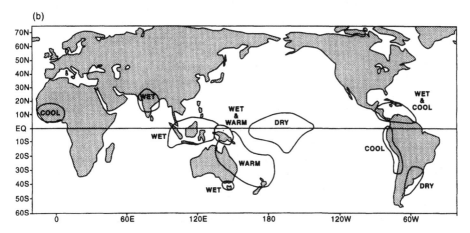

Figure 6.9   Major global climatic anomalies for La Nina (cold) events for (a) December to February; (b) June to August. La Nina events are an exaggeration of the average climate pattern. Reproduced with the permission of Climate Prediction Center (CPC), USA

hemisphere winter, especially over the north Pacific, storms tend to be more intense and take a more northward track than normal. This results in warm moist air being pumped northwards into areas such as the Gulf of Alaska and the northern regions of the United States, as well as over eastern Asia, resulting in warmer and wetter conditions than normal in these areas.

La Nina, the antithesis of ENSO warm events, also has associated with it climatic anomalies that are consistent from one cold event to another. For many of the ENSO-sensitive regions, there is a complete reversal of the ENSO climatic anomaly in La Nina years. Generally for these areas the cold event climatic anomaly is an accentuation of the normal climatic pattern. For example, during La Nina, the normal central Pacific dryness is exacerbated due to cooler than normal sea surface temperatures suppressing rainfall formation processes through their effect on atmospheric stability (Figure 6.9).

Table 6.1 Impacts of the 1983 ENSO event (modified from Canby, 1984)

Region	Impacts	Cost (US$ millions)
Australia	Worst drought this century resulting in enormous livestock losses. Bushfires fuelled by winds caused property loss and 75 deaths.	2500
Indonesia/Philippines	Drought destroyed crops; 340 people in Indonesia starved.	750
Pacific Islands	Hawaii suffered damage from a rare hurricane. French Polynesia struck by six tropical cyclones; 25 000 people made homeless in Tahiti.	280
Pacific Ocean	High ocean temperatures cause coral reef dieback and disruption of ocean food chain; 17 million nesting birds evicted from Christmas Island due to lack of food; over-fished anchoveta stocks depleted further because of disrupted food chain.	
India/Sri Lanka	Crops fail due to drought; widespread occurrence of jaundice and other epidemics as a result of water shortages in India.	150
US West Coast	Crops, homes and roads destroyed by mudslides and floods as a result of Pacific storms and enhanced snowmelt.	1100
Mexico/Central America	Crop production reduced due to dry conditions; grazing lands and water supplies severely affected.	600
US Gulf Coast	Five months of incessant and intense rain caused flooding, crop and property damage. Deaths reached 65.	1270
Ecuador, Peru	Intense rains; floods and landslides; 600 deaths. Failure of anchoveta fishery.	1000
Southern Africa	Two-year drought exacerbated by lack of rain which reduced crop production by 40–70%. Disease and malnutrition widespread.	1000

ENSO events also have an ecological and human dimension, in terms of a range of socio-economic impacts and impacts on flora and fauna (Glynn, 1990). The 1982–83 ENSO event, so far the most intense this century, offers a clear illustration of the type and geographical extent of these impacts. Total damage resulting from the anomalous weather patterns associated with the 1983 ENSO has been estimated as US$8.65 billion. A regional breakdown of this total and the associated damage is given in Table 6.1.

It remains to be seen whether the ENSO event that has developed throughout 1997, which in its early stages is as intense if not more intense than the 1982–83 event, produces an environmental and socio-economic toll as great as the 1982–83 event. In many ways this could very much depend on the preparedness of ENSO-sensitive countries.

DIURNAL VARIATIONS OF THE TROPICAL CIRCULATION

Diurnal wind systems are only of climatological importance where they occur frequently and regularly. This is the case in many tropical areas, where diurnal temperature variations are generally more conspicuous than in the mid-latitudes. The thermal changes between day and night, so typical of the tropics, are the main driving force of the diurnal wind systems, because they differ in intensity over land and water surfaces, and over highlands and lowlands. Other favourable circumstances are the generally small pressure gradients and low wind velocities of the general circulation, which reduce large-scale turbulence and allow the rapid formation of local pressure differences. The absence of fronts and strong depressions makes the development of diurnal wind systems a rather regular phenomenon in the tropics, and they are an important feature of many tropical climates.

Because of their limited duration, diurnal wind systems usually are effective only over relatively small areas. They can, but rarely do, extend far from their regions of origin and by their very nature show many local variations.

Generalizing broadly, there are two main types of location of diurnal winds: coastal regions, both along the sea and near large lakes, where systems of land and sea (or lake) breezes occur frequently, and areas of variable relief, where different types of valley and mountain winds can develop.

SEA AND LAND BREEZES

Coastal wind systems with a clear diurnal cycle are not limited to tropical locations, but they generally show their most regular occurrence and strongest development here. Thermal differences between land and water surfaces are their main cause. During the day, the land heats up rather quickly under the influence of solar radiation, while water surfaces remain cooler, because the heat is dissipated over thicker layers of water by turbulence and waves, and by direct penetration and absorption. As a result, a small convectional cell develops, with winds near the earth's surface blowing towards the land – the sea breeze (Figure 6.10). At night, the land cools off rapidly due to longwave radiation loss. The water, because of its thermal inertia, remains at about the same temperature as during the day. Consequently the daily pressure pattern is reversed and a land breeze is formed as relatively cool land air moves down the small local pressure gradient to the area of lower pressure over the sea.

The sea breeze is usually the stronger of the two winds. It can, under favourable conditions, reach speeds of 4–8 m s^{-1}, and the thickness of the air layer involved can be as much as 1000 m. The sea breeze in the tropics can reach inland as far as 100 km, up to 80 km further than its mid-latitude counterpart. In some locations, the sea breeze may be so well developed it may push over coastal topographic barriers and penetrate inland. Distinguishing sea breezes at distances beyond 50 km from the coast is, however, difficult as they may interact with other local circulations. At some distance inland, air rises in the ascending arm of the sea breeze convection cell and returns towards the sea at about 1500–3000 m.

The sea breeze usually starts, near the coast, a few hours after sunrise, typically mid

(a) Day

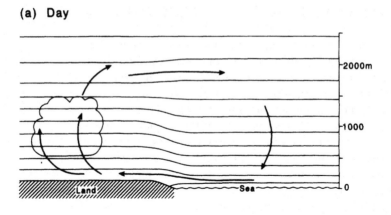

(b) Night

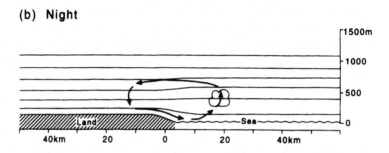

Figure 6.10   The basic pattern of sea and land breezes: (a) daytime sea breeze and (b) night-time land breeze. Horizontal lines indicate isobaric surfaces. Note the contrasting vertical and horizontal scales of development.

to late morning, but expands landward and seaward during the day. It attains its maximum development when sea–land temperature contrasts are at about their maximum. In some places this can be in the morning, as opposed to the afternoon, as land temperatures in the afternoon may be depressed due to cloud development related to the sea breeze. The sea breeze usually continues until shortly after sunset, but its circulation at higher levels may persist a few hours longer.

Seasonally, the sea breeze and its lake breeze equivalent are strongest when insolation is intense. They are therefore best developed during the dry season. In the outer tropics, the summer is also a season of strong sea breezes because the low wind velocities of the general circulation and the unstable air masses create propitious conditions for their development.

As with all diurnal winds, the actual strength and direction of the sea breeze are controlled by local factors. Low surface water temperatures, caused by cold ocean currents or upwelling of water from below, increase the strength of the sea breeze. Factors which increase the daytime temperatures over land, such as the lack of vegetation and dry surfaces have the same effect. A dense vegetation cover, swamps or flooded ricefields generally lead to reduced sea breezes as these decrease the land–sea temperature contrasts. The presence of mountains near a coast often creates a combined sea breeze–valley-wind system (Riehl, 1954).

The sea breeze rarely brings much precipitation, because its air masses have experienced a stabilizing downward movement over the sea to begin with. But when it converges with winds from a different direction, often a sea breeze front is formed and this can cause local cloud development and rainfall (Ramage, 1964) (Plate 3). Sea breezes, for example in the Hawaiian Islands, interact with the trade winds. Cloud develops in the zone of convergence between these opposing local and synoptic-scale systems (Hastenrath, 1985). Over islands and peninsulas, systems of converging seas breezes from opposite coasts can cause a regular afternoon rainfall maximum (Byers and Rodebush, 1948; Nieuwolt, 1968; Skinner and Tapper, 1994). During the early morning, the beginning sea breeze can carry disturbances developed over the sea during the night, to coastal areas, thereby producing rainfall.

The land breeze is weaker than the sea breeze in most tropical climates. This is because in the tropics the land–sea temperature difference due to daytime heating is much greater than that due to night-time cooling. Its main cause is the rapid cooling of the land surface during the night. This cooling influence is limited to a thin surface layer of air. Moreover, this layer is strongly affected by friction (Figure 6.10). Therefore the land breeze rarely exceeds 3 m s^{-1}; however, it may be enhanced by katabatic flows (see next section). The thickness of the moving air layer in a land breeze is usually only a few hundred metres. The land breeze does not normally reach more than 15–20 km seaward. It generally starts about three hours after sunset, increasing in strength until sunrise and, at times, continuing beyond sunrise.

Land breezes are best developed in areas where the water surface is relatively warm: in equatorial regions, near warm ocean currents and relatively shallow lakes, such as Lake Victoria in East Africa. Long and clear nights, occurring during dry seasons and, in the outer tropics, during the winter, also favour land breezes.

All local circulations are influenced by winds of the general circulation and sea and land breezes are no exception. Where synoptic-scale winds are strong, no breezes develop at all, since turbulence prevents the establishment of local temperature and pressure differences between water and land surfaces. With weaker general winds, the breezes are often limited to changes in the direction and speed of these winds (Nieuwolt, 1973). In doldrum areas and near the equator, where synoptic-scale winds are extremely weak, local circulations may dominate.

Other variations of breezes are related to the general form of the coastline which can cause local convergence or divergence (McAlpine, Keig and Falls, 1983). Convergence and cloud development is favoured over headlands while divergence and broken lines of cloud development is favoured over bays. Breeze systems develop over islands, when these are not too small (a diameter of about 15 km seems to be the minimum). Over seas, such as the Straits of Malacca, convergence of opposing land breezes may occur at night creating precipitation (Ramage, 1964).

As with most local winds, sea and land breezes are not affected by the Coriolis force, unless they prevail over large distances in extra-equatorial latitudes. In the outer tropics this factor can cause a slight deviation of the breezes, which can become parallel to the coast, but this is rarely the situation in the low latitudes. Land and sea breezes, however, can interact with synoptic-scale gradient winds to produce resultant winds which flow obliquely to the coastline. For example, a gradient wind blowing parallel with the coast over the land can interact with a sea breeze wind blowing at right angles to the coast. The interaction of these two wind directions will produce a resultant onshore wind flow angled at about 45° to the coastline.

Sea and land breezes are of great practical importance. In many tropical land areas they bring welcome relief from the oppressive heat during the hottest hours of the day (Nieuwolt, 1973). This is caused by both the advection of cooler air and by improved ventilation. Traditionally, fishermen also use the breezes: they sail out to sea with the land breeze in the early morning and return to land with the afternoon sea breeze. However, local coastal circulations do have their disadvantages as they are essentially closed circulation cells. For this reason the location of air polluting activities in tropical coastal locations where sea and land breezes are climatologically important should be avoided. This is because pollutants emitted during the day, although diffused vertically in the rising landward limb of the sea breeze cell, will be returned to land in the descending seaward and landward lower branches. At night subsidence over the land may also bring pollutants back down to the surface.

MOUNTAIN AND VALLEY WINDS

Over areas with large differences in relief, diurnal wind systems often develop. These winds are particularly regular and strong in the tropics. Their basic origins are heating and cooling of air on slopes. During a sunny day mountain slopes heat up rapidly due to large radiation receipts. The free atmosphere over the lowlands remains less affected by these large insolation inputs and is slightly cooler than air over the mountain slopes. Mountain slope air therefore becomes unstable and tends to rise up the slope. This type of upslope flow is called "valley" wind or "*anabatic*" flow (Figure 6.11). It can be easily recognized as it is often accompanied by the formation of cumulus clouds near mountain tops or over escarpments and slopes. At night, a reverse temperature difference develops, as the highlands cool off rapidly because of longwave radiation loss. This cooler dense air then moves downslope under the influence of gravity and is called a "mountain" wind or "*katabatic*" flow (Figure 6.11).

These wind systems can develop in very different dimensions: over a single mountain or valley, or even an individual slope; along mountain ranges or escarpments, and between extensive highlands and lowlands such as the Tibetan–Himalayan massif and the Ganges Plain in northern India. Obviously, large regional variations of the general pattern occur, but upslope winds during the day and downslope winds during the night prevail in most cases.

Anabatic flows are usually stronger and more persistent than katabatic ones. They frequently continue well after sunset and this tendency is particularly strong in the outer tropics during the summer, when insolation is very intense and nights rather short. Under these circumstances the anabatic winds, if developed on a large scale, can continue throughout the night. This happens, for instance, at the foothills of the Himalayan mountain range (Flohn, 1960). For the Highland regions of Papua New Guinea where large mountains surround open basins, afternoon steady anabatic flows have been measured at 12–13 m s^{-1} (McAlpine, Keig and Falls, 1983).

Where winds of the general circulation prevail from one direction, as in the case of

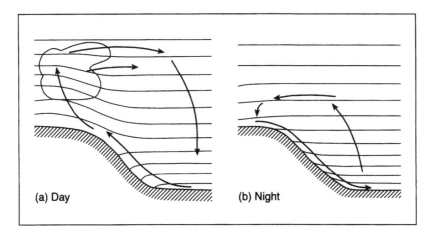

Figure 6.11   The basic pattern of valley and mountain winds: (a) daytime valley wind or anabatic flows and (b) night-time mountain wind or katabatic flows. Horizontal lines indicate isobaric surfaces

the trade winds or the monsoons in some areas, anabatic winds generally reinforce the prevailing wind on the windward side of the mountains. Here, they can contribute to orographic rainfall and these areas frequently exhibit a clear afternoon rainfall maximum (Chapter 10). However, on leeward slopes the anabatic winds are usually suppressed by the winds of the general circulation.

Katabatic winds are normally weaker than daytime anabatic winds because thermal differences are usually smaller and friction reduces wind speeds near the earth's surface. Katabatic winds, however, can be just as strong if not stronger than anabatic winds in some cases. This is especially true for high-elevation tropical mountain environments where, because of elevation effects, night-time cooling can be extremely rapid under clear sky situations. Given this set of conditions, downslope topographically channelled katabatic flows can be very strong: gusts can be in excess of $15\,\mathrm{m\,s^{-1}}$ on Mount Wilhelm in Papua New Guinea. The main visible effect of katabatic winds is the rapid dissolution of clouds near the mountain tops or over slopes such as Mount Kenya (Hastenrath, 1985). Cool descending air may result in the formation of valley and basin fogs as katabatic flows cool valley air to its dew point. In anomalous climatic conditions such as occur in the Papua New Guinea Highlands during El Nino events, gentle katabatic flows can enhance the potential for frost formation. Katabatic flows and land breezes may also combine in areas of steep coastal topography to enhance the night-time offshore flow of air. These may converge with directionally opposed synoptic-scale seasonal flows to produce night-time offshore convection zones (Figure 6.12).

DIURNAL PRESSURE VARIATION

An interesting feature of the low latitudes is a quite regular semi-diurnal (12 hourly) cycle in air pressure: maxima of different intensity are observed around 1000 and 2200 hours and minima around 0400 and 1600 hours local time (Figure 6.13). Similar

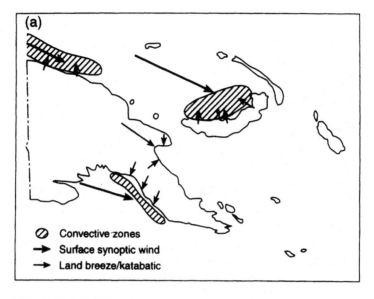

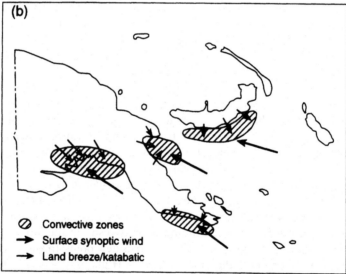

Figure 6.12   Contrasting areas of offshore convective zones around Papua New Guinea during (a) the north-west ("wet season") monsoon and (d) the south-east ("dry season") monsoon. Source; McAlpine, Keig and Falls (1983)

pressure cycles occur at all latitudes, but their amplitudes decrease with distance from the equator where it may reach 3–4 mb to about 1.0 mb in mid-latitudes and 0 mb at the poles. The 12 hourly semi-diurnal wave is a progressive wave as it moves from east to west and follows the progression of the sun. Other diurnal changes in pressure occur over the globe – namely 24 hourly and 8 hourly cycles – but these are of subordinate importance in the low latitudes. In the low latitudes, where irregular changes of

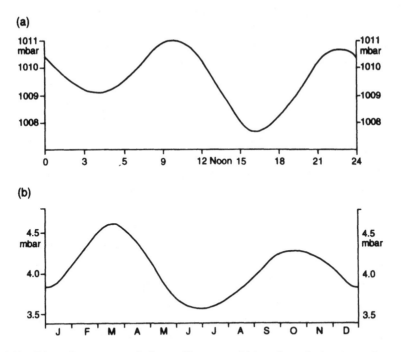

Figure 6.13   Diurnal pressure variation at Singapore: (a) hourly variation (annual means); (b) seasonal variation of the semi-diurnal amplitude. Observation period is 18 years

pressure are generally smaller and less frequent than in higher latitudes, the semi-diurnal cycle must be taken into account in short-term forecasting and the calibration of altimeters.

Various explanations have been suggested for the existence of diurnal pressure cycles. These include resonance oscillations in the atmosphere, diurnal temperature variations and tidal effects (Frost, 1960). The thermal and gravitational nature of the various diurnal cycles or atmospheric tides, as they are often referred to, have been discussed extensively by Chapman and Lindzen (1970). The seasonal variation of the semi-diurnal cycle's amplitude near the equator, which shows maxima around the equinoxes, seems to suggest a relation with direct absorption of solar radiation, most likely in the upper levels of the atmosphere (Figure 6.13). However, tidal influences of the sun and moon, amplified by the natural 12-hour resonance of the atmosphere, undoubtedly contribute to the pressure cycle.

SUMMARY

The tropical circulation is characterized by a number of non-seasonal variations which vary in their frequency. Belonging to the group of high frequency variations are semi-diurnal pressure oscillations and diurnal local circulation changes. The former, often referred to as atmospheric tides, have their greatest amplitudes at the equator and appear to be related to direct solar heating of the upper atmosphere. Diurnal changes in local circulation systems manifest themselves in the form of sea/lake and land breezes and mountain and valley winds. Sea/lake and land breezes develop because of day–night thermal contrasts between the land and sea. Valley and mountain winds are a product of daytime heating and night-time cooling of mountain slopes. Daytime upslope flows are referred to as anabatic flows while night-time downslope winds are called katabatic flows. Although described in this chapter as non-seasonal variations it should be noted that both atmospheric tides and local circulation systems do display seasonal variations in their intensity with more often than not the greatest intensities being displayed during the months of maximum solar heating.

In ascending order of periodicity are the slowly varying 40–50 day tropical oscillation (Madden–Julian Oscillation – MJO) in surface pressure, upper level winds and temperature at a variety of levels, the quasi-biennial oscillation (QBO) in equatorial stratospheric winds between easterly and westerly modes, and the El Nino southern oscillation (ENSO). The MJO has been shown to play an important role in explaining short-term variations in weather at equatorial locations and accounting for active and break phases of the monsoons. The weather and climate impacts of the QBO remain uncertain. This is in stark contrast to the coupled atmosphere–ocean phenomenon called ENSO that occurs in the Pacific Basin on a time scale of 3–7 years. Although the main ocean and atmospheric signals, in the form of sea surface temperature and atmospheric pressure reversals, are largely confined to the Pacific (especially the central and eastern Pacific), the climate impacts of ENSO are far reaching, often extending beyond the tropical Pacific to mid-latitude northern and southern hemisphere locations. In the Pacific the occurrence of ENSO results in a complete reversal of the "normal" climate pattern, especially rainfall, which can have disastrous environmental and socio-economic consequences for those countries bordering the Pacific Basin.

CHAPTER 7

Seasonal Variations in Regional Circulation Systems: The Monsoons

The term "monsoon" appears to have originated from the Arabian Sea region where the word *mausim* means "season". The main characteristics of the monsoon regions are as follows:

1. the prevailing wind direction shifts by at least 120° between January and July;
2. the average frequency of prevailing directions in January and July exceeds 40%;
3. the mean resultant winds in at least one of the months exceeds 3 m s^{-1}, and less than one cyclone anticyclone alteration occurs on average every two years in any one month in a 5° latitude–longitude rectangle (Ramage, 1971).

The monsoon regions of the world according to this set of criteria are shown in Figure 7.1.

Not only are the monsoon areas important in terms of their geographical coverage, but also in demographic terms, as they are home to over 55% of the globe's population. Because the monsoons are such a dominant feature, the social and economic welfare of many tropical countries is intimately linked to the vagaries of the annual monsoon cycle. For this reason we will describe the general climatic factors that give rise to monsoons and control their variability as well as present a general model of the annual monsoon cycle. Following this, a systematic description of the structural and variability characteristics of the regional monsoon systems will be presented.

THE MONSOON MAKERS

Three general factors account for the existence of the monsoons. They are (1) the differential seasonal heating of the oceans and continents, (2) moisture processes in the atmosphere and (3) the earth's rotation (Webster, 1987).

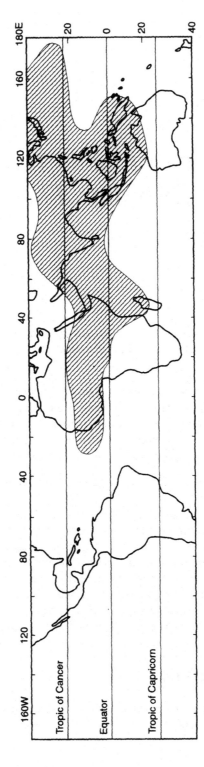

Figure 7.1   Areas with monsoon circulations according to the criteria of Ramage (after Ramage 1971, p. 4)

SEASONAL HEATING

Over the large ocean basins, seasonal changes in the tropical circulation are limited to minor latitudinal shifts and small variations in intensity of the main components, but the general pattern remains virtually the same throughout the year. However, the picture is entirely different over the tropical continents and adjacent seas. Here important seasonal temperature and pressure changes take place. Seasonal contrasts in land surface temperatures produce atmospheric pressure changes which produce seasonal reversals of the pressure gradient force, the basic driving force of the winds. As a result there are major seasonal wind reversals. It is these which are often referred to as "the monsoons".

MOISTURE PROCESSES

As moist warm air rises over summertime heated land surfaces, the moisture eventually condenses, releasing energy in the form of latent heat of condensation. This extra heating raises summer land–ocean pressure differences to a point higher than they would be in the absence of moisture in the atmosphere. Moisture processes therefore add to the vigour of the monsoon.

THE EARTH'S ROTATION

Because the earth rotates on its axis and produces a rotation force called the Coriolis force, air in the monsoon currents moves in curved paths. This is because air from high pressure areas tends to spiral into low pressure areas. Inter-hemispheric differences in the direction of the Coriolis force also cause winds to change direction as they cross the equator.

These three mechanisms collectively cause the monsoons. We will now consider their interplay in the annual cycle of the monsoon.

THE ANNUAL MONSOON CYCLE

The relationship between the general mechanisms that generate the monsoons, the seasonal climate cycle and the annual monsoon cycle are shown in Figure 7.2. In the transitional months between the southern and northern hemisphere summers, the ITCZ is located in the equatorial regions (Figure 7.2(a)), where maximum surface heating can be found. At this stage of the seasonal cycle, the northern hemisphere tropical–subtropical latitudes are beginning to warm up. Vertical motion is present, but weak. The northern hemisphere Hadley cell still predominates at this stage, as does the offshore flow of air. With the northward movement of the sun in May to June, the heating of northern tropical land masses intensifies, as does the vertical motion over these land masses (Figure 7.2(b)). Northern tropical atmospheric moisture contents also increase as the southern hemisphere Hadley cell intensifies; the predominant wind direction is onshore. By May to June, belts of precipitation associated with the ITCZ have moved well north of the equator, signalling the onset of the summer wet monsoon

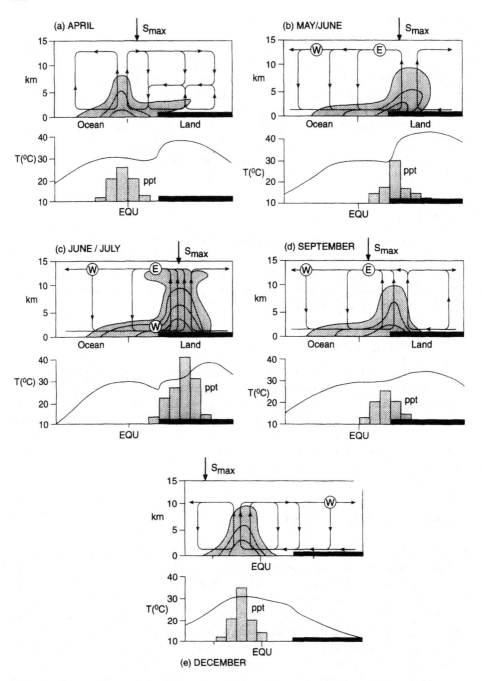

Figure 7.2   The annual monsoon cycle. From Webster (1987). Reproduced by permission of John Wiley and Sons, Inc.

for many areas (Figure 7.2(b)). By June to July, sensible heat input at the surface is close to a maximum, as is the vertical motion and atmospheric moisture over the northern hemisphere tropical land masses (Figure 7.2(c)). Maximum values of the pressure gradient force have also been attained by this stage and the monsoon reaches its maximum intensity. The rush of air northwards over the equator is subject to a strong Coriolis force, which in some monsoon regions, generates a jet-like wind, e.g. the East African jet (Chapter 5). Outflows of air in the upper atmosphere, above the surface thermal lows when subject to the Coriolis force, also produce strong upper winds. At this stage the amount of precipitation and its northward extent have reached a maximum. By September, surface heating has decreased markedly with maximum insolation now positioned close to its April position. The structure of the monsoon at this time of the year is therefore very similar to that of April (Figure 7.2(d)). September heralds the cessation of the northern hemisphere monsoon wet season and the onset of the dry season. By December, the southern hemisphere wet season is well under way as precipitation belts associated with the ITCZ have moved well south of the equator, accompanying the strengthening of the northern hemisphere Hadley cell and the southern position of the zone of maximum heating (Webster, 1987).

VARIABILITY OF THE MONSOONS

The schematic model of the annual cycle presented in the section above is for mean annual conditions. However, in reality, there is considerable variability in the onset, duration and magnitude of the monsoons. The mechanisms responsible for monsoon inter-annual variability may be categorized into two groups. These are "internal dynamics" and "boundary forcing" (Shukla, 1987).

A variety of aperiodic variations in the atmospheric circulation such as travelling disturbances, thermal and orographic forcing, non-linear associations between different scales of atmospheric motion and tropical–extratropical interactions constitute some of the internal dynamic controls on the monsoon (Shukla, 1987).

Boundary forcing refers to changes in surface conditions. The areal extent of snow cover, surface hydrological effects and sea surface temperatures are important boundary forcing factors through their influence on the surface energy balance and thus tropospheric thermodynamics. Changes in boundary conditions will affect the geographical distribution of heat and moisture sources and sinks in the atmosphere, and thus the pattern of moisture-bearing tropospheric winds.

It is highly likely that internal dynamical and boundary forcing factors interact to produce variations in the monsoons. The specific mechanisms believed to account for the inter-annual variability of the various regional monsoon systems will be discussed below. In addition to inter-annual variation, the various monsoon systems experience intra-seasonal variations in the form of *active* and *break* phases. The former is a period of marked rainfall increase and the latter a period of diminished or absent rainfall.

REGIONAL MONSOON SYSTEMS

Traditionally, three main monsoon systems have been recognized. These are the

African, Asian and Australian monsoon systems. However, as more atmospheric data have become available over the last 20 years, mostly through monsoon field experiment programmes, there appears to be a strong case for dividing the Asian monsoon into two separate subsystems, namely an Indian monsoon and an East Asian monsoon, and for recognizing another monsoon region in the eastern north Pacific.

THE INDIAN MONSOON

The Indian monsoon is made up of a number of components:

1. the monsoon trough over northern India;
2. the Mascarene anticyclonic system;
3. the low level cross-equatorial jet;
4. the Tibetan high pressure system;
5. the tropical easterly jet;
6. monsoon cloudiness;
7. rainfall (Krishnamurti and Bhalme, 1976) (Figures 7.3 and 7.4).

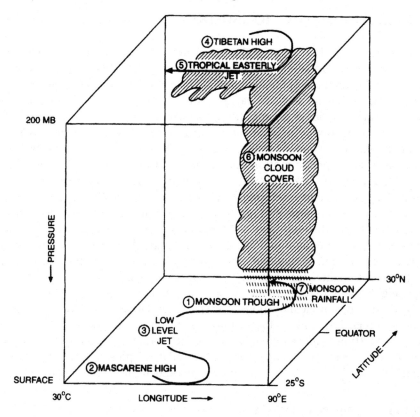

Figure 7.3   Main structural elements of the Indian monsoon. From Krishnamurti and Bhalme (1976). Reproduced by permission of the American Meteorological Society

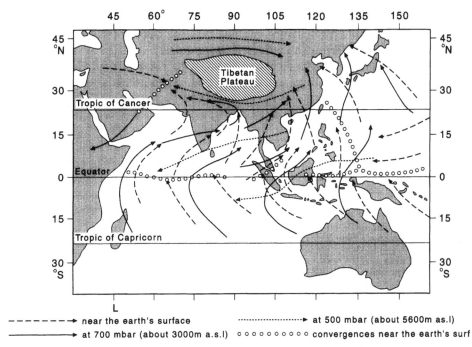

Figure 7.4   The Asian summer monsoon. Winds (arrows) and convergence zones (dots) from about June to September, near the earth's surface. Prevailing winds at 700 mb (3000 m) and 500 mb (5800 m) levels

The monsoon trough is formed over northern India in the northern hemisphere summer as part of the global ITCZ. It differs slightly, however, in its dynamic structure in that to the south of the trough, south-westerly winds exist, whereas to the north, easterly winds prevail. The trough is a zone of intense horizontal wind shear. The monsoon trough is associated with an area of surface low pressure. The Mascarene high is situated over the south-east Indian Ocean at around 30°S, 50°E (Krishnamurti and Bhalme, 1976). From this area there is a large outflow of air. This moves north over the equator where it becomes a south-westerly flow known as the cross-equatorial East African or Somali jet (Chapter 5). Reaching its maximum intensity in the months of June to August, the jet splits into two branches at around 10°N, 60°E at this time. These cross the southern Arabian Sea and arrive over the central west and southern coasts of India (Figure 7.4). Variations in the intensity of this jet are believed to be important for determining rainfall amounts over western India.

The Tibetan high (unlike the Mascarene high) is an upper level anticyclone. It is found above the surface monsoon trough located over northern India as low level convergence is matched by upper level divergence. It is best developed at the 200 mb level. By July the Tibetan high is well established over the Tibetan highlands where it remains until around September, after which, it moves in a south-south-eastward direction as the zone of maximum surface heating and low pressure move south with the onset of the southern hemisphere summer. The outflow of air from the southern flanks of the Tibetan high gives rise to the Tropical easterly jet as described in Chapter

5. This jet lasts from around June until September when the Tibetan anticyclone diminishes in strength. With these developments the upper atmosphere easterlies associated with the easterly jet are replaced by westerlies.

Cloud cover is an important component of the Indian monsoon and is the manifestation of moist convective processes over the Indian subcontinent. Cloud cover varies in both space and time. During active phases of the monsoon, cloud cover maximums can be found in a belt running from the western shores of the Bay of Bengal to the northern shores of the Arabian Sea. At the same time, cloud cover minima are found over the foothills of the Himalayas and southern India and Sri Lanka. For break phases, there is a reversal of this pattern. Rainfall distribution is similar to the cloud distribution.

A distinct structural feature of the Indian monsoon is the development of a cyclonic vortex at the time of monsoon onset. This forms on the cyclonic shear side of the low-level jet in the lower troposphere over the east Arabian Sea in the region of 10°N, 65°E (Krishnamurti, 1985). Although not a prerequisite for monsoon onset, its development appears to play a role in the rapid changes that occur in cloudiness and rainfall patterns at the time of monsoon onset. Associated with onset vortex formation are increases in the intensity of the monsoonal flow and storm development over the Arabian Sea. Formation of the onset vortex is due to the instability arising from a rapid increase of the horizontal shear of the large-scale monsoonal flow (Krishnamurti, 1985). A consequence of its explosive evolution is the influx of large amounts of water vapour from the southern hemisphere. This influx accounts for the rapid changes in the vertical humidity structure over the Indian peninsula at the time of monsoon onset.

Variability of the Indian monsoon has been historically important as at times failure or an oversupply of monsoon rains has brought much human suffering to regions affected by this monsoon system.

Recently much attention has been focused on Eurasian snow cover as a possible factor in Indian monsoon variability, as pre-summer-monsoon snow cover can alter the surface energy balance of the Tibetan Plateau. Extensive snow cover remaining in pre-monsoon months will mean that large amounts of energy are consumed in snow and ice melt and evaporation of water from melt-saturated ground. Pre-monsoon surface heating will, therefore, be much reduced. As a result, the thermal lows that eventually develop will be less intense compared to those developed following a normal snow year, as will land–ocean temperature contrasts. Empirical evidence supports such a contention as Eurasian snow cover has been found to be matched with a delayed monsoon onset over India (Dey and Bhanu Kamar, 1983; Barnett et al., 1989) and low rainfall amounts (Dickson, 1984; Yang, 1996).

Sea surface temperature (SST) variations in the Arabian Sea, Indian Ocean and the Equatorial Pacific may also exert an influence on the behaviour of the Indian monsoon. For example, a relationship between warm (cool) Arabian SST and above (below) average rainfall over the Indian continent has been found (Shukla and Misra, 1977). However, the SST anomalies are small when compared to other oceanic areas. This is because of complex feedback effects between the ocean and the atmosphere in this region. Stronger circulation over the Arabian Sea, in warmer and wetter years, enhances the mixing and upwelling in the Arabian Sea. This results in a suppression of what should otherwise be strong SST anomalies. Despite suppressed positive SST anomalies, evaporation is enhanced due to the increased circulation intensity.

Although the annual cycle of SST in the Indian Ocean plays an important role in the establishment of the Indian monsoon and the Indian Ocean is an area of net heat gain, the extent to which Indian Ocean SST anomalies control the inter-annual variability of the monsoon remains unclear. This uncertainty appears to be a product of the lack of oceanic data for this region.

During the 1980s strong associations between the inter-annual behaviour of equatorial eastern Pacific SST and a range of tropical climatic elements including Indian summer monsoon rainfall were established. Below-average monsoon rains for both the Asian and African monsoon areas were often found to be related, but not exclusively, to above-average equatorial SST in the eastern Pacific (ENSO events, Chapter 6). The simple fact that 21 out of 25 ENSO events were associated with Indian summer monsoon rainfall below the long-term median (Rasmusson and Carpenter, 1982) convinced many that equatorial Pacific SST, through their control on the southern oscillation, were the pacemakers of the Asian monsoons (Angell, 1981; Ropelewski and Halpert, 1989) and that the southern oscillation could be used as a predictor of monsoon variability (Shukla and Paolino, 1983). It was postulated that ENSO was the active system which controlled the inter-annual variability of the Asian summer monsoon. Recent research, however, suggests that ENSO and the monsoon over Asia are mutually interactive systems, and what is more, the monsoon possibly plays an active role, as opposed to a passive role, in tropical and indeed global climate variability (Yasunari, 1990; Webster and Yang, 1992). This is very much a return to the original beliefs of Sir Gilbert Walker after whom the Walker Circulation is named.

Although the Asian monsoon and ENSO systems may be interactive, it must also be remembered, as pointed out by Webster and Yang (1992), that the anomalous monsoon develops as part of a larger scale and longer period anomalous system which itself may be controlled, for example, by the extent of Eurasian snow cover. To unravel the exact nature of this complicated interacting system will require extensive research on climate sensitivity studies using numerical climate models. Such studies will, it is hoped, shed light on how the ENSO system, Eurasian snow cover and monsoon intensity do in fact interact.

On the intra-seasonal time scale, a number of variations in the intensity of the Indian monsoon system occur (Yasunari, 1980). These include variations at the synoptic scale of 5–7 days, and the lower frequencies of 10–20 and 30–60 days which match those found by Madden and Julian (1971) for the equatorial regions. A characteristic of the low frequency modes is their northward and eastward propagations (Krishnamurti and Bhalme, 1976; Gueremy, 1990). Krishnamurti and Ardanuy (1980) have also identified westward-propagating systems on a time scale of 10–20 days.

For surface pressure, rainfall, cross-equatorial flow and monsoon cloudiness, a bi-weekly mode of 13.1 days is dominant. This falls within the 10–20 day low-frequency mode. There also appears to be a clear pattern of development. Normally the monsoon trough intensifies first. This is followed by an intensification of (1) cloud development, (2) the Mascarene high, (3) the Tibetan high, and (4) the tropical easterly jet. Rainfall maxima over central India move steadily westward with the monsoon depression (Krishnamurti and Bhalme, 1976). This sequence of events describes the evolutionary cycle of the active or pronounced rainfall phase of the monsoon as the monsoon trough intensifies over central India.

Active–break phases are typified by northward propagations of cloudiness, rainfall and heating patterns. These most likely have their origins in the equatorial Indian Ocean and move northward at about 1° per day and eastward at about 5° per day on a 40–50 day time scale (Madden and Julian, 1994). Mechanisms suggested to explain the northward propagation of the monsoon trough and related convergence zone, and thus the occurrence of active and break phases, include the 30–60 (40–50) day tropical oscillation (Chapter 6) and land–atmosphere feedbacks (Webster, 1983; Gueremy, 1994).

THE EAST ASIAN MONSOON

Over the last 10–15 years it has become clear that the monsoon over East Asia is not simply an eastward extension of the Indian monsoon, but a separate component of the large Asian monsoon system. Facts that have led researchers to this conclusion include the frequent opposite behaviour of monsoon activity over East Asia and India; the existence of heat source and sink regions over the South China Sea and Australian regions (these are important for driving the East Asian monsoons and contrast with those of the Indian monsoon); intra-seasonal changes in the origins of the air masses involved in the respective monsoons; and contrasts in onset and cessation times of the two monsoons. These contrasting features are all evidence of the different structures of the East Asian and Indian monsoon subsystems. Additionally, the East Asian monsoon has a very strong cold winter signature, a characteristic not possessed by its Indian counterpart or any of the other monsoon systems. Because the East Asian monsoon can be divided clearly into summer and winter monsoons these will be discussed separately.

EAST ASIAN SUMMER MONSOON

The East Asian summer monsoon can be divided into seven major components: (1) the Australian high; (2) the cross-equatorial jet at about 110°E; (3) the monsoon trough; (4) its associated zones of convection; (5) the tropical easterly jet, part of the upper level north-easterly return flow; (6) the western Pacific High; (7) the Mei-Yu Front; and (8) mid-latitude disturbances (Figures 7.4 and 7.5). These structural components differ in detail to those described above for the Indian monsoon.

In addition to the cross-equatorial flow from the southern hemisphere, the East Asian monsoon also originates in air flows from two other air mass regions (Figure 7.5). These are (1) the Indian Ocean, the source of the Indian summer monsoonal airflow, and (2) the western Pacific from where the south-east monsoon flow emanates from the western flanks of the western Pacific high.

The cross-equatorial flow from the southern hemisphere is an extremely important component of the East Asian monsoon as it carries with it large amounts of moisture. It is therefore extremely important for inter-hemispheric mass exchange. When the pressure difference between the Australian high and the thermal low over China is

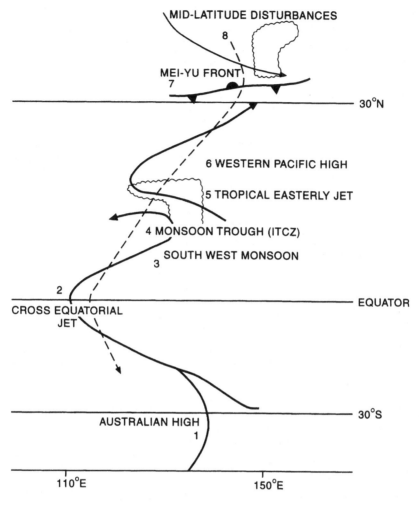

Figure 7.5   The main structural components of the East Asian monsoon. From Domroes and Peng (1988). Reproduced by permission of Springer-Verlag GmbH & Co.

enhanced, the cross-equatorial flow intensifies. The air in this flow is originally stable and dry over the Australian continent and it keeps these characteristics over the south-eastern islands of Indonesia. However, as it moves north-eastward and recurves to become a south-westerly flow under the influence of the Coriolis force as it crosses the equator, it becomes more humid and unstable. By the time it has reached the South China Sea it is extremely unstable and moist. Over the Malaysian Peninsula the cross-equatorial flow meets the westerly Indian monsoonal flows (Figure 7.5). In this zone of confluence large amounts of rainfall are produced.

   The South-East Asian summer monsoonal flows emanate from the region occupied by the western Pacific subtropical high. The subtropical high is especially important in the period from July to the end of August. During this period the western Pacific high moves rapidly northward, as does the ITCZ to the south-west. At its origin, the air in the subtropical high region is dry and stable and the atmospheric structure is dominated by a trade inversion. As the subtropical high air moves westward, over warm ocean surfaces, it is rapidly modified and becomes unstable as it turns north-westward. Consequently, moist warm air from the south-east is advected over the Philippines and into north-eastern, south-eastern and southern parts of Indo-China, China and Japan respectively. The zone of confluence with the cross-equatorial air from the Australian region is generally to the east of the Philippines but can reach southern Japan and even Korea. Generally, the south-easterly component of the East Asian monsoon is greatest over the South China Sea and southern China and Japan. For this reason the summer monsoon is often referred to in these areas as the south-east monsoon.

An extremely important structural feature of the summer monsoon over East Asia is the monsoon trough which stretches from the western Pacific across Indo-China. For most of the time this trough remains separate from the monsoonal trough over the Bay of Bengal and India associated with the Indian monsoon. Occasionally, these troughs may become connected. Associated with the monsoon trough are distinct zones of convergence, cloudiness and rainfall. The trough region is therefore characterized by wet, cloudy and often windy weather. The movement of this cloudiness and rainfall into a region often heralds the onset of the wet season, or the initiation of an active monsoon phase, which is one form of monsoon variability.

The East Asian monsoon varies in its activity on a variety of time scales (as does the Indian monsoon). The shortest of these time-scales is around 4–5 days while the clearest and longest is at the inter-annual time scale. Oscillations of monsoon rainfall activity on a time scale of 4–5 days are due mainly to tropical disturbances in the form of tropical cyclones which move westward from the Pacific Basin (Cheang, 1991). These can deliver large amounts of rainfall in very short periods. Variations in convective activity at a time scale of 10–20 days over South-East Asia (Zangvil, 1975) are most likely due to the same factors that have been suggested to explain similar oscillations in Indian monsoonal rainfall.

Low-frequency oscillations in cloudiness of 30–50 days in the East Asian region have been related to active and break periods of the East Asian monsoon. Chinese meteorologists have identified four stages in the active and break periods of the East Asian monsoon (Ding, 1994). Each stage lasts for around 10 days. Over the four stages there is a northward advance of the monsoon trough associated with a strengthening of the equatorial westerlies and a subsequent southward retreat as the westerlies weaken. Concurrent with this, the south-west monsoon over the South China Sea, with its origins to the west, is replaced by the south-easterly monsoon from over the western Pacific.

The first of the four stages is characterized by westerly winds over southern China and northern Indo-China. At this time the western Pacific subtropical high extends into and dominates the South China Sea region. Consequently, there is no ITCZ in this region. Over the Bay of Bengal the monsoon trough is oriented almost north–south with south-westerly flows to the east of the trough pushing across Indo-China and into southern China. For these areas the monsoon is at its peak activity, but the south-west monsoon over India is inactive. Over Malaysia, rainfall is at a minimum.

In the second stage, the westerly winds over southern China and northern Indo-China weaken. Over India, the equatorial westerlies enhance significantly and push eastwards in the direction of southern Indo-China. Over the western Pacific the subtropical high moves northward thus relieving southern China and northern Indo-China of the westerlies that dominated in the previous 10 days. For these regions the westerlies are replaced by easterlies emanating from the southern flanks of the western Pacific anticyclone. Convergence in the south-easterly flows over the South China Sea precipitates the development of the monsoon trough here. This subsequently moves northward and eastward as the equatorial westerlies increase in strength.

Stage three of the active–break cycle is characterized by a strong easterly flow over southern China. This coincides with the most active stage of the monsoon trough over the South China Sea. This contrasts with the situation for this region in stage one. Associated with the peak development of the monsoon trough is tropical cyclone activity. Over Indo-China the equatorial westerlies dominate and convective activity over northern Indo-China and southern China is suppressed. Rainfall over Malaysia, however, is at a maximum at this stage. The development of the active monsoon trough over the South China Sea as well as the south-west monsoon is enhanced by the cross-equatorial flow from the Australian region. To the north of the monsoon trough, easterlies are at a maximum and a long monsoon trough extends from the Bay of Bengal to Guam in the western Pacific (Ding, 1994).

In the fourth and final stage the easterly wind that dominated southern China for the previous 10 days begins to wane. At about the same time the monsoon over the southern parts of the Bay of Bengal begins to weaken significantly. Over the southern part of the South China Sea the cross-equatorial flow remains active. This helps to preserve the south-west monsoon over the southern South China Sea. Concurrently, the subtropical high pushes eastward to attain a position farther east than normal over southern China and northern Indo-China where convective activity is virtually absent. Here the equatorial westerlies are extremely weak or replaced by easterlies. At this time a double monsoon trough develops north and south of the equator in the western Pacific. This now becomes the site of maximum activity with the active formation of tropical disturbances. Disturbances in these troughs bring short but intense spells of rain to Malaysia, Borneo, Sumatra and the Java Islands (Cheang, 1991). Over the Indian region the monsoon is not generally active at this East Asian monsoon break stage.

Inter-annual variability of monsoon rainfall, referred to as the Meiyu ("plum rains") in southern China, is a product of variations in the behaviour of several of the East Asian monsoon circulation components. These components are especially affected when ENSO onset times are in the spring and summer. In such a situation the subtropical anticyclonic ridge over the western Pacific is displaced north-eastwards with a concomitant weakening of the anticyclonic system. Such a situation is unfavourable for the northward transport of moisture from the western flanks of the anticyclonic system. This leads to less moisture advection over eastern subtropical China and below normal rainfall amounts. This meridional behaviour of the subtropical anticyclone appears to be related to convective activity over the Indo-China–South China–Philippines region. When convective activity and thus rainfall is pronounced (subdued) in this region, which is usually associated with above (below) average SST in the western Pacific, anomalous atmospheric heating (cooling) causes the western Pacific subtropical high to move northwards (southwards) such that rainfall in eastern China is below (above) average. ENSO events have a pronounced effect on the patterns of convective activity from Indonesia to the Philippines. High SST in the western Pacific occur in non-ENSO years. Because of the teleconnections between SST/convective activity in the western Pacific/Indo-China region and western Pacific subtropical anticyclonic activity described above, below-average rainfalls over sub-tropical eastern China result in ENSO years (Ding, 1994).

ENSO causes other changes to the monsoon circulation components, including a southward displacement of the Australian high; a south-eastward extension of the ITCZ into the central Pacific, such that the cross-equatorial flow from the southern hemisphere is more easterly than usual and thus flows into the western Pacific instead of the South China Sea; the summer monsoon over India is at a break stage and the south-westerly monsoon flow from the Bay of Bengal cannot easily penetrate into southern and eastern China. For years in which the ENSO onset is in the preceding autumn and winter, the position and intensity of these circulation components in the spring and summer are closer to normal or further westward, thus producing near- and above-normal Meiyu rainfall (Ding, 1994).

EAST ASIAN WINTER MONSOON

With the onset of the northern winter the predominantly cyclonic circulation pattern over eastern Asia is replaced by an anticyclonic pattern (Figure 7.6). The associated high pressure centre reaches a great intensity at latitudes of about 40–60° over Mongolia and middle Siberia. Regions under the influence of this high pressure experience very low temperatures due to longwave radiation losses under minimal cloud cover conditions and minimal shortwave gains due to the high albedo of snow cover. Air moves out from the anticyclonic centre over Mongolia and middle Siberia in a southerly direction over Korea, China, Japan, Indo-China and the western Pacific. At about 15–20°N, over the South China Sea, the northerly airflow converges with the north-easterlies from the Pacific (Figure 7.6). The two air currents merge gradually on their way to the south-west, where they form the north-east monsoon of the Malaysian Peninsula.

The onset of the winter monsoon in this area in September is related to three major features of the circulation (Cheang, 1991). The first is the monsoon trough. By September a large-scale monsoon trough from the northern South China Sea has moved south and establishes a quasi-stationary position over the equatorial South China Sea. When in this area, the monsoon trough is often referred to as the "northern near-equatorial trough" (Cheang, 1991). North of this trough north-easterly trade winds prevail. The second feature is a surge of cold air within the northerly monsoon current from the northern Asian anticyclonic system. This frequently reaches the Malaysian Peninsula area and interacts with the Pacific north-east trades, often producing heavy rainfall (Cheang, 1977; Chiyu, 1979). The third feature is the reversal of upper level winds from easterly to westerly over southern China. This occurs when the north–south temperature gradient across the Asian continent reverses (Cheang, 1991). Over the Malaysian Peninsula the onset of the monsoon brings rain to the east

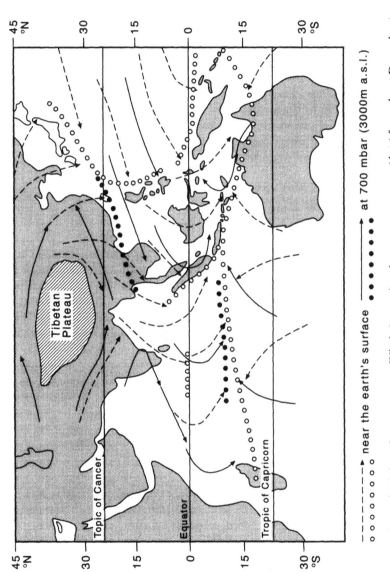

Figure 7.6   The Asian winter monsoon. Winds (arrows) and convergence zones (dots) from about December to March, near the earth's surface and at 700 mb (3000 m) and 500 mb (5800 m) levels

– – – – ▶ near the earth's surface ▬▬▬▶ at 700 mbar (3000m a.s.l.)
○ ○ ○ ○ ○ near the earth's surface ●●●●●

coast of this area. A distinct feature of the winter monsoon are cold surge events. These will be discussed in Chapter 8.

Compared to the summer monsoon circulation, the winter monsoon circulation is shallow. Near its origin the cold anticyclonic high does not reach to great heights confirming its thermal origin. Over most of the Asian continent, winds at the 700 mb level are westerly which means that the northerly winter monsoon current is confined to levels below 700 mb. Westerly winds flow mainly on the southern side of the Tibetan–Himalayan Plateau but a westerly stream is also maintained on the northern side. Both westerly streams meet in a zone of confluence on the eastern side of the Tibetan–Himalayan Plateau. The confluence zone may reach as far as southern Japan. Within this zone depressions may develop. Further to the south, where the winter monsoon is reinforced by the north Pacific trades, it reaches to the 700 mb level. Over northern Australia the winter monsoon circulation is shallow as easterlies prevail at the 700 mb level.

THE AUSTRALIAN MONSOON

The Australian summer monsoon is very much a southern hemisphere extension of the larger Asian winter monsoon system (Figure 7.6). The main northern hemisphere components of this circulation have been discussed in the East Asian monsoon section above. In addition to these are a couple of southern hemisphere components. These are the monsoon shear line or southern near-equatorial trough and the Pilbara and Cloncurry heat lows of north-western and north-eastern Australia respectively. These structural elements are season-long features of the Australian monsoon. They have associated with them distinct cloudiness and rainfall patterns. The monsoon shear line and its associated horizontal wind and vertical velocity patterns also display distinct spatial behaviour during active and break phases of the Australian monsoon.

The heat lows of northern Australia have both thermal and dynamic origins and are perhaps the most important factors in the establishment of the Australian monsoon (Suppiah, 1992). Once established, cyclonic inflow into the area of the heat lows follows. This dynamic process helps to augment the developing surface low pressures through convergence and ascent. Of the two heat lows, the Pilbara heat low is the most intense because of the greater land–sea temperature gradient on its north-western and western flanks (Suppiah, 1992).

Because the Pilbara and Cloncurry heat lows are very sensitive to changes in surface heating processes, the position of the monsoon shear line, and its associated trough, moves on daily and longer time scales, especially during active and break periods. Davidson, McBride and McAvaney (1983) have summarized the main structural properties of the active and break monsoon.

The active and break phases of the Australian monsoon coincide with the peaks and troughs of the 40–50 day tropical oscillation. This is clearly evident in near-surface and upper level winds (McBride, 1987) and cloudiness patterns (Kuhnel, 1989). Although clear 40-day peaks in rainfall, as found in the Indian monsoon area, have not been found for northern Australia, Hendon and Leibmann (1990) have observed a close correspondence between the low level wind patterns and rainfall, such that the strengthening of westerly winds, which occurs on the time scale of the 40–50 day oscillation, often brings increases in rainfall.

Inter-annual variations of the monsoon are manifest by variations in the nature of the 40–50 day oscillation, the onset and withdrawal of the monsoon and the total monsoon season rainfall. Much emphasis has been placed on the role of ENSO in determining inter-annual variability (Holland, 1986). The effect of this coupled ocean–atmosphere system on the 40–50 day oscillation appears to be clear in the Australian region as during ENSO (non-ENSO) years the eastward propagation of cloudiness associated with the 40–50 day oscillation is slow (fast), the cyclic variations of the oscillation are irregular (regular) and the time scale of the oscillation is longer (shorter) than the normal 40–50 day period. However, the extent to which ENSO forces the 40–50 oscillatory behaviour of cloudiness is uncertain given the contention that this oscillation itself may trigger the onset of ENSO events (Lau and Chan, 1986).

Leaving aside the triggering mechanisms, it is clear that there are ENSO signals in rainfall and temperature throughout the Australasian–Indonesian monsoon region (Quinn et al., 1978; Allan, 1988; Allen, Brodrfield and Bryan, 1989; McGregor, 1989, 1992). Although there are some discrepancies (McBride and Nicholls, 1983), generally drought in the Australian monsoon region is associated with ENSO events and negative SST anomalies over the eastern Indian Ocean and western Pacific. The correlation of ENSO with tropical drought is strongest for intense ENSO events when strong negative rainfall anomalies are recorded throughout the Australian monsoon region (Drosdowsky, 1993). However, the strength of the link between the SO and Australian summer rainfall may be variable on time scales from months to decades (Allan, 1991). At the decade time scale, the correlation between the SO and climate has been strongest since the mid-1940s and prior to the First World War. At the monthly time scale, correlations between the SOI and rainfall are weakest in the summer months. The link between the SO and climate is strongest for the winter and spring months and this appears to persist into the following summer, although at a much weaker level. Inter-annual variations in monsoon onset also appear to be related to ENSO events. Early onset precedes ENSO and cooler SSTs in northern Australian waters, while in ENSO years themselves, monsoon onset is delayed (McBride and Nicholls, 1983).

THE AFRICAN MONSOON

The monsoonal circulations over Africa differ from the Indian and East Asian systems in their magnitude, thickness of flow and geographical coverage. Futhermore, high latitude air masses are not involved. However, because of the shape of the African continent there are basic differences in the structure and physical properties of the monsoon systems between West and East Africa.

In West Africa, a large continental area north of the equator contrasts with the oceanic regions of the South Atlantic Ocean. Consequently, the two West African monsoonal winds demonstrate a good deal of difference in their physical properties. In East Africa, the continent stretches on both sides of the equator, though it is broader in the northern hemisphere. The two East African monsoonal winds therefore differ

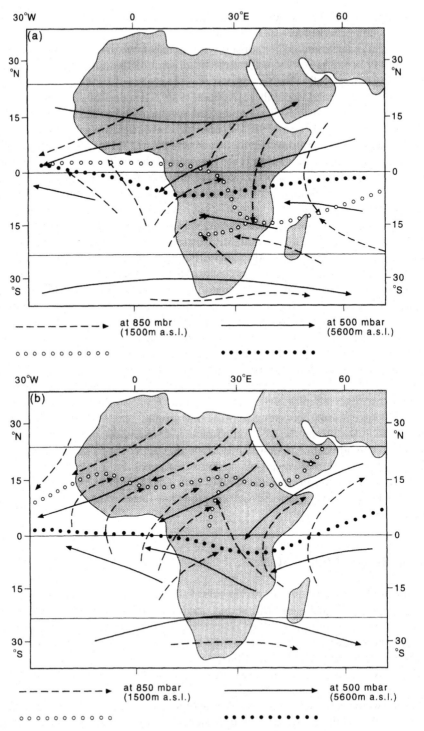

Figure 7.7   The circulation over Africa during (a) January and (b) July at the 850 mb (1500 m) and 500 mb (5800 m) levels. Arrows are winds, dots are convergence zones

mainly in their direction as there is little physical contrast between their source regions. Other factors in East Africa which accentuate the east–west differences are large meridional mountain ranges on either side of the Great Rift Valley, large inland lakes which may modify monsoon flows, and the influence of the Indian monsoon.

In West Africa the main structural elements of the monsoon are the surface pressure trough at the zone of maximum surface heating, the zone of confluence between north-easterly and south-westerly winds as represented by the ITCZ with its associated wind flow discontinuity, the linear bands of cloudiness and rain called "squall" or "disturbance lines" (Chapter 8) and the South Atlantic (St Helena) and North African (Saharan) subtropical high pressure systems (Figure 7.7).

In the northern hemisphere winter, West Africa is largely under the influence of north-easterly trade winds (Figure 7.7(a)). These emanate from the semi-permanent subtropical high pressure system which establishes itself over northern Africa during the northern winter. The north-easterlies prevail to an elevation of about 3000 m and bring dry and stable air masses, which often carry dust particles from the desert regions over which they originate. These winds are locally known as the "Harmattan" in many parts of West Africa. At this time of the year the ITCZ is located to the south over the southern West African coast where the rainfall is confined. For most of West Africa the northern hemisphere winter is the dry season.

During the northern hemisphere summer a thermal low pressure area builds up over the African continent at about 20°N. Consequently the ITCZ, often referred to locally as the Intertropical Discontinuity or the Intertropical Front, slowly moves northward to reach a position at around 15°N (Figure 7.7(b)). Similar heat troughs also develop over the North Atlantic Ocean. Concomitantly, over the South Atlantic the semi-permanent St Helena anticyclonic system builds up. The development of this is reinforced by the northward flow along the south-west African coast of cold water in the Bengula current. As a result a considerable pressure gradient develops between the South Atlantic Ocean and northern central Africa. In response, south-easterly flows make their way northward where they recurve to become the south-westerly monsoon flow over West Africa (Figure 7.7(b)). These cool and moist flows push north into sub-Saharan Africa.

Associated with the northward progression of the cool south-westerly moist air masses are zonally oriented disturbance lines, which bring much needed rain to the sub-Saharan area. These do not represent the air mass discontinuity between the dry hot north-easterly and cool moist south-westerly flows. This occurs further south. At the sub-Saharan latitude, northerly hot dry continental air is less dense than the cool moist south-westerly monsoonal flow and is therefore uplifted. As only the oceanic air masses produce the precipitation, the zone of maximum rainfall is where the air masses are thicker, i.e. further south (Figure 7.8). Because of these meridional air mass differences, distinct weather zones exist. North of the ITCZ, in the southern Sahara area, conditions are generally cloudless as this area remains under the influence of subsiding north-easterly flows. In the vicinity of the flow discontinuity, there is limited cloud development and little rain, as north-easterly flows cap the leading edge of the south-westerly monsoonal flow. South of this zone cloud development reaches a maximum, as does rainfall. This zone also possesses the most changeable weather, especially in its equatorward parts. In this zone the depth of the monsoon flow is

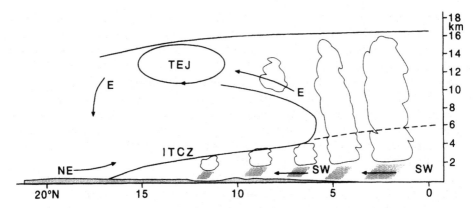

Figure 7.8   Meridional cross-section of the troposphere over West Africa during August (after Leroux, 1973). Vertical exaggeration is about 50 times

around 3000 m, above which north-easterly air dominates. Despite the presence of an inversion at around 6000 m, produced by the warm north-easterly air capping the cool moist south-westerly air, the vertical development of the cumulonimbus, associated with the zonally oriented disturbance lines, reaches the upper troposphere. In the upper troposphere, to the north of the disturbance lines, the tropical easterly jet prevails. This separates a northern zone of upper tropospheric convergence and subsidence associated with the subtropical high pressure belt over the Sahara and a southern upper tropospheric zone of divergence which represents outflow from the top of the ITCZ where vertical movements of air are strong (Figure 7.8).

In East Africa several major structural elements of the monsoon can be identified. These include the ITCZ, the Arabian and Mascarene high pressure systems, the East African low level jet stream, West African mid-tropospheric jet stream, the tropical easterly jet stream and the two subtropical westerly jet streams. Two major monsoon wind systems dominate over East Africa. These are the north-easterlies, which originate from the Arabian high over north-eastern Africa and the Arabian peninsula, and the south-easterlies which have their origins in the south-western regions of the Indian Ocean in the area of the Mascarene anticyclonic system (Figure 7.7). Both these anticyclonic systems, through variations in their intensity, location and orientation, play an important role in determining the strength of the monsoonal flows and thus the rate of moisture transport over East Africa. The St Helena anticyclonic system over the South Atlantic also exerts an influence in East Africa. It does this through its control on the rate of moisture flux from the Atlantic into central southern Africa in the area of Zaire and Congo. In this region south-easterly flows from the Mascarene high converge with the westerly flows from the St Helena high in an area known as the Zaire air boundary. This has a meridional orientation and is a major convergence zone.

Because the West African south-west monsoon has its origins over the east Atlantic ocean, climatologists have looked for, and found tenable ocean and atmosphere explanations in this region for monsoon variability. For dry years these include the following:

1. a retardation of the northward movement of the south-west monsoon by up to 200–300 km short of its long-term mean position (20°N in August);
2. a southward displacement of the intertropical discontinuity, the region where the northern and southern hemisphere Atlantic trades meet;
3. southerly displacements of the near-equatorial trough and the zone of maximum surface pressure north and south of the equator (the Azores and St Helena highs respectively); and
4. anomalously cool (warm) water to the north-west (south-east) of a line linking south-western West Africa and north-east Brazil (Lamb and Peppler, 1991).

In addition to these surface or boundary controls, internal dynamic controls in the form of the upper tropospheric easterly jets north and south of the equator may also play a role in determining the variable timing of monsoon onset. This is because jet-related upper atmosphere divergence patterns influence the position of lower level convergence (Newell and Kidson, 1984).

Physical and dynamic changes occur to the south-west monsoon in dry (normal and wet) years. These include: (1) smaller (higher) lower tropospheric humidities or mixing ratios; (2) less (greater) water flux convergence; (3) a more northerly (southerly) and stronger (weaker) 700 mb sub-Saharan easterly wind maximum; (4) moisture supply from north (south) or west; (5) shallow (deep) invariant (variable) south-westerly flow depths; (6) smaller (greater) size and degree of disturbance-line organization and less (more) intense disturbance-line rainfall (Lamb and Peppler, 1991; Bell and Lamb, 1994).

The regional ocean–atmosphere mechanisms described above do not always account for monsoon variability and the occasional failure of the monsoon, as occurred in 1983. As a result, climatologists have looked further afield for possible global ocean and atmosphere explanations for variations in West African monsoon dynamics at the decadal and inter-annual time scales. On the interdecadal scale, it appears that dry monsoon years are associated with SST warmer than normal in the southern and equatorial oceans, and in the whole Indian Ocean, and with colder than normal SST in the northern oceans (Palmer, Folland and Parker, 1986; Folland et al., 1991). However, at the inter-annual time scale this global pattern is not well established (Adedoyin, 1989; Ward, 1992).

Although global scale patterns may not be able to explain inter-annual monsoon variability, it seems likely that eastern Pacific SST anomalies may play a role through creating anomalous upper tropospheric equatorial westerlies which extend eastwards over the Atlantic Ocean to Africa (Palmer, 1986). That these may well explain the extreme rainfall anomaly patterns associated with the 1983 drought has been partially confirmed by empirical and climate modelling studies (Janicot, 1992, 1994). These have revealed that the responses of West African monsoon dynamics to eastern Atlantic and eastern Pacific SST variations are quite different. It appears that West African rainfall may well be influenced by the displacement of a zonal circulation between the eastern Atlantic and Pacific Oceans or variations in its intensity. This zonal east to west circulation is characterized by an ascending branch over Central America (western dipole) and a descending one over Africa (eastern dipole) at about 30°W (see Figure 6.3), and is responsive to SST variations in the eastern Pacific and Atlantic Oceans. For

El Nino or warm eastern Pacific (La Nina – cold) years the east to west circulation is stronger (weaker) than normal, thus producing abnormal subsidence (ascent) over West Africa and thus below (above) average convectional activity. In years when these teleconnection patterns are weak and SST and atmospheric pressure in the dipoles of the zonal circulation vary in the same way, then coupling between the eastern Pacific and eastern Atlantic is weak. In such years eastern Atlantic SST appear to be predominant in determining West African monsoon dynamics largely through their control on the location of the ITCZ as suggested by Lamb and Peppler (1991).

Compared to West Africa, little work has been done on understanding the mechanisms that may control the inter-annual variability of the East African monsoon. Although ENSO signals are present in East African seasonal rainfall, these are not region-wide. For example, equatorial eastern Africa experiences above-normal rainfall from October of an ENSO year through to the following April. Drier than normal conditions are experienced from March to May of an ENSO year. This behaviour is opposite to that for northern parts of eastern Africa which experience wet conditions when equatorial regions are dry. The ENSO-related features for abnormal monsoon behaviour in this region are as follows:

1. anomalous upper westerly flows;
2. the movement of a mid-tropospheric trough eastward over the Mozambique channel, and the related building of a mid-tropospheric anticyclone over south-eastern Africa; and
3. a weakening of the tropical easterlies over southern Africa, as well as lower level convergence.

Associated with these developments is a regime of subsidence and low atmospheric moisture contents. These changes led to a weakening of the monsoon flows and the development of a very diffuse ITCZ (Atheru, 1994). SST patterns in the Arabian Sea and Indian Ocean are also important. Generally, anomalously warm SST in these regions lead to a weakening of the Arabian and Mascarene high pressure systems. Consequently, the related monsoon flows are weaker.

EASTERN NORTH PACIFIC MONSOON

Mainly due to the lack of observations of the annual climate cycle in the eastern Pacific, this region has not been traditionally considered a monsoon region. Analyses of climatological data for the Pacific Ocean coast of Mexico (Douglas et al., 1993) and over the eastern north Pacific (Wang, 1994) have, however, led to the conclusion that a monsoon system exists in this region because (1) seasonal rainfall distributions along the Pacific coast of Mexico are similar to those of monsoon Asia; (2) a distinct ocean summer rainfall season exists, as reflected by low outgoing longwave radiation amounts in July; (3) temperatures peak prior to the onset of the rains; (4) an annual reversal of winds up to the 200 mb level to the south of the Mexican anticyclonic system occurs; and (5) an annual reversal of the surface wind system from easterly in January to westerly in July occurs (Douglas et al., 1993; Wang and Murakami, 1994). The triangular region for which these monsoon characteristics exist is defined by the northern boundary of the equatorial cold SST tongue in the south; to the west a straight line between 5°N, 120°W and 20°N, 100°W; and to the east the Central American land bridge (Wang and Murakami, 1994). This system has been called the Eastern North Pacific monsoon as opposed to the Mexican monsoon because stronger monsoonal characteristics exist over the ocean compared to the Mexican coast; the eastern north Pacific ocean is the moisture source of the summertime rainfall peaks over land; the mechanisms that drive the land-based "Mexican monsoon" are mainly oceanic as opposed to land-based ones.

SUMMARY

The monsoons are characterized by the seasonal reversal of the prevailing wind systems and seasonal contrasts in regimes of cloudiness, precipitation and temperature. Although three main monsoon systems were traditionally recognized, it is now considered that the Asian monsoon comprises two subsystems in the form of the Indian and East Asian monsoons. A distinct feature of the latter is a very strong winter signature, a characteristic not possessed by its Indian counterpart or any of the other monsoon systems. Recent studies have also revealed clear differences in the monsoon systems between western and eastern Africa, while a further monsoon system has possibly been identified over the eastern Pacific.

Differential seasonal heating of the oceans and continents, moisture processes in the atmosphere and the earth's rotation are the main factors which explain the existence of the monsoons which are economically important as they bring with them rainfall on which the economies of many tropical regions depend. Variations in the monsoons occur at the intra-seasonal time scale in the form of active and break phases. These are closely tied to the MJO discussed in the previous chapter. At the inter-annual time scale, internal dynamic and boundary forcing mechanisms are responsible for the variations in the onset, duration, intensity and cessation of the monsoons. Although there are clear ENSO signals in monsoon rainfall, the extent to which this phenomenon alone "controls" monsoon rainfall, especially in the Indian, East Asian and northern Australian regions, is not entirely clear as recent research suggests that a complex interaction between Eurasian snow cover, Indian monsoon rainfall and the zonal atmospheric circulation over the Pacific appears to be at play. The nature of this interactive behaviour will have to be unravelled before the long-range forecasting of monsoon rainfall can be done with a great deal of accuracy.

CHAPTER 8

Tropical Disturbances

Superimposed on the low frequency seasonal and non-seasonal variations of the tropical atmosphere which were discussed in Chapters 5–7 are higher frequency variations due to tropical disturbances. These arise from temporary aberrations in the pressure and subsequently the wind and moisture fields in the atmosphere. Such disturbances, which are of widely different duration, size and intensity, give rise to the short-term weather changes in the tropics and are therefore of prime interest to the meteorologist and forecaster. Climatologically, tropical disturbances are of importance where they are so frequent or intense that they have a clear impact on average or extreme conditions. This is the case in many parts of the tropics, because despite their general irregularity, many tropical disturbances exhibit clear maxima in their frequencies of occurrence, with regard to both their seasonal and diurnal distribution.

The main condition for the formation of tropical disturbances is the presence of warm and humid air masses, in which no inversion exists. This type of air mass can easily become unstable. In this state any upward movement of air will be reinforced by the release of large amounts of latent heat of condensation, which is the main source of energy of tropical disturbances. The rising air movements can therefore reach high levels. This requirement of instability explains why tropical disturbances are less numerous in dry air masses than in humid ones, and why they hardly ever develop over cool ocean currents and areas where the trade wind inversion is strong.

There are several factors which can create instability in suitable air masses. The most common is simple heating from the earth's surface, which increases the lapse rate of the lower layers of the troposphere. The resulting upward movement of air, called convection, takes place predominantly in relatively small cells, with a diameter of the order of a few kilometres. These convection cells are easily identified by their typical cumulus or cumulonimbus clouds.

Instability may also be an indirect product of convergence, confluence or orographic forcing. Convergence occurs where air flows decelerate as occurs in the trades as they approach equatorial latitudes (Chapter 5). Convergence at the surface is matched by divergence aloft, which helps maintain the upward movement of air. It should, however, be noted that some disturbances may well be the product of divergence aloft as these may initiate surface convergence (Chapter 2). Convergence may also be a product of increased surface friction. This often occurs over land as a result of friction reducing the Coriolis force. Confluence also results in vertical air movements and thus instability indirectly. This finds its best expression in the ITCZ (Chapter 5) where the moist humid trade winds from either hemisphere meet. As the two major wind systems meet, air is forced to rise. Orographic forcing of flows of air by mountains, hills and coastlines also causes lifting of air.

Convergence, confluence and orographic forcing cause instability indirectly because they provide the opportunity for vertical movements of air. If these movements result in warm moist air in the lower levels being forced to upper levels, where the air is drier, convective or potential instability will result. This occurs because when moist air is forced upwards into drier air, it will cool at a lapse rate greater than the saturated adiabatic lapse rate. Consequently, the lapse rate is steepened and the air becomes unstable, often leading to an overturning of the layer of air which initiates further vertical air movements.

Vertical movements of warm moist air lead to a variety of tropical disturbances which occur over a range of space and time scales. These will be discussed in the following sections and include convection cells occurring individually or in groups with a quasi-circular or linear form, cyclonic vortices in the form of tropical cyclones and wave disturbances or easterly waves. All these disturbances originate within the tropical latitudes. However, some disturbances have an extra-tropical origin such as cold surges and subtropical cyclones. These will also be discussed.

THUNDERSTORMS

It is open to question whether thunderstorms can be properly classified among disturbances, not only because of their small size but also because of their short duration. However, as they are the basic spatial elements of a number of tropical disturbances, ranging from mesoscale convective complexes to tropical cyclones, and as they contribute to the majority of the vertical fluxes of heat and momentum within the tropical atmosphere, they warrant consideration.

Thunderstorms are basically individual convection cells. Their origins are mainly due to instabilities in the atmosphere. Such instabilities come about due to free convection (surface heating) or forced convection (orographic forcing). The outcome of these convectional processes is the steepening of lower level lapse rates and the realization of conditional instability in the case of moist adiabatic processes, or convectional/potential instability in the case of upper level dry air.

Individual thunderstorm cells are usually localized affairs as they often form and decay *in situ*. They rarely reach diameters of more than 10 km and have on average durations of 1–2 hours. Where thunderstorms prevail over longer periods it is merely a repetition of the short process. Well-developed thunderstorms composed of cumulonimbus, reach well up into the troposphere and beyond, attaining heights of 20 000 m with vertical wind velocities (updraughts) of 10–14 m s^{-1}. Such well-developed deep convective systems are often referred to as "hot towers" because of the large releases of latent heat of condensation associated with their development.

It has been estimated that at any one time about 3000 thunderstorms are taking

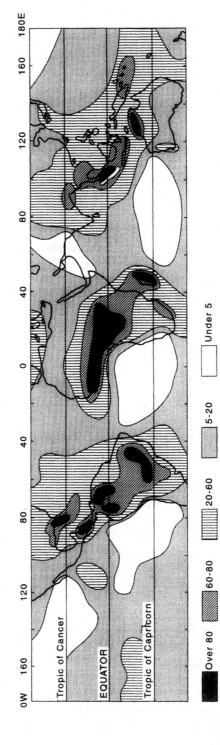

Figure 8.1   Mean number of days with thunderstorms per year

place near the earth's surface. Most of these occur in the tropics, where their intensities are also generally higher than their mid-latitude counterparts. Many parts of the tropics have, on average, well over 40 thunderstorm days throughout the summer months (Figure 8.1). For Africa, well over 20% of the continent records more than 100 thunderstorm days a year. Some locations even have thunderstorm days in excess of 200 per year: 242 at Kampala, Uganda; 221 at Bukavu, Zaire Republic; 216 at Calabar, Nigeria. In the equatorial region of Asia and over insular and peninsula Asia, thunderstorm days may reach over 180 per year.

The basic conditions for the development of thunderstorms over land are warm and humid air masses, which can become unstable over considerable vertical layers. Areas where cool or dry air masses prevail, like the eastern parts of the ocean basins and the deserts, have few thunderstorms. Although Figure 8.1 shows greatest thunderstorm frequencies over the land, this is a little misleading and related to the lack of data for the ocean basins, as here thunderstorm activity for some areas, such as the ITCZ, can occur almost on a daily basis. Over the oceans, convection and thus the potential for thunderstorm activity, is greatest where sea surface temperatures exceed 26 °C.

It is generally estimated that a thunderstorm will develop when the instability reaches levels around 8000 m. Whether this condition will be met depends largely on local conditions, which trigger off the process by causing an initial upward air movement. The most common factor is convection, resulting in widely scattered thunderstorms developing over places where surface heating is most intense. Nighttime convection over water bodies and advection of cold air aloft are other factors which can cause scattered thunderstorms.

The thunderstorm process itself can be divided into three stages, each lasting between 20 and 40 minutes (Figure 8.2). In the initial or cumulus stage, strong updrafts prevail in the thunderstorm cell and the cumulus cloud grows rapidly upward, reaching levels around 8000 m. In this stage little or no precipitation takes place and lightning is rare.

In the second or mature stage, the thunderstorm reaches its highest intensity. Updraughts continue to be strong, but in some parts of the cell they are replaced by downdraughts. These are accompanied by precipitation. This precipitation is rather intense and strongly localized: it is very high over a small, well-defined area, but outside that area no rainfall is received. In this stage, the cumulonimbus cloud frequently reaches levels of 18 000–20 000 m, often developing an anvil head which is caused by upper tropospheric wind shear.

In the final or dissipating stage downdraughts prevail in most of the cell and lighter rain falls. At this stage the convection cell is deprived of a source of supersaturated air so that cloud droplets can no longer grow and the self-fuelling mechanism of condensation and latent heat release ceases. Consequently, precipitation stops and the cloud mass decays by evaporation.

So far our discussion has been concerned with single convective elements. However, these often occur in complexes or groups referred to as mesoscale convective complexes with dimensions of around 100 km across. These in turn may aggregate into large cloud clusters which attain dimensions of 100–1000 km in diameter. Although outnumbered by individual cumulus and cumulonimbus, these clusters dominate the cloudiness and rainfall of the tropics. Typically their lifetimes are longer than a few

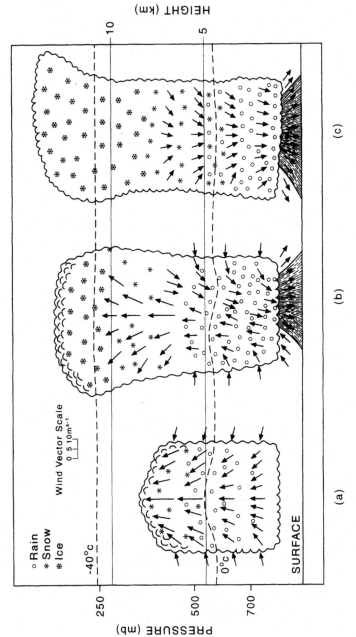

Figure 8.2 Three-stage development of a thunderstorm. From Wallace and Hobbs (1977). Reproduced by permission of Academic Press, London

hours and as much as several days. They may form parts of much longer lived weather systems and may develop into tropical storms or linear systems such as squall lines (Hastenrath, 1985).

LINEAR SYSTEMS

Linear systems consist mainly of numerous thunderstorms organized into mesoscale convective complexes or cloud clusters, aligned in bands, which because of a common origin, develop and move more or less as an organized system. These systems, also referred to as squall lines or linear disturbance lines, can be hundreds of kilometres long, while their width is in the order of 10–30 km. They normally do not consist of uninterrupted cumulonimbus clouds, but exhibit zones of stronger and weaker thunderstorm activity. They are characterized by their explosive growth, rapid propagation and in plan view their convex leading edge. Although the individual convective elements may only last for an hour or so, the whole squall line system can last from 3 to 15 hours. Occasionally durations of several days may occur. Squall lines are of climatological importance as they contribute significantly to the annual rainfall total where they occur. Squall lines have been documented for a large number of tropical locations including northern Australia, north-western India, Bangladesh, Malaysia, Indonesia, South America–Caribbean and West Africa. Of these regions the squall lines of West Africa have been studied the most.

Several factors appear to be important for the generation of the West African squall lines. These include the diurnal solar cycle, topographic features, easterly waves, the large-scale flow and the atmospheric moisture content (Rowell and Milford, 1993). Regionally specific factors such as land and sea breezes are also likely to be important.

Because the frequency of occurrence of squall lines demonstrates a clear mid- to late afternoon peak, this indicates the importance of surface heating for the release of conditional instability in the boundary layer. Release of the instability may also be enhanced by orographic effects, as African squall lines show a tendency to generate immediately west of the Air plateau between 16 and 19°N and 7 to 9°E; just west of the Adrar des Iforas mountains (20°N, 2°E), and over the Jos Plateau around 9–12°N and 7–10°E (Rowell and Milford, 1993). The role of African easterly waves (next section) is unclear as some researchers have suggested that thermodynamic conditions just ahead of the waves present ideal conditions for squall line growth (Reed, Norquist and Recker, 1977) while others have found no association between the passage of these waves and squall line generation (Bolton, 1984). However, it is possible that there may be some geographical variation in the relationship between easterly waves and squall line development such that towards the West African coast, where the easterly wave disturbances grow more intense, squall line development could be associated with these disturbances. However, further to the east in the area of easterly wave generation there appears to be little evidence of a relationship between these two weather systems (Rowell and Milford, 1993). In the case of West Africa, the component of the large-scale flow believed to be important for squall line initiation is the African easterly jet, especially its influence on low level vertical wind shear.

Atmospheric moisture is a prime ingredient for squall line development and mainte-

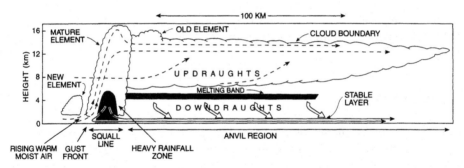

Figure 8.3   Model of squall line structure. Modified from Gamache and Houze (1982)

nance. The generation of a squall line is associated with moist lower levels and drier air at middle levels. The latter, because of its effect on steepening lower level lapse rates, is an important prerequisite for rapid squall line development. Dry middle air is also partly responsible for the generation of gust fronts at the leading edge of squall lines. Evaporation of precipitation in the squall line environment cools the mid-level dry air. Consequently, this descends in the form of downdraughts. These in turn undercut the warm moist low level air which enhances the development of convection and thus further rainfall production. This positive feedback mechanism is important for sustaining convective cell growth at the squall line's leading edge and thus its propagation. Variations in atmospheric moisture at the intrastorm and seasonal time scale are also important for squall line development. Throughout the squall line's life, convection operates to overturn the atmosphere such that the ideal vertical moisture structure for squall line development is destroyed. As a result there is often a period of quiescence following squall line passage as low level moisture levels rebuild (Rowell and Milford, 1993). At the seasonal time scale, the depth of the moist layer within the south-west monsoon varies. At the height of the monsoon the middle level dry layer is often absent, so that squall lines often display peaks in frequency at the beginning and end of the monsoon season, especially on the southern coast of West Africa, where the monsoon reaches its greatest depth during mid-summer.

Squall lines have a distinct structure which makes them different to mid-latitude squall lines. The most distinctive and contrasting feature is the trailing anvil composed of decaying convective elements (Figure 8.3). In mid-latitude squall lines this anvil is positioned forward of the squall line. Although the precipitation that falls from the trailing anvil may be important in terms of total squall line rainfall, it is not as intense as that which falls from the mature convective elements in the squall line. At the rear of the mature convective elements cool divergent downdraught air occurs. This cool air outflows to the rear to produce a stable layer in the anvil region as well as a gust front at the leading edge. The gust front air subsequently meets warm moist air at the leading edge and aids in the development of new convective elements which eventually form the new leading edge.

While the West African squall lines are basically a part of the large-scale south-west summer monsoon wind system, local wind systems can also generate squall lines by confluence and convergence. Convergence of sea breezes is known from many islands, and even from large peninsulars such as Florida and Malaya. The convergence can

Plate 1   Bangkok, Thailand during a pollution episode. Note the mixture of building styles typical of rapidly growing tropical mega-cities; squatter to high rise. Such cities may well lose their utility as places to live as air quality and human thermal comfort conditions deteriorate as the population grows and climate changes.

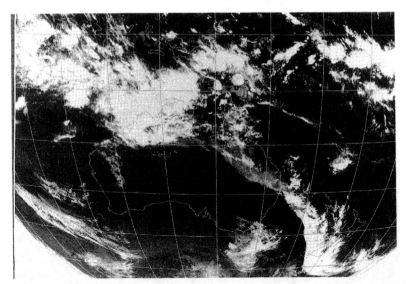

Plate 2   An active ITCZ spans the Southeast Asian–Pacific region. Note the line of intense convective clusters extending southeast from Indochina, and passing north of New Guinea and out towards the central Pacific. Supplied by Australian Bureau of Meterology.

Plate 3   Sea breeze front advancing over the Fly River delta, Papua New Guinea.

Plate 4   Typhoon Isa on 19 April 1997. Note spiralling cloud bands and the well defined eye.
Japan Meteorological Agency GMS image supplied by Royal Observatory of Hong Kong.

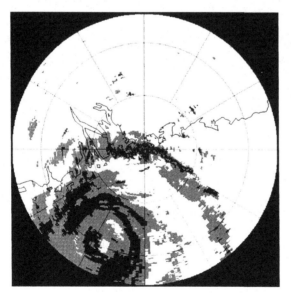

Plate 5   Radar image of typhoon Sally lying over the south China coast in the vicinity of Hong
Kong on 9 September, 1996. Note clear rain band structure; dark tones indicate intense rain
bands. Supplied by Royal Hong Kong Observatory.

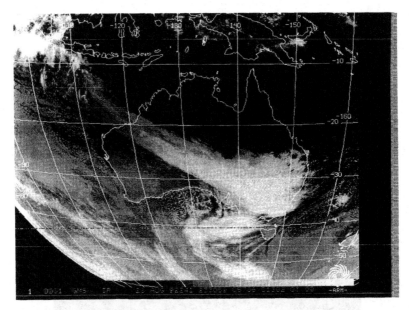

Plate 6   Tropical extra-tropical cloud band stretching from southeast Australia to the Indonesian archipelago. Supplied by Australian Bureau of Meteorology.

Plate 7   Cumulus developed over islands south of Ponape Island, central Pacific as a result of orographic forcing of the sea breeze. Such cloud development can lead to afternoon showers.

Plate 8   Soil erosion as a result of intense rainfall on devegetated tropical soils.

Plate 9   Heavy cumulus cloud cover associated with the ITCZ positioned over the Papua New Guinea Highlands during the wet season. Note rainstorm and associated thundercloud in middle distance.

Plate 10   Cleared forest patch for subsistence gardening.

Plate 11   Strip logging of tropical forest. This is one of the main causes of deforestation in the Amazon Basin. Supplied by C.A. Nobre, Centro de Provisio de Tempo e Estudos Climatico, Brazil.

Plate 12   Satellite image of deforestation in the Amazon Basin. Note the herring bone pattern due to strip logging (Plate 11). Abracos Forest is the site of deforestation and climate studies (see Gash et al., 1996). Supplied by C. A. Nobre, Centro de Provisio de Tempo e Estudos Climatico, Brazil.

Plate 13   Mangrove communities. Just one of the tropical coastal ecosystems that will come under threat from sea level rise due to climate change.

Plate 14 Kaafu Atoll in the Maldives which is not more than two metres above sea level is just one of the many atolls on which human habitation would be rendered impossible as a result of an enhanced Greenhouse Effect related sea level rise. © Andy Crump/Still Pictures.

also be between land breezes over a narrow strait, such as the Straits of Malacca, or between a local wind and a major air current of the general circulation. Sea breezes sometimes develop against the mean general wind direction, creating a linear system known as a sea breeze front (Nieuwolt, 1973).

As an example of a linear system caused by many factors and occurring outside the West African region, the "sumatras" of south-west Malaya may be described. They consist of a band of cumulus and cumulonimbus clouds with a length of 200–300 km, which forms over the Straits of Malacca. These disturbances develop only during the south-west monsoon season (May–September), usually during the night. They move slowly with the main monsoon current towards the coast of Malaya where they arrive during the late hours of the night or in the early morning. They are most frequent between Singapore and Port Swettenham, where the Straits are only 40–80 km wide, and occur less often further north.

The origin of the sumatras is not related to a major air boundary, because most of them are formed within one airstream. Three main factors can usually be recognized in their development. Firstly, the air masses of the rather sluggish south-west monsoon are heated, during the day, over the extensive lowlands of Sumatra. Consequently, they often become convectively unstable. During the night, when they are situated over the Straits of Malacca, this instability is reinforced by radiation losses from the cloud tops.

Secondly, when those air masses reach the Malayan south-west coast, they are uplifted orographically. They can also be undercut by the land breeze. Though this breeze is usually rather weak when cloudy conditions prevail, it is locally intensified where the form of the coastline is concave, resulting in convergent land breezes.

Thirdly, where the Straits of Malacca are narrow, land breezes from Malaya and Sumatra may converge. This last factor explains why sumatras are most frequent in the southern parts of the Straits.

The duration of a sumatra is about 1–2 hours. Wind gusts accompanying the squall can reach speeds up to $20 \, \text{m h}^{-1}$. The real importance of the sumatras, however, is their rainfall, as an individual sumatra may bring up to 80 mm. Due to their high frequency they make a significant contribution to total monsoon rainfall. They are also responsible for the diurnal rainfall maximum (Chapter 10) during the early hours of the morning which prevails at Malacca and other stations along the south-west coast of Malaya during the south-west monsoon (Nieuwolt, 1968).

Because of the importance of linear systems for the rainfall climatology of the areas in which they occur, any inter-annual variability in the requisite atmospheric conditions for their formation will have dramatic effects on annual rainfall totals. For example, it has been suggested that the dramatic decrease in Sahelian rainfall since the 1950s may be attributed to a progressive decrease (increase) in the number of large (small) squall lines and also decreases in the monsoon season average squall line size and intensity (Bell and Lamb, 1994).

EASTERLY WAVES

Although the trade winds are generally very steady and regular winds, which bring stable weather conditions over most areas where they prevail, in some parts of the tropics this tranquil situation is at times interrupted by westward travelling disturbances. These show a number of variations in size and intensity, but they have one characteristic in common: their main centre of low pressure is not circular or elliptic, but in the form of a wave in the isobaric pattern. This trough of low pressure extends in a south-west to north-east direction in the northern hemisphere. As these waves move with the easterlies they are generally called easterly waves.

Easterly waves are best known in the Atlantic–Caribbean area, but similar waves have been described for the Pacific Ocean. In the Atlantic region easterly waves generally occur at about 15°N, have a wavelength of between 2000 and 4000 m and move westwards at 6–7° per day longitude or roughly 8 m s^{-1}. Typically they have a life span of around 1–2 weeks. They are most common over the western parts of the Atlantic; however, they have their maximum amplitude over the West African region where they originate in the region of the African tropospheric easterly jet. An instability in this jet, known as baroclinic–barotropic instability (potential vorticity begins to decrease north), is known to generate the easterly waves. Two preferred regions of development either side of the easterly jet have been identified. These are downwind of the Hoggar mountains in the Sahara between 18°N and 25°N and 10°W and 5°E, and in the main West African rainbelt between 8°N and 15°N and 0° and 10°E (Thorncroft and Hoskins, 1994). Disturbances originating in these regions often merge downwind off the west coast of Africa (Reed, Norquist and Recker, 1977). As the easterly waves pass over the cool eastern Atlantic waters they often weaken considerably but may reach the western Atlantic, Caribbean and as far as the Pacific.

The formation, maintenance and development of easterly waves involves a complicated interaction between adiabatic dynamics, boundary layer, moist and radiative processes. For their maintenance it appears that barotropic processes are more important over the oceans while baroclinic processes are more important over land (Thorncroft and Hoskins, 1994).

A typical easterly wave covers a large area (Figure 8.4). It is usually relatively weak near the surface, reaching its strongest development at around the 700 mb level where vorticity is at a maximum. In their lower levels, easterly waves lag behind the main environmental flow; but at higher levels where the trades are weaker, they move ahead of the main flow. Ahead of the easterly wave, flow is divergent with a marked eastwards component (Figure 8.5). Behind the wave, however, air is strongly convergent. As air behind the trough axis has a northward component it gains both absolute and relative vorticity as it approaches the trough axis. Consequently there is vertical stretching of the atmospheric column which encourages low level convergence. The opposite is true ahead of the wave as air here has an equatorward flow so the vorticity components decrease. Consequently there is vertical shrinkage of the atmospheric column and thus horizontal divergence. These contrasting patterns of divergence and convergence, westward and eastward of the trough axis respectively, produce contrasting weather patterns such that ahead of the easterly wave the weather is particularly fair as the trade wind inversion is temporarily reinforced and moved to a lower level by the subsident air. At the rear of the main axis, in the zone of convergence, squall lines prevail and the trade wind inversion is destroyed.

Easterly waves are climatologically important as they bring large amounts of rainfall to areas which are generally dry as long as the trades are undisturbed. Easterly

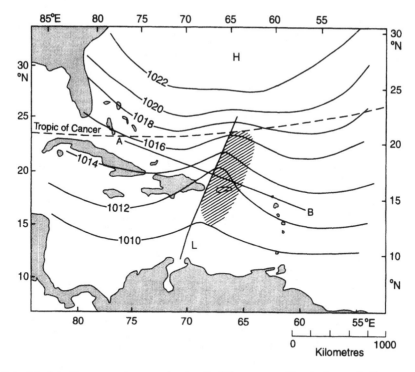

Figure 8.4   Model of an easterly wave in the Caribbean area. Hatched area indicates main rainfall zone

waves are the main reason for the late summer rainfall maximum observed in many islands of the Caribbean and in the western parts of the Pacific Ocean. In these areas the rainfall is strongly concentrated during only a few days per year. For instance, at Oahu on the Hawaiian Islands, over a period of 10 years, 66% of the annual rainfall total was normally received over about 10 days during the passage of easterly waves (Mink, 1960).

Fully developed or remnant easterly waves often provide the initial disturbance required for the development of tropical cyclones.

TROPICAL CYCLONES

A tropical cyclone is the generic term for a synoptic-scale low pressure system with no fronts, occurring over tropical or subtropical waters with organized thunderstorm activity (Holland, 1993) (Plate 4). Tropical cyclones possess a cyclonic surface wind circulation and are characterized on the surface weather map by almost circular isobars around the centre of low pressure. Tropical cyclones are called hurricanes in the north Atlantic Ocean, the north-east Pacific east of the dateline, or the south Pacific Ocean east of 160°E; typhoons in the north-west Pacific Ocean west of the dateline; a severe tropical cyclone in the south-west Pacific Ocean west of 160°E or south-east Indian Ocean east of 90°E; a severe cyclonic storm in the north Indian

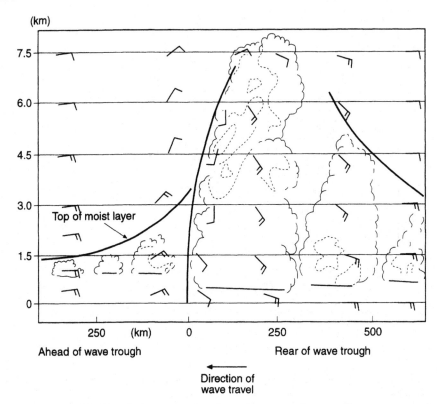

Figure 8.5 Model of vertical cross-section through an easterly wave. Convergence and vorticity increase to the rear of the wave in poleward-flowing air while there is divergence and decreasing vorticity to the fore of the wave in equatorward-flowing air (see Figure 2.7). From Sumner (1988).

Ocean; and tropical cyclones in the south-west Indian Ocean (Neumann, 1993). These weather systems have diameters of around 500–800 km, but some typhoons have been larger. Pressure gradients within tropical cyclones are steep, with the centre of low pressure around 950 mb, but as low as 870 mb as recorded for Typhoon Tip in the north-west Pacific in 1979. Because very strong pressure gradients prevail near the centre of the cyclone, wind speeds can be high: they usually exceed $33 \, m \, s^{-1}$ and winds over $55 \, m \, s^{-1}$ have been observed. In many cases higher wind speeds could not be recorded because instruments had been damaged or destroyed. Maximum sustained wind speeds are used to classify tropical cyclones in terms of their intensity in the following way: tropical depression, winds less than $17 \, m \, s^{-1}$; tropical storm, winds $17–33 \, m \, s^{-1}$; hurricane, typhoon, severe tropical cyclone or tropical cyclone, winds $33 \, m \, s^{-1}$ and above.

Tropical cyclones are different from extratropical cyclones in a number of ways. Tropical cyclones do not possess fronts whereas extratropical cyclones do. This indicates a basic difference in their formation. The centre of a tropical cyclone is warmer than the surrounding air whereas the opposite is true for extratropical storms. Tropical cyclones derive most of their energy from evaporation of water over warm tropical oceans with subsequent release of latent energy of condensation, whereas extratropical storms, which are often associated with fronts, derive their energy from horizontal temperature gradients (baroclinic instability) within the atmosphere. Structurally there are also differences. In the case of extratropical storms, strongest wind speeds are in the upper atmosphere, whereas surface winds are the strongest for tropical cyclones. Tropical cyclones are also well known for possessing a distinct eye at the centre of the cyclonic vortex whereas extratropical storms do not. Finally, tropical cyclones tend to be smaller in size (300–800 km) than extratropical storms (1000–1600 km).

Tropical cyclones possess distinct geographical (Figure 8.6) and temporal (Figure 8.7) distributions which reflect the factors necessary for their formation. These include the following:

1. the rising air masses in the core of the depression must be warmer than the surrounding air masses up to a level of about 10 000–12 000 m. As the main driving force of tropical cyclones is latent heat of condensation, the rising air masses must also be humid. These disturbances will therefore form only over large ocean areas, where the surface water temperature is above the critical value of 27°C. This condition explains why tropical cyclones form mainly over the western parts of the large ocean basins where no cold currents occur. It also accounts for the fact that the main season of the tropical cyclones is towards the end of the summer, when sea surface temperatures are highest (Figure 8.7).
2. Unlike the situation in other tropical disturbances, vorticity is an essential part of the circulation in tropical cyclones. Therefore they do not develop at latitudes below about 5°. Closer to the equator, the Coriolis force is too weak to divert the inflowing airstreams and even a strong surface low will fill rapidly.
3. The basic air current in which the tropical cyclone is formed should have only a weak vertical wind shear, since vertical shear inhibits the development of a vortex – a situation very different to the case of tornadoes. This is the main reason why no tropical cyclones develop in the Asian summer monsoon when it is at full strength. Both over the Bay of Bengal and in the Arabian Sea, tropical cyclones have their highest frequencies at the beginning and towards the end of the summer monsoon season (Figure 8.7).
4. A small low pressure centre is required to trigger off cyclonic development. The initiative can come from small vortices near the ITCZ or from wave disturbances such as easterly waves. However, only about 85% of intense Atlantic hurricanes have their origins as easterly waves (Landsea, 1993). Because the ITCZ is very rarely positioned far south of the equator in the Atlantic and south-eastern Pacific, tropical cyclones rarely form here (Figure 8.6).
5. Combined with an area of low level convergence an area of upper level divergence should be present between 9000 and 15 000 m, so that the outflow of air in the upper atmosphere is maintained. The absence of this condition often accounts for why many incipient cyclones fail to develop in intensity.

A remarkable feature of the tropical cyclone is the warm core, the evolution and maintenance of which is vital for the development and longevity of the tropical cyclone. Before the warm core develops, an opportunity for cyclone growth must be

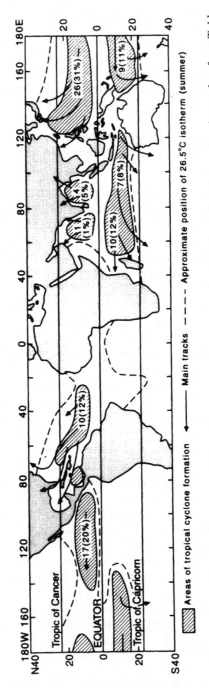

Figure 8.6 Main zones of tropical cyclone formation and predominant tracks. Annual numbers (percentages) are taken from Table 8.1 and are for tropical storms or stronger (sustained wind speeds greater than 17 m s^{-1})

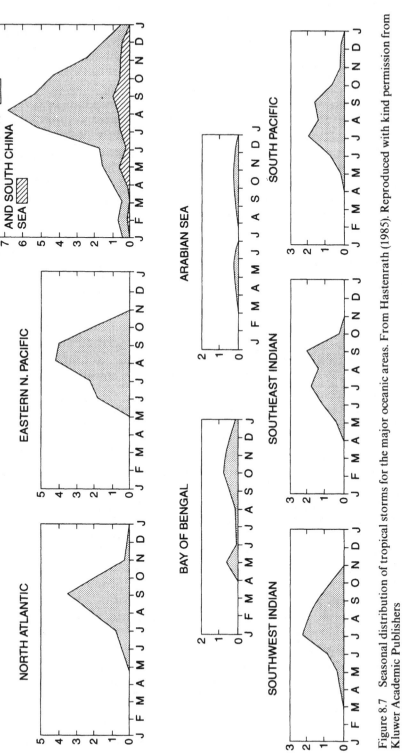

Figure 8.7 Seasonal distribution of tropical storms for the major oceanic areas. From Hastenrath (1985). Reproduced with kind permission from Kluwer Academic Publishers

provided in the form of a pre-existing disturbance which encourages vertical air movements. Once initiated, the surface disturbance encourages the inflow of moist air which fuels convection and the development of cumulonimbus towers which release large amounts of latent heat of condensation. It is this heat which helps to produce the initial warm core. However, a problem exists in terms of maintaining the warm core.

Warm core maintenance is achieved by a combination of outflow of air in the upper levels and inflow of moist air in the lower levels of the tropical cyclone. This provides the opportunity for further convective activity. This happens via a positive feedback mechanism referred to as conditional instability of the second kind. This works in the following manner: large releases of latent heat in the upper atmosphere enhance outflow in the upper levels of the tropical cyclone which encourages convergence in the lower moist layers. The low level influx of moist air provides convective fuel to the bases of the "hot" cumulonimbus towers, promoting further growth of these. These in turn produce more latent heat of condensation and upper level outflow so that the self-fuelling cycle is completed. Upper outflow may also be enhanced by the presence of a tropical upper tropospheric trough.

A further mechanism also aids the maintenance of the warm core. This mechanism usually comes into play when the central core pressure has fallen substantially and air is brought to the centre from considerable distances over warm sea surfaces. As air spirals into the vortex, isothermal warming of the air occurs. This leads to an enhanced buoyancy of the core air compared to the surrounding air. This in turn aids cumulonimbus growth, with more heat being added to the core of the tropical cyclone. The net effect of this, of course, will be to lower surface pressure further by positive feedback. In very mature tropical cyclones, very low surface pressures ensure that this positive feedback effect is very large, several degrees in some cases, thus contributing to the high efficiency of the tropical cyclone as a thermodynamic engine (McIlveen, 1992). This efficiency can also produce rapidly deepening tropical cyclones with central pressure drops of more than 42 mb in 24 hours. Such rapid deepening appears to be dependent on the presence of warm surface waters in excess of 28 °C to a depth of 30 m and an eye dimension of around 40 km.

Once formed, tropical cyclones move with the mid- to upper tropospheric winds or environmental flow. Winds in the mid-troposphere (700–500 mb), at a 5–7° latitude radius from the cyclone centre, have the best correlation with cyclone movement. In the northern hemisphere, tropical cyclones move 10–20° to the left of their surrounding flow while in the southern hemisphere they move around 10° to the right. Generally tropical cyclones move approximately 1 m s^{-1} faster than their surrounding flow (Chan and Gray, 1982). Although their movement is generally westward, at a speed usually of the order of 15–30 km/h^{-1}, many tropical cyclones show a tendency to recurve and move polewards as they approach the western margins of the ocean basins. The reasons for this are not completely understood but seem to be related to the position and intensity of the upper (200 mb) westerly flows relative to the tropical cyclone centre. For recurving storms in the north-west Pacific, strong deep westerlies have been found to lie at only 10° north of the cyclone centre, in contrast to non-recurving storms for which the westerlies are much further north, shallower and weaker (George and Gray, 1977).

On making landfall, tropical cyclones tend to decay. This is not because of the

Table 8.1 Tropical cyclone statistics for the various ocean basins (1968–1989 northern hemisphere; 1968/69–1989/90 southern hemisphere) (Neumann, 1993)

Basin	Tropical storm or stronger ($> 17 \text{ m s}^{-1}$ sustained wind)			Hurricane/typhoon/severe tropical cyclone ($> 33 \text{ m s}^{-1}$ sustained wind)		
	Max.	Min.	Mean	Max.	Min.	Mean
Atlantic	18	4	9.7	12	2	5.4
NE Pacific	23	8	16.5	14	4	8.9
NW Pacific	35	19	25.7	24	11	16
N Indian	10	1	5.4	6	0	2.5
SW Indian	15	6	10.4	10	0	4.4
SE Indian/Australia	11	1	6.9	7	0	3.4
Australia/SW Pacific	16	2	9	11	2	4.3
Global	103	75	83.7	65	34	44.9

increased surface friction, as this would act to increase the low level inflow and theoretically enhance the positive feedback mechanisms for maintaining the warm core. Tropical cyclones tend to die over land because of the large moisture flux available from over warm ocean surface, which provides the energy to drive the tropical cyclone, is no longer available. Without this moisture supply, deep convection and maintenance of the warm core is not possible. Central pressure therefore rises and the tropical cyclone tends to fill. For example, as tropical cyclones cross the Philippines and Taiwan their intensities decrease by 33% and 40% respectively (Brand and Belloch, 1973, 1974). Perhaps the only situation in which tropical cyclones may be sustained over land is if they move over extensive swampy regions such as the Florida Everglades. In the Bay of Bengal and the Arabian Sea, tropical cyclones usually move towards the continent soon after formation. Their life spans are therefore shorter than those of oceanic cyclones. However, both in their movements and their development, cyclones are notoriously erratic and the prediction of their tracks is one of the main problems in tropical forecasting.

Statistics summarizing tropical cyclone occurrence for the various ocean basins are presented in Table 8.1. The maxima and minima information in Table 8.1 indicates that the frequency of occurrence of tropical cyclones exhibits large variations from year to year in each of the ocean basins. The cause of these variations appears to be non-seasonal variations in the ocean–atmosphere system (Chapter 6). Of these it appears that ENSO and the QBO play an important role in determining the interannual variability of cyclone frequency for some of the ocean basins. In the Atlantic Basin, tropical cyclone activity is suppressed in ENSO years due to increased tropospheric vertical wind shear. Wind shear also appears to vary with the phase of the QBO. Consequently this non-seasonal feature of the tropical circulation also influences hurricane activity. When the QBO is in its east phase for 12–15 months, Atlantic basin hurricane activity is reduced, while in the following west phase, which lasts on average for 13–16 months, there is an increase in hurricane activity. Interestingly, ENSO and QBO operate at different frequencies such that only occasionally do strong ENSO events coincide with the height of east phase QBO events. Such a configuration as occurred in the years of 1972 and 1983 results in very low numbers of Atlantic

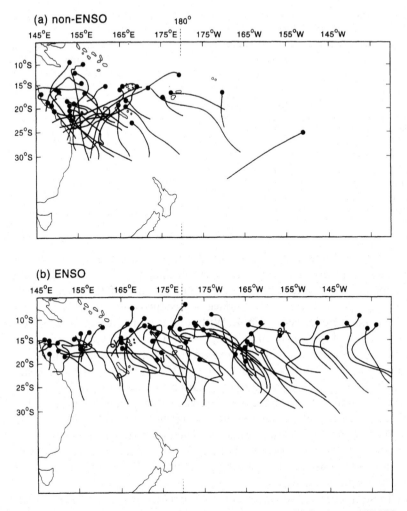

Figure 8.8   Distribution of tropical cyclogenesis points and tracks for (a) non-ENSO and (b) ENSO periods in the south-west Pacific. From Hastings (1990). Reproduced by permission of the *International Journal of Climatology*, Royal Meteorological Society

hurricanes (Gray and Schaeffer, 1991). For the Atlantic basin indices of atmosphere and ocean behaviour have been used to develop statistically based hurricane forecasting models (Gray et al., 1994).

In the Australian/South Pacific region ENSO has a marked spatial impact on tropical cyclone occurrence. During non-ENSO or La Nina events, tropical cyclone activity is very much confined to the south-western Pacific between 145°E and 165°E (Figure 8.8(a)). A diffuse pattern is, however, evident for warm ENSO conditions such that cyclone activity spreads eastwards across the Pacific to 145°W with numbers decreasing in the usual area of occurrence between 145°E and 165°E (Figure 8.8(b)). There also appears to be a tendency for the cyclogenesis points to be closer to the equator in ENSO episodes (Hastings, 1990). For the north-west Pacific, ENSO events result in no change in frequency but a geographical shift in cyclone location from west of 16°E to 16°E to the dateline. For the South China Sea there are also clear ENSO signals in tropical cyclone occurrence (McGregor, 1995b). As yet, no clear ENSO signals in cyclone activity have been found for the eastern portion of the north-east Pacific, the south-west Indian, the south-east Indian/Australian and north Indian basins.

Although at the global scale there is no evidence of changes in tropical cyclone climatology, for some ocean basins on the decadal time scale, some changes are evident. Whether these represent shifts in cyclone climatology, or whether they are a part of a cyclic change is difficult to establish due to lack of long records of tropical cyclone occurrence. However, analyses of historical data stretching back to the 15th century for the Caribbean, indicate significant variations in the favoured cyclone tracks and levels of activity for this area; notably, high levels of activity during the 1770s to 1780s, 1810s, and 1930s to 1950s and a shift eastwards in the favoured tracks from the mid-20th century onwards (Reading, 1990). For the Atlantic basin as a whole, although there has been no significant change in the total frequency of tropical storms and hurricanes over the period 1944–95, the number of intense hurricanes dropped during the 1970s and 1980s. This was also matched by a moderate decrease in the maximum intensity reached by all storms over a season (Landsea, 1993). Notably the extended 1991–1994 ENSO event resulted in the quietest hurricane season for the Atlantic since 1944. However, this was followed in 1995 by a very active hurricane season. For the Australian region it appears that the high frequency of ENSO events during the 1980s to early 1990s has had a dramatic effect on the number of tropical cyclones with an overall reduction since the mid-1980s (Nicholls, 1984, 1992).

Tropical cyclones have received much attention because of their destructive power and the consequent damage they cause; a direct result of strong winds and an indirect result of large rainfall amounts and storm surges. The geographical pattern of damage is often related to the physical structure of the tropical cyclone. Strongest winds occur in that part of the cyclone where the winds in the vortex are moving in the same direction as the steering environmental flow. This is because the cyclone and environmental flow wind effects are additive. In the northern (southern) hemisphere this is usually on the right (left) hand side of the cyclone. Rainfall in tropical cyclones is usually concentrated in a rather narrow zone around the core, but the actual amounts received vary largely depending on size, intensity and movement of the disturbance (Plate 5). In a moving cyclone, rainfall in a circle with a diameter of 100 km around the centre has been calculated at 863 mm per day. Rainfall also tends to decrease in a linear fashion away from the eye wall region (Riehl, 1954). How much of this is actually received at a location depends on the time that this zone is situated over a particular location. The amounts of rainfall can also be increased markedly when orographic lifting takes place. Values of over 2000 mm per day have been recorded in the Philippines under these circumstances. La Reunion Island in the Indian Ocean is also well known for receiving record tropical cyclone related rainfalls. For example, cyclone Hyacinth delivered 3240 mm of rainfall during 24–27 January (72 hours) in 1980. This location also holds a number of other tropical cyclone rainfall records such as 1144 mm in 12 hours (tropical cyclone Denise, 7–8 January 1966).

Storm surges can also cause much damage through coastal flooding. These occur when strong winds push the ocean water ahead of the storm on the right (left) hand side of the storm in the northern (southern) hemisphere in the direction of coastlines. As the wind strength is related to the surface pressure gradient within the storm, the central pressure can be used to give a rough estimate of the height of the storm surge. For example, a storm with a central pressure of about 900 mb may produce a storm surge of 6–10 m. As well as wind strength, the storm surge height will also depend on the coastal topography, speed of the tropical cyclone, the storm's angle of incidence and the state of the tide.

MONSOON DEPRESSIONS

Monsoon depressions are visible on synoptic weather charts as quasi-circular or elliptical isobars around a core of low pressure with a diameter of anywhere between 500 and 1300 km. The central pressure near the earth's surface is about 3–10 mb below the normal pressure. These disturbances, which are best developed at 2–4 km, but characterized by a deep cyclonic flow up to about 7–9 km and an upper easterly anticyclonic outflow, last from a few days to about one week. Although they account for a small proportion of the total monsoon rainfall they can deliver average daily falls of 10–20 cm over a widespread area, mainly in their south-western quarter. Disturbances of this kind occur on average about three times per month in the Asian summer monsoon season. Consequently they are usually referred to as monsoon depressions.

The most important area of origin is in the Bay of Bengal, but occasionally monsoon depressions also develop over the Arabian Sea. At both locations their development takes place when the main monsoon trough is nearby. Over the Bay of Bengal monsoon depressions generally develop in the following manner: a typhoon or tropical storm arrives at the Asian coast near 20°N, and at the same time the monsoon trough is active over north-west India around 80°E; surface pressure falls across southern China and Vietnam; pressure gradually rises across Indonesia and Burma over about 7 days; 4–5 days later pressure falls and a monsoon depression forms over the Bay of Bengal with high pressure to the east over Burma and Indonesia (Krishnamurti et al., 1977). Convergence in the westerly monsoon current near the earth's surface, combined with divergent easterlies aloft, initiates vertical air movements. This is intensified by the release of large amounts of latent heat from the rapidly condensing moisture in the deep convective cloud masses.

Once formed, the monsoon depressions move slowly in a west-north-west direction with the upper air flow. Over land these depressions usually weaken rapidly. They almost never develop into strong tropical cyclones as the strong wind shear in the monsoon flows prevents the development of a vortex.

TROPICAL DISTURBANCES WITH EXTRATROPICAL ORIGINS

The origin of some tropical disturbances lies outside the tropics. Cyclones from the Polar Front region occasionally venture toward the equator. On meeting warm and

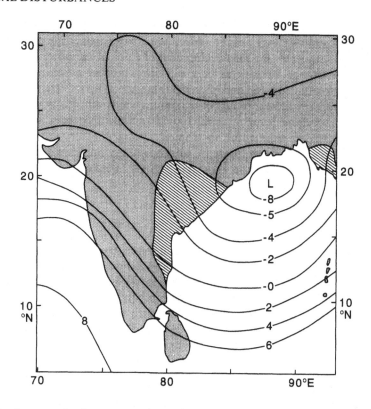

Figure 8.9   An example of a monsoon depression over the Bay of Bengal for 1200 hours GMT, 20 August 1986. 1000 mb contours are departures in decametres from the average pressure. After Koteswaram (1958)

humid air masses they may regenerate into true tropical disturbances but without their prior mid-latitude frontal characteristics. Other tropical disturbances are caused by waves in the upper westerlies, from which cold air masses are sheared off towards the equator. The upper air divergence related to this cold core can create subtropical cyclones. Surges of cold air that often invade tropical latitudes also have a mid-latitude origin. Known as cold surges, these find their best expression in the East Asian monsoon area.

SUBTROPICAL CYCLONES

Subtropical cyclones represent hybrid systems of both the low and mid-latitudes as they possess the basic characteristics of tropical cyclones and mid-latitude cyclones. These are low pressure systems and may be found in either tropical or subtropical latitudes, but have a tendency to occur in the outer tropics such as over the Pacific Ocean to the north-east of Hawaii, western India, the north-eastern parts of the Arabian Sea and in the south-west Pacific in the vicinity of New Zealand and Australia.

Their origins are both a result of horizontal temperature gradients in the atmosphere (baroclinic atmosphere), as for mid-latitude cyclones, and deep convection, akin to that which is important for the development of tropical cyclones. Physically, they may resemble a tropical cyclone as they can often possess a broad eye of around 150 km width, surrounded by a band of precipitating cloud up to 300 km in width. Like tropical cyclones, they have a radius of maximum winds but this is farther out than that of a tropical cyclone, being at about 100–200 km distance from the centre. These winds are also weaker than those in tropical cyclones, rarely attaining velocities greater than 33 m s^{-1}. Climatologically, subtropical cyclones are important as they bring considerable rainfall to the Hawaiian Islands at times, as well as north-west India.

COLD SURGES

A distinct characteristic of the winter monsoon over East Asia are spells of very cold weather, related to cold surges within the winter monsoon flow (Ramage, 1955). Cheang (1991) and Ding (1994) have described the stages leading to the onset of a cold surge. In the initiation stage baroclinic processes related to large north–south temperature gradients and the pooling of cold air to the north of the Tibetan Plateau appear to be important primers for surge onset. In this stage the role of the Tibetan Plateau is recognized as important as it restricts the movement of polar continental air to the south, effectively trapping it, which enhances the intensity of the Siberian high through positive feedback mechanisms. This accentuates the intensity of the baroclinic zone between cold northerly air and warm southerly tropical air. Atmospheric dynamic processes appear to be responsible for causing cold air outbreak from the Siberian high area. To the east of the Tibetan Plateau, over the western Pacific, between Japan and the Philippines at about 130°E, a quasi-stationary upper level westerly longwave trough with a length of around 500 km is maintained during the winter. Through this trough, shortwave disturbances in the form of cyclonic vortices travel. The effect of these is to spark off intense periods of anticyclogenesis over northern China and cyclogenesis over the eastern parts of the South China Sea (Hastenrath, 1985; Cheang, 1991). Associated with these changes is a downstream acceleration of the upper westerly jet over East Asia (Cheang, 1991). This produces a steep meridional pressure gradient over East Asia with the result that there is a burst of cold air from northern China over the South China Sea. This is the equatorial propagation phase which occurs in two stages (Ding, 1990, 1994).

The first stage is characterized by a rise in pressure over the South China Sea. Propagation speeds in this stage have been measured at between 40 m s^{-1} and 27 m s^{-1}(Ding, 1994). The second stage is associated with a sharp decrease in humidity as cold dry air streams out from northern China. Wind speeds may increase in either of the stages. A southward-moving cold front, however, is only seen in the second stage. This moves at an average speed of 11 m s^{-1}. Cold surges also affect areas south of the equator.

The maximum frequency of cold surge occurrence occurs over the South China Sea between 10 and 20°N and 110 and 118°E. A second high frequency area to the east of the Philippines also occurs but this has only about half the occurrences of the South

China Sea area (Ding, 1990). Analyses of upper air data have also revealed that cold surges are very much shallow surface layer events, not being present above 700 mb.

Cold surges occur most often in December with secondary peaks in the months of November and January. Events are characterized by below normal temperatures and above normal wind speeds and may last between 4 and 6 days. Oscillations in cold surge activity and associated atmospheric pressure and rainfall have been noted at time scales of 4–5 and 10–20 days (Murakami, 1979). Oscillations at 4–5 days appear to be related to disturbances moving into the East Asian area from the western Pacific (Cheang, 1991). Rainfall oscillations at the 10–20 day time scale are at the same time scale as that of cold surges, suggesting that 10–20 day rainfall oscillations are related to the interaction of bursts of northerly monsoon activity with the moist Pacific easterlies. Surges at the 10–20 day time scale also appear to be important for enhancing convective activity in the near-equatorial trough region thus giving impetus to the heat engine that drives the winter monsoon (Chang, Erickson and Lau, 1979).

Cold surge events are economically important as they may disrupt sea transport services because of their windiness, bring heavy rains and floods to southern South-East Asia areas lying under confluence zones such as Malaysia and Borneo (Ramage, 1971; Cheang, 1977), and cause damage to agricultural crops due to frost occurrence or chilling injury in northern Indo-China and southern China. For example, in tropical Hainan Island off the north-eastern Indo-China coast (18–20°N), severe cold injury to rubber trees was reported on five occasions between 1950 and 1979. In January 1955, 10–30% of the juvenile rubber trees were killed because of cold injury as temperature dropped to 0.4°C, almost 17°C below the January average for this area (Domroes and Peng, 1988). Because of their importance to agriculture and other industries, attempts at forecasting cold surges abound. Most forecast models pay particular attention to atmospheric dynamics in the three cold surge source areas over the Arctic seas, Russian and Siberian lowlands, especially the Lake Baikal area at 90°E. In addition to their importance for the East Asian region, cold surges may play a role in the modulation of the Australian summer monsoon (Suppiah, 1992).

Cold surges also occur over the South American continent. Here these events appear to be associated with an intensifying surface anticyclone west of Chile over the south-east Pacific which moves east and then northwards. Such a configuration results in cold air being advected from the south with an associated cold front moving northwards into the western Amazonian basin. At its most frequent occurrence during the southern hemisphere winter these cold surges often produce frost and freezing conditions in subtropical agricultural areas with the effects being identical to those of cold surges in the South-East Asian region.

SUMMARY

Tropical disturbances arise from short-term perturbations in the pressure, wind and moisture fields of the atmosphere which give rise to atmospheric instability, the main prerequisite for their formation. Occurring on a variety of time and space scales they are generally characterized by their ferocity in terms of rainfall intensities and winds. The main energy source which fuels tropical disturbances such as thunderstorms, squall lines, tropical cyclones and monsoon depressions is latent heat which is released when condensation occurs in the large volumes of ascending moist air in these systems.

Although often destructive, tropical disturbances are of climatological importance because they contribute significantly to the annual rainfall in a large number of tropical locations. Despite the majority of tropical disturbances having their origins in the low latitudes, subtropical cyclones and cold surges have an extratropical origin. The former is especially important for the rainfall climates of subtropical locations while the latter often causes severe frost and cold injury to tropical agricultural crops in South-East Asia and South America.

CHAPTER 9

Water in the Tropical Atmosphere

The climatological importance of atmospheric water is related not only to its capacity to control precipitation, but also to its strong influence on both reflection and absorption of terrestrial and solar radiation. It therefore has a large effect on atmospheric temperature conditions (Chapter 4). A further relevant fact is that as water changes between its gas, liquid and solid states energy is either produced or consumed. Such phase changes are therefore very important sources or sinks of energy, especially in the tropical atmosphere where the condensation of large amounts of water vapour leads to the release of large amounts of energy called the latent heat of condensation (Chapter 2).

In the tropics, where the atmospheric water content is comparatively high, its role in climate is especially important. As has been seen in the previous chapter, stability conditions of tropical air masses and the latent heat of condensation are the main driving forces of tropical disturbances. Both factors are closely related to atmospheric moisture.

This chapter follows the atmospheric part of the hydrological cycle, starting with evapotranspiration, the origin of atmospheric water vapour. The horizontal and vertical distribution of humidity in the atmosphere is then described, including a consideration of the different measures of humidity. Following this, the process of condensation is briefly considered. A discussion of the climatic importance of the products of the condensation process – clouds – concludes the chapter. The remaining stage of the hydrological cycle, the return of atmospheric water to the earth's surface in the form of precipitation, is the subject of the next chapter.

EVAPOTRANSPIRATION

All atmospheric moisture originates from the earth's surface, where water in its liquid and solid phase is transformed into water vapour, which can be carried both vertically

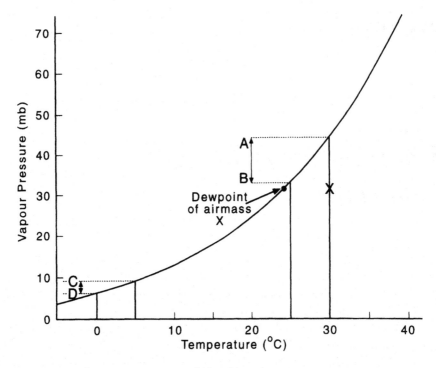

Figure 9.1 Saturation vapour pressure relationships

and horizontally by air movements. The water vapour originates from two processes. The first is evaporation, which takes place from water and ice surfaces, and over land areas when soils, rocks or vegetation cover are wet after recent precipitation or when groundwater is close to the surface. The second source is transpiration, physically the same process as evaporation, but carried out by organisms, mainly plants. The combined effects of these two processes are often called *evapotranspiration*, which indicates the total flow of water into the atmosphere (Thornthwaite, 1948).

Evapotranspiration is controlled by three conditions: the capacity of the air to take up water vapour, the amount of energy available for use in the evaporation and transpiration processes, and the degree of turbulence of the lower layers of the atmosphere, necessary to replace the saturated air layers near the earth's surface by unsaturated air from higher levels. In respect of these three requirements the conditions in the tropics are rather favourable. In the first place, the capacity of the air to retain moisture increases rapidly with temperature (Figure 9.1). Warm tropical air masses can therefore take up more water vapour than cold ones. The actual amount that can be absorbed into the atmosphere also depends on the humidity of the air – the lower this is, the more favourable are the conditions for further evaporation. The dry tropics therefore have very high rates of evapotranspiration.

Secondly, the energy for evapotranspiration is mainly provided by solar radiation, which is available in large quantities in the tropics. Thirdly, turbulence is caused by winds or by convectional currents. While winds are not particularly strong in the tropics, compared to other climates, convection is very frequent.

It is for these reasons that the latitudinal distribution of evapotranspiration shows a clear maximum in the low latitudes (Figure 9.2). However, within this broad geographical region there are variations in the evaporation rates. Highest values of around 2000 mm year^{-1} occur over the subtropical oceans where the subtropical anticyclones dominate in both hemispheres. Because evaporation exceeds precipitation on a mean annual basis here, these oceanic regions are often referred to as ocean deserts. Evaporation rates are also high where warm ocean currents exist, especially in the region of the Gulf Stream and Kuroshiro currents off the east coasts of North America and southern Japan respectively. Here winds on the south-western flanks of subtropical anticyclones recurve northward pushing warm tropical water to higher latitudes. These warm sea surface temperatures allow evaporation rates to reach around 2000 mm year^{-1}. In contrast, the areas off the west coasts of northern and southern Africa and North and South America where cold ocean currents predominate have evaporation rates of less than 1000 mm yr^{-1} (Figure 9.2).

For the low latitude oceans equatorward of 20°, lowest evaporation rates occur in the equatorial regions (Figures 9.2 and 9.3). Here precipitation is plentiful, cloud cover is high, air masses are often at or near saturation, wind velocities are low and in some locations oceanic upwelling of cold water occurs. All these factors operate to reduce evaporation rates. Although the oceanic equatorial regions show little seasonal variation in evaporation rates, the subtropical oceans have a winter (summer) maximum (minimum) (Figure 9.3). This seasonal contrast is due to the changing intensity of the trade winds such that, in the winter hemisphere, the trade winds are stronger; increased turbulence offsets lower winter sea surface temperatures to produce higher evaporation rates.

In contrast to the oceans, land evaporation rates are highest for the equatorial regions, reaching around 1200 mm year^{-1}. These high rates are because of a combination of high precipitation and temperatures, the latter perhaps being more important. This relates to the surface energy balance such that over the oceans, because moisture supply is unlimited, a large amount of energy goes into evaporation (Chapter 3), leaving little for sensible heating of the ocean surface and overlying air. Conversely, over the land, water supply is limited, compared to the oceans, with the result that much more energy goes into sensible heating of the surface; air temperatures are therefore greater here. These facilitate higher evaporation rates, especially in areas where there is plentiful precipitation.

Despite high energy inputs, the subtropical desert regions have the lowest mean evaporation rates. Rates as low as 200 mm year^{-1} occur here because moisture is severely limited; a result of negligible precipitation amounts.

Although the atmospheric factors of energy, humidity and turbulence are important, the ultimate determinant of evaporation is moisture availability. Over water surfaces, where this supply is plentiful, evaporation is not restricted and the values reached are called the *potential evapotranspiration*. Over land areas, water supply is often a limiting factor so that the *actual evaporation* remains lower than the potential evaporation. Over large land areas its maximum is set by the total amount of precipitation received. This is the main reason why globally the oceans have a higher mean actual evaporation rate of 1176 mm year^{-1} compared to 480 mm year^{-1} for land (Piexoto and Oort, 1992). However, in some parts of the humid tropics the continental rate of actual

ANNUAL EVAPORATION (cm)

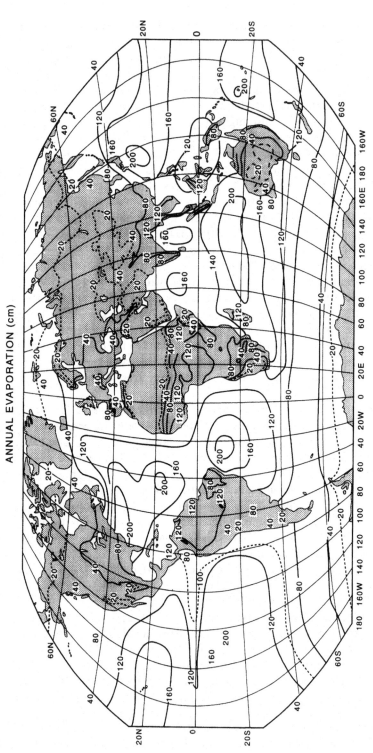

Figure 9.2   Global distribution of annual evaporation (cm). From Piexoto and Oort (1992). Reproduced by permission of American Institute of Physics

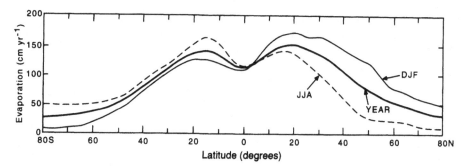

Figure 9.3 Meridional distribution of evaporation over the oceans. JJA is June to August and DJF is December to February. Note low (high) evaporation rates in the equatorial (subtropical) latitudes. From Piexoto and Oort (1992). Reproduced by permission of American Institute of Physics

evapotranspiration may occasionally exceed the potential value recorded over the oceans in the same climatic area. This is caused by the enormous increase of the transpiration surface of the luxuriant rainforests with their multiple canopies and thus transpiring surfaces.

MEASUREMENT AND ESTIMATION OF EVAPORATION

Actual measurements of evaporation over the oceans are rare compared to those over land, therefore the methods outlined below mainly refer to the measurement of terrestrial evaporation. A number of methods exist and include the gravimetric method, the use of drainage lysimeters and the water budget method. The first of these involves taking a number of samples of soil to a depth which includes the root zone. The samples are weighed, dried and reweighed in order to determine the average water content, which is equivalent to the change of mass after drying. This process is repeated with more samples after a period of time. Any differences in weight of these samples are attributed to changes in average water content which must be due to either precipitation in the case of weight gain, or evaporation in the case of weight loss once any precipitation gains have been accounted for.

Drainage lysimeters are used to determine evaporation based on measuring the difference between the amount of water applied to a vegetated surface and the amount of water withdrawn from beneath that surface. With this method a known amount of water is applied daily to the soil in a large container, such as an oil drum, which has been set in the ground. A sump beneath the drum collects the water that manages to drain through the soil. Water is then pumped from the sump the next day. The difference in the amount of water applied to the soil and that pumped from the sump is assumed to have evaporated. This process is repeated on a daily basis.

The water budget method is usually used for determining the evaporation from a reservoir or lake. The method relies on application of the water balance equation which states that evaporation is equivalent to inflow plus precipitation, minus outflow, leakage and rise in water level. In this case inflow and outflow are to or from the lake or reservoir respectively. This method therefore relies on determining the difference between inputs and output plus or minus the change in storage.

The three methods described above allow an estimation of the actual evaporation rate, which is different from the potential rate which is that amount of evaporation occurring when water is freely available. This is usually determined using evaporation pans. These, in simple terms, are large dishes of water. The widely used US Class A evaporation pan is 250 mm in depth and 1.22 m in diameter. It is usually located on a slightly elevated platform at about 150 mm above the ground. Daily changes in the pan water level are attributed to evaporation once precipitation inputs have been accounted for.

Although evaporation pans are widely used they often indicate inflated evaporation values. This is due to the "oasis effect" (Oke, 1987) caused by the relatively small evaporating surface. Saturated air layers over the pan are very frequently replaced by relatively dry air from the surrounding areas, where normally no water is available for evaporation. The pan is thus analogous to an oasis in a dry desert. Over large water bodies, such as lakes, or over densely vegetated surfaces, the replacing air layers come mainly from other parts of the evaporating or transpiring surface. This is the main reason why figures from evaporation pans are often corrected by a factor of around 0.7 to make them comparable to evaporation data from lakes and reservoirs (Weisner, 1970). Evaporation pans also suffer from other inaccuracies, due to the direct absorption of solar radiation or heat from the ground, or splashing and interference by animals. Pans with different designs also give different evaporation rates.

Given the problems of measurement, potential evapotranspiration, also called the evaporation from an open water surface, is often estimated. Such estimates come from calculations which use the meteorological factors that control evaporation rates. Estimation techniques are usually used in preference to measurement techniques because the meteorological variables used are often available from climate stations.

Perhaps one of the most widely applied estimation techniques is Penman's method (Penman, 1948). This is essentially a combination of two other estimation methods: the mass or turbulent transfer method and the energy balance method (Oke, 1987). The mass transfer component of the Penman estimation method takes into account wind speed and vapour pressure, both of which are important for controlling the removal of evaporated water vapour from above an evaporating surface. The energy balance component of Penman's formula estimates the amount of energy available for evaporation. This is the difference between the energy gained and lost from a particular site and is called the net radiation (Chapter 3). Estimates of energy gain are based on values of radiation (measured or calculated), average cloudiness and surface albedo, while those of energy loss are based on values of air temperature and vapour pressure.

An often used alternative method for estimating evaporation is Thornthwaite's method (Thornthwaite, 1948). This method uses monthly temperature data and a day length factor. It is only suitable for estimating monthly evaporation, in contrast to that of Penman's which can be used to estimate daily evaporation rates. Evaluations of the applicability of Thornthwaite's method in the tropics have shown, however, that it is less satisfactory here compared to the temperate climate environments for which it was originally developed. A useful comparison of the evaporation estimates from a number of different estimation methods is given in Jackson (1989). This analysis shows that for

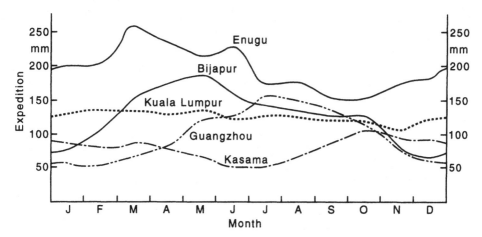

Figure 9.4 Mean annual evapotranspiration at Kuala Lumpur, Malaysia, 3°7'N, 101°42'E; Enugu, Nigeria, 6°28'N, 7°33'E (evaporation pan data only); Kasama, Zambia, 10°12'S, 31°11'E; Bijapur, India, 16°53'N, 75°42'E; Guangzhou, China, 23°N, 113°15'E. Data at all stations except Enugu are based on estimates according to Penman's formula

the three tropical locations analysed there is reasonably close agreement between US Class A pan measurements and Penman estimates.

Penman evaporation estimates for a number of tropical locations are shown in Figure 9.4. Although estimates show generally small seasonal variations for the tropics they show a gradual increase with latitude, mainly because of seasonal differences in solar radiation and temperature. It should be borne in mind that the high values at Enugu are possibly caused by the fact that they have been recorded at evaporation pans. A notable feature of the seasonal patterns is their variability, which reflects the local and regional controls on evaporation rates.

In contrast to the variety of seasonal patterns, the diurnal pattern of evaporation is simple: a clear maximum during the day is followed by a very low minimum at night. In the tropics, where the nights are long, the humidity high and turbulence limited, the potential evapotranspiration comes to an almost complete standstill at night for most locations.

HUMIDITY

An outcome of the evaporation process is the vertical transfer of moisture away from the earth's surface into the atmosphere with the result that atmospheric water vapour levels build. This process not only is a *mass transfer process*, as water is removed from the earth's surface, but is also an *energy transfer process*. This is because the energy consumed in the evaporation process is moved, along with the water vapour, into the atmosphere. This is often referred to as *latent heat transfer*.

The term humidity refers to the content of water vapour in the air. At the earth's surface, atmospheric water vapour is always strongly concentrated in the lower layers and normally about half of its total is found below the level of 2000 m.

Humidity may be measured and expressed in a variety of ways. The simplest form of measurement is by dry and wet bulb thermometry. A wet bulb thermometer is simply an ordinary thermometer with a sleeve of wet fabric around the bulb. Dry and wet bulb thermometers are often combined into one instrument called a psychrometer which is attached to a handle so that it can be swung around in the air; a whirling pyschrometer. The way in which this instrument works is straightforward. As the psychrometer is swung, water evaporates from the wet bulb thermometer. The amount that evaporates will depend on the water vapour content or humidity of the air. The lower the humidity, the greater the rate of evaporation from the wet bulb. As evaporation is a heat-consuming process, the temperature of the wet bulb will fall while that of the dry bulb will remain at the air temperature. Eventually the wet bulb temperature will stabilize at a level below that of the dry bulb. The difference between the dry and wet bulb temperature is called the wet bulb depression. This is used with the aid of Regnault's equation (Linacre, 1992) in the calculation of vapour pressure, which is one indirect measure of atmospheric moisture content.

Vapour pressure is expressed in millibars (mb) and is a measure of the total pressure exerted by water molecules in the atmosphere, remembering that the global average surface pressure exerted by all atmospheric gases is around 1013 mb. Saturation vapour pressure varies with temperature (Figure 9.1).

Other indirect measures of humidity or atmospheric water content include relative humidity, saturation deficit and dew point. Absolute humidity and the mixing ratio are two direct measures of humidity. All indirect and direct measures have different units of measurement of which relative humidity, expressed as a percentage, is probably the most widely known measure of atmospheric moisture.

Relative humidity is the ratio of the amount of water in the atmosphere to that which could possibly be present in the atmosphere for a given temperature; the saturation vapour pressure. Although relative humidity shows the relationship to saturation, it is an inadequate indication of the amount of atmospheric moisture because it is dependent on temperature (Linacre, 1992). Relative humidity varies considerably throughout the day such that when the temperature rises, relative humidity falls and vice versa. This behaviour occurs irrespective of whether there are actual changes in atmospheric moisture levels. Relative humidity can be measured using dry and wet bulb thermometers or hygrographs.

Saturation deficit (measured in millibars) is the difference between the possible and actual levels of moisture for a given temperature. The larger (smaller) the saturation deficit, the drier (wetter) the atmosphere.

Dew point is another indirect measure of atmospheric moisture. It is defined as the temperature to which air must be cooled for saturation to occur. This can be easily understood with reference to air mass X in Figure 9.1. This air mass has a temperature of 30 °C and a vapour pressure of around 30 mb. As the saturation vapour pressure for air at 30 °C is approximately 42 mb (this value can be found in standard meteorological tables) air mass X is unsaturated with a saturation deficit of 12 mb (42 mb–30 mb). For saturation to occur, the temperature of air mass X must be lowered to a point where the saturation vapour pressure is 30 mb, which is approximately 24 °C; the dew point of air mass X. Further cooling below 24 °C of air mass X would result in oversaturation which means that "too much" water is present in the air. Consequently air must shed some of its moisture. This is achieved through the process of condensation, onto either

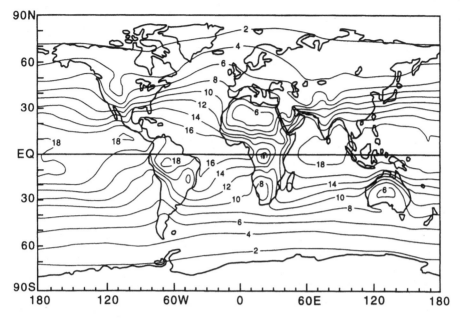

Figure 9.5   Mean annual atmospheric moisture content (specific humidity in g kg^{-1}). From Piexoto and Oort(1992). Reproduced by permission of American Institute of Physics, Boston

cool surfaces at the earth's surface (dew formation), or onto microscopic particles in the atmosphere (condensation nuclei – the starting points of raindrop growth).

Absolute humidity is the mass of vapour in a given volume of moist air. Its unit of measurement is grams per cubic metre. Specific humidity is the ratio of the mass of water vapour to that of humid air. It is measured in grams per kilogram which makes it a density. Although a direct measure, one disadvantage of absolute humidity is that its value is altered with a change of volume as occurs as an air parcel rises or falls. For this reason a preferred measure of humidity is the mixing ratio expressed in grams of water vapour per kilogram. It is defined as the ratio of the mass of water vapour to the mass of dry air containing it. At sea level its value (in grams per kilogram) is approximately 62% of the actual vapour pressure (Linacre, 1992).

One further measure of moisture in the atmosphere is precipitable water; usually expressed in grams per centimetre or centimetres. As the name suggests, it is the depth of water that would result if all the water vapour in a unit column of air was condensed and precipitated on the ground. Specific humidity is used in the determination of perceptible water (Piexoto and Oort, 1992).

DISTRIBUTION OF HUMIDITY IN THE TROPICAL ATMOSPHERE

The latitudinal distribution of atmospheric humidity shows a very simple pattern with a clear maximum near the equator (Figure 9.5). This maximum is the effect of the high evapotranspiration rates in the low latitudes and the prevalence of warm air masses, which can contain large amounts of water vapour. Although humidity decreases polewards, within the tropics themselves there are large differences (Figure 9.5). Over the oceans, humidity is generally controlled by the sea surface temperature and it is normally close to the saturation vapour pressure. Therefore, the western parts of the ocean basins and the equatorial oceans have generally high humidity values. This effect is transferred to the adjacent continents, though in eastern Africa this influence is limited to a narrow coastal strip owing to topographic and dynamic effects. The eastern ocean basins, where cold currents prevail, have much lower humidities. Humidity levels here are also low because of the regime of subsidence found on the eastern flanks of the subtropical anticyclones.

Over the continents, atmospheric humidity depends on the characteristics of the prevailing air masses. Generally the high ocean humidities in equatorial regions are replicated over the continents. Here large rainfall amounts over, and evapotranspiration rates from, equatorial forests ensure high humidities. This is particularly true for the African and South American equatorial regions (Figure 9.5).

Rapid meridional decreases in humidity levels over low latitude land masses are seen in northern and southern Africa in the areas of the Sahara and Kalahari deserts respectively, and over northern and central Australia. In these areas, despite high temperatures and thus potentially high saturation vapour pressures, humidity levels are low because surface water is limited.

To illustrate some regional and local variations, Table 9.1 provides mean humidity data expressed as vapour pressure. The effects of elevation are shown by the station pair Singapore–Tanah Rata, in equatorial conditions where seasonal differences in atmospheric humidity are small, and by the Zimbabwe stations where the effects are limited during the dry season, but considerable in the rainy part of the year. Seasonal differences in monsoon climates are illustrated by the two stations in southern and eastern Asia. The Tanzanian station pair shows the combined effects of elevation and distance from the ocean, Tabora being about 750 km from the Indian Ocean coast, where Dar es Salaam is situated. The reduction in mean vapour pressure remains the same in the dry and wet seasons.

The Ugandan station near Lake Victoria is compared with a Kenyan station at the same elevation to illustrate the influence of the lake on humidity conditions. This is, of course, a purely local effect, experienced by many stations near water surfaces. Over low latitude ocean surfaces, humidity levels change very little throughout the year, as shown for the central Pacific locations of Tarawa and Funafuti where the January–July contrasts in vapour pressure are negligible (Table 9.1). The two stations on Malagasy are used to indicate that the low level humidity conditions are not affected by their situations on either side of the large mountain ranges of Madagascar. Despite their large differences in rainfall (Tamatave has an annual rainfall of 3526 mm, while Maintirano receives only 998 mm per year) the two stations differ very little in humidity. Finally, two extratropical stations are given to illustrate the quite different magnitudes of vapour pressure in comparison to the tropics.

As well as considerable spatial variability, humidity levels in the tropics demonstrate considerable temporal variability. This is clearly seen in Figure 9.6 which shows the January minus July specific humidity. Negative (positive) values indicate areas where January humidity levels are lower (higher) than July levels. Immediately obvious are

Table 9.1   Mean atmospheric vapour pressure (mb) for low latitude locations

Station	Lat.	Long.	Elevation (m)	January	July
Singapore	1°N	194°E	8	27	30
Tanah Rata, Malaysia	4°N	101°E	1448	17	17
Calcutta, India	23°N	88°E	6	15	34
Koshun, Taiwan	23°N	120°E	10	15	31
Dar es Salam, Tanzania	7°S	39°E	14	30	23
Tabora, Tanzania	5°S	33°E	1190	20	13
Entebbe, Uganda	0	32°E	1146	20	20
Moyale, Kenya	4°N	39°E	1113	15	16
Bulawayo, Zimbabwe	20°S	29°E	1344	17	9
Beitbridge, Zimbabwe	22°S	30°E	456	22	10
Tamatave, Malagasy	18°S	49°E	5	29	21
Maintirano, Malagasy	18°S	44°E	23	30	20
Tarawa, Pacific Ocean	1°N	172°E	4	28	29
Funafuti, Pacific Ocean	8°S	179°E	2	31	31
London, England	52°N	0	33	7	14
Cairo, Egypt	30°N	31°E	30	8	17

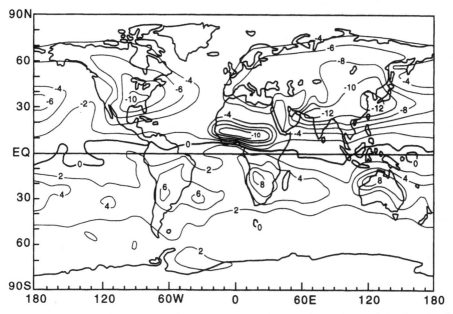

Figure 9.6   January minus July atmospheric moisture content (specific humidity in g kg^{-1}). Negative (positive) values in the northern (southern) hemisphere indicate that winter (summer) moisture is less (greater) than summer (winter) levels. Note the large differences over India, West Africa and northern Australia where seasonal moisture differences are a product of the monsoon systems in these areas (Chapter 7). From Piexoto and Oort (1992). Reproduced by permission of American Institute of Physics, Boston

the large variations that occur in the monsoon regions of West Africa, Asia and northern Australia which show specific humidity contrasts between the summer and winter of $8-12\,g\,kg^{-1}$. Of these areas the Asian monsoon region has the largest annual variation of humidity. Such large variations are due mainly to advection of moisture into these regions by strong monsoonal flows over warm oceanic waters (Chapter 7). Although the dry tropical areas have low atmospheric humidities throughout the year (Figure 9.5), values during the winter of their respective hemispheres are lower than during the summer, as shown by northern and southern Africa and Australia (Figure 9.6).

While the seasonal variations in humidity follow a simple pattern in most parts of the tropics, being controlled mainly by temperature and air mass characteristics, the diurnal variations are more complicated. At most tropical stations, the daily minimum of vapour pressure is recorded at night, shortly before sunrise. Temperatures are then at a minimum, so that the lower layers of air frequently become saturated and dew is formed. This process abstracts moisture from the air. During the early morning hours, most of the dew is readily available for evaporation and transpiration is high, since the plants have abundant water available; vapour pressure builds to a maximum. Later in the morning, when the dew has disappeared and transpiration and evaporation are limited by water supply, and when convection currents develop, which transfer water vapour to higher elevations, the atmospheric humidity decreases to a secondary minimum. Unless precipitation occurs, this minimum persists throughout the afternoon. After sunset, when the temperatures fall and convectional activity dies down, atmospheric humidity increases again. This increase is particularly strong after rainfall. This is clearly seen for the case of Singapore where day length hardly varies because of its equatorial position (Figure 9.7). The diagram for Singapore, which is based on annual means, illustrates the small increase during the afternoon, caused by frequent rainfall in this part of the day. For Lusaka (Figure 9.7), where day length varies with the season (earlier morning and later evening maxima in July), there are clear seasonal differences in the diurnal cycle of vapour pressure. January in the middle of the rainy season, brings not only much higher vapour pressure values, but also a modest afternoon minimum. In July the minimum is better developed and persists for a longer period. The very clear morning maximum illustrates the importance of dew formation during the rather cold winter nights.

Over the oceans the vapour pressure shows almost no diurnal variations, because the sea surface temperatures remain practically unchanged throughout the day.

Although humidity levels are high in the low latitudes not all of the atmospheric water vapour is derived from the same latitude as significant equatorward transfers occur mainly in the lower branches of the Hadley cells. As well as considerable imports of moisture into the low latitudes, export of water vapour also occurs to the mid-latitudes, especially on the western sides of the ocean basins (Figure 9.8).

Because there are large movements of water vapour in the atmosphere and evaporation is rarely equal to precipitation for a location, some areas experience moisture divergence while others experience moisture convergence. Areas of divergence represent areas where evaporation is greater than precipitation. They are therefore source areas of moisture. Moisture convergence areas are moisture sink areas; precipitation exceeds evaporation.

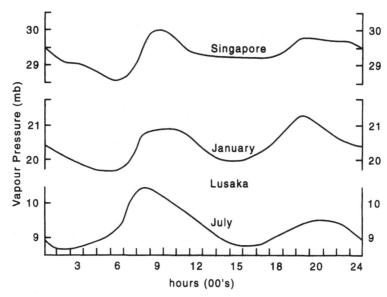

Figure 9.7   Diurnal variation of mean vapour pressure at Singapore (1°21'N, 103°54'E, 8 m a.s.l.), based on annual means, and at Lusaka (15°19'S, 28°15'E, 1154 m a.s.l.) for January and July

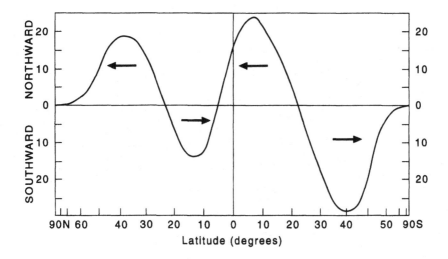

Figure 9.8   The mean annual meridional transfer of water vapour in the atmosphere (in 10^{15} kg). Note that the south-east trades transport a greater mass of water vapour compared to the north-east trades. After Sellers (1965, p. 94)

In the tropics, moisture convergence generally prevails over the equatorial regions in the area of the ITCZ, while in the subtropics divergence prevails. Several centres of water vapour convergence may be identified over land (Figure 9.9). These occur over the headwaters of large river systems such as the Amazon in South America, the

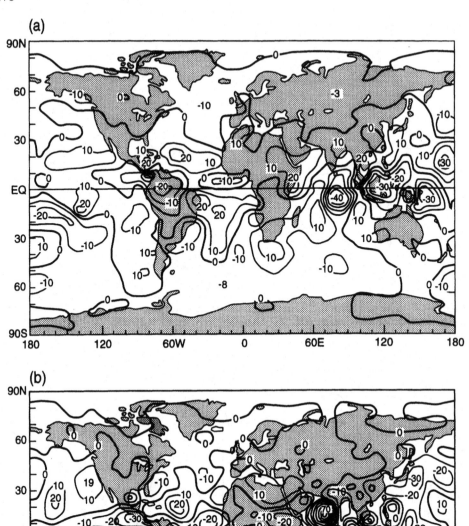

Ubangi, Congo, Senegal and Blue Nile in Africa, and the Indus, Ganges, Mekong and Yangtze in Asia. Areas of moisture convergence demonstrate seasonal shifts in accord with the seasonal shifts of the rising branches of the Hadley cells (Piexoto and Oort, 1992). For example, in the Asian–Australian area, major areas of convergence demonstrate significant shifts from equatorial regions between 70°E and 160°E in the southern hemisphere months of December to February (Figure 9.9(a)) to around 15°N to 20°N and 65°E just west of India at the height of the northern summer (Figure 9.9(b)).

Geographical shifts in the areas of divergence also occur in association with the seasonal shifts of the subtropical anticyclones. This is very clear over the southern oceans where not only geographical shifts are seen, but also changes in the magnitude of water vapour divergence (Figure 9.9).

CONDENSATION

In the atmosphere, condensation takes place when the air is cooled beyond its dew point. Cooling can be the result of radiation losses during the night which may lead to the development of dew or fog. Perhaps more important and widespread is the cooling of air by expansion, when it moves upward and reaches levels of lower pressure; an adiabatic process (Chapter 2). This form of cooling results in condensation and the formation of clouds.

There is a quantitative difference in the amount of liquid water produced by these processes between the tropical and extra-tropical climates, because warm air masses change their capacity to retain water vapour much faster than cooler ones (Figure 9.1). Actual amounts of water produced by cooling of $1\,m^3$ of air by 5°C decreases from 6.6 g for air starting at 30°C, through 4.5 g for air at 20°C to only 2.0 g at 5°C. Warm airmasses therefore generally produce more water by condensation than cooler ones.

The level where condensation starts depends on the humidity of rising air. It is generally higher in warmer air masses. Above the condensation level, latent heat set free maintains the buoyancy of warm air masses over deeper layers than in cold air. Therefore in the tropics, condensation is not only more vigorous, but also extends over thicker layers of air than in cooler climates.

Dew and morning fog are frequent occurrences in many parts of the tropics. During rainy periods, the humidity of the air at night close to the earth's surface is usually near the saturation point, so that only a slight amount of cooling is sufficient to produce dew. In dry periods, when the humidity of the air at night is much lower, cooling during clear nights is much stronger and even then dew will form quite often.

Morning fog results from the cooling of thicker layers of air. It is therefore usually related to air drainage. For this reason, morning fogs occur mainly in basins and

Figure 9.9 Global distribution of horizontal divergence of total water vapour transport in units of $0.1\,m\ year^{-1}$ for mean conditions in (a) December to February and (b) June to August. Note the strong negative values over monsoon areas in the respective summer hemispheres where there is moisture convergence: excess of precipitation over evaporation. Positive values represent evaporation excess. From Piexoto and Oort (1992). Reproduced by permission of American Institute of Physics, Boston

valleys. Near coasts, where land breezes often produce some turbulence during the night, morning fogs are rare. Morning fogs usually disappear soon after sunrise, as solar heating and increased turbulence cause warming of the lowest air layers.

Fogs caused by the cooling of moist air over cold ocean waters are particularly frequent in dry tropical areas on the western coasts of the continents, as in Namibia, Morocco, northern Chile and Peru. These fogs persist for long periods, as daytime heating by insolation is prevented from reaching the earth's surface. Surface heating over ocean surfaces is also much reduced because the majority of heat received goes into evaporation and not surface heating. In many instances the only precipitation source in coastal deserts such as the Namib is that of fog precipitation, often referred to as occult precipitation as it collects on both flora and fauna.

TROPICAL CLOUDS AND CLIMATE

The main products of condensation in the atmosphere are clouds. Clouds are climatologically significant, not only because they are the source of precipitation and control the intensity of solar radiation during the day and the extent of radiation losses at night, but also because they are the main controlling factor of the otherwise rather uniform temperatures in many parts of the tropics (Nieuwolt, 1968).

It has long been recognized that there is a close two-way link between tropical sea surface temperatures and clouds: sea surface temperature controls the local humidity and thermodynamic processes within the marine boundary layer which is the moisture source for clouds; cloud development influences the radiative and mass exchange, convection and evaporation over the sea surface and also mesoscale convergence and subsidence patterns in the atmosphere.

Of the various tropical ocean areas the western Pacific is of interest because sea surface temperatures here rarely exceed 30°C. This phenomena may be related to a built-in control or feedback mechanism which operates between the ocean and the atmosphere at time scales of around 3–6 days – the time scale over which clouds and sea surface temperatures seem to respond to each other (Zhang, Ramanathan and McPhaden, 1995). This response is in terms of a change in cloud physical properties due to a change of surface evaporation caused by variation of the sea's surface temperature. Changes in cloud physical properties, in turn, feedback on sea surface temperatures through the cloud's effect on radiation transfer to the ocean surface, thus limiting the surface temperature. Increases of wind speed with increasing surface temperatures may also play a role in mixing of the warm surface water with cooler water below the surface (Webster, Clayson and Curry, 1996).

Evaporation also appears to be controlled by a cloud-related feedback mechanism such that evaporation is temperature-limited; a relationship contrary to the commonly held belief that evaporation increases with surface temperature. This self-regulating mechanism appears to operate in such a way that evaporation increases to a point around 28–29°C; however, beyond this temperature, evaporation levels decrease despite a sharp increase in the saturation deficit between the ocean surface and the air. This apparent physical dilemma may be explained with reference to the behaviour of wind speed as sea surface temperature rises. Surprisingly, as temperatures increase, wind speed falls. Consequently there is decreased turbulence over the sea's surface with the result that less water vapour can be transported away from the surface by turbulent transfer, although surface to atmosphere vapour pressure gradients are high. Evaporation is thus wind-speed-limited. Changes in the wind speed regime appear to be related to interactions between convection and the large-scale low level circulation as at the centre of large-scale convergence zones, which are products of convection, wind speeds are low (Zhang, Ramanathan and McPhaden, 1995). Such low wind speeds limit evaporation.

CLOUD TYPES AND CLOUDINESS

In the tropics, the various cloud types occur at higher levels than in the mid-latitudes. This is especially true for high and medium level clouds which are related to sub-zero temperatures. Cirrus clouds, in higher latitudes, often found at altitudes of 3500 m, rarely occur below 6000 m in the tropics. Medium-high clouds, which in cooler climates are limited to elevations below 5000 m, often reach levels as high as 7500 m in the tropics.

Although for almost all cloud types there is a peak of occurrence in the equatorial regions, both over land and ocean, there are some considerable differences between the tropics and the higher latitudes in the amounts (product of frequency and coverage) of certain cloud types. The main difference is a result of the paucity of stratiform clouds in the tropics. In the mid-latitudes this is the prevailing type, especially in the area of the polar fronts (Figure 9.10).

Over the oceans, cumulus (Cu) generally decreases polewards, with the largest amounts found in the general regions of the trade winds on the upstream side of the ITCZ (Figure 9.10(a) and (b)). Cumulonimbus (Cb) is also a feature of the ITCZ and the South Pacific convergence zone. Similar to Cu, Cb shows marked seasonal shifts north and south of the equator especially in the maritime continent region. Cb also tends to decrease from west to east in the Pacific and Atlantic equatorial regions in accordance with decreases in sea surface temperatures.

Of all the cloud types, stratus (St) and stratocumulus (Sc) demonstrate the largest differences between low and mid-latitude environments (Figure 9.10). Stratus clouds in the tropics are limited to areas of widespread convergence, in easterly waves and other disturbances, and to low level fog caused by air drainage at night or over cold ocean surfaces. Stratus is especially prevalent over the western coasts of southern Africa and South America where the cold Benguela and Humbolt currents can be found. At the equatorial limits of these cold currents considerable seasonal variation in St amounts occur. Smallest St amounts in the low latitudes occur where there are local Cu maxima (Warren et al., 1986, 1988).

Altostratus (As) varies from very low amounts of around 10% in both winter and summer just south of the equator in east central Pacific and Atlantic, to well over 30% in the Indonesian area which has large amounts of clouds at all levels. Cirrus (Ci) type clouds have their largest amounts in the equatorial zone where they are associated with the anvil tops of towering Cb.

Discussions to this point have dealt mainly with cloud amount which has been

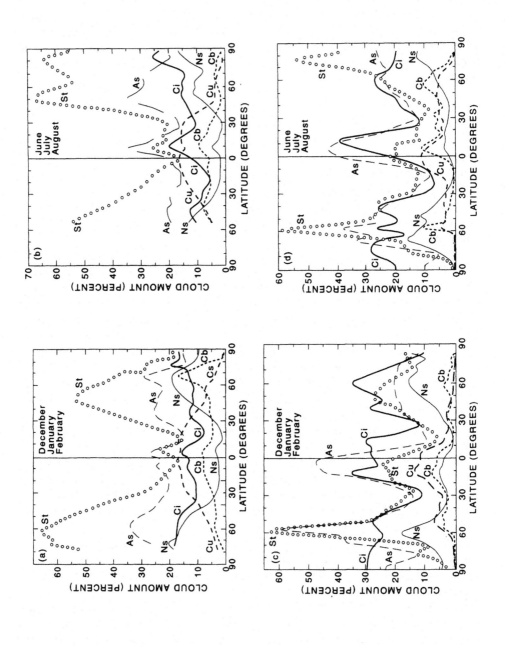

defined as the product of frequency of occurrence of a particular cloud type and its coverage. It is the latter of these two measures that is routinely reported at most climate stations and is often referred to as cloudiness. This is the proportion of sky in eighths (oktas) or tenths which is covered by clouds, independent of type. Compared to tropical land areas, the tropical oceans on average have greater cloudiness mainly because of the contrasts in the subtropical zones (Figure 9.11). In the equatorial zones, however, cloudiness tends to be higher over land for all seasons, especially in the southern hemisphere for the summer months (Figure 9.11(b)). The oceans are also much more conservative with regard to the seasonal changes in cloudiness, showing inter-seasonal differences of less than around 10%. Peak cloudiness for the oceans is also conservative in geographical terms, as the maximum zone of low latitude cloudiness does not shift south of the equator over the oceans with only a limited geographical range of less than 5° (Figure 9.11(a)). This contrasts markedly with tropical land areas which can show up to 40% range in cloudiness and a geographical range of around 18° between the summer and winter months (Figure 9.11(b)). This reflects the greater seasonal movement of the ITCZ over land areas in comparison to the thermally conservative oceans (Chapter 5).

In terms of cloud levels, high and low clouds dominate the inter-annual variation of total cloudiness over the western and central tropical oceans. Variations in these cloud types are due to changes in large-scale motions, resulting from changes of large-scale sea surface temperature gradients which contribute to the change of convective stability. Variations in low cloud play a major role in variability of total cloudiness in the subtropical eastern Pacific and to the west of Africa and Australia. This is consistent with the strength of the trade inversion in these areas which prevents deep convection (Fu, Liu and Dickinson, 1996).

Three major cloud patterns prevail in the tropics: amorphous cloud masses, showing no distinct structure, vortex patterns and cloud bands. Amorphous cloud masses are often formed in trade wind areas, their basic units are the trade wind cumuli. Amorphous cloud masses can also be identified for the ITCZ area as these represent the extensive anvils that sometimes form above mesoscale convective systems and cloud clusters. Vortex patterns are usually associated with cyclonic disturbances (Chapter 8). Cloud bands are found in zones of convergence or horizontal shear, or near-linear systems (Conover, Lanterman and Schaefer, 1969). Some cloud bands can stretch from the tropics to the extratropics. These are called *tropical–extratropical cloud bands* and are a clear expression of how the tropics and extra tropics interact (Plate 6).

Tropical–extratropical cloud bands are important channels for the exchange of latent heat and moisture with subtropical and mid-latitude areas. Such cloud bands seem to form when areas of active tropical convection connect with bands of middle to high level cloudiness which are a product of the polar front jet and the subtropical jet connecting. This occurs at times of intense mid-latitude cyclonic storm development. Typically, tropical–extratropical cloud bands are associated with incursions of mid-latitude troughs which bring upper tropospheric westerlies into the low latitudes or

Figure 9.10   Meridional profile of seasonal cloud amounts (product of cloud cover and frequency) for (a, b) oceans and (c, d) land. St is stratus; As is altostratus; Ci is cirrus; Ns is nimbostratus; Cu is cumulus; Cb is cumulonimus. From Warren et al. (1986, 1988). Reproduced by permission of NCAR/UCAR, USA

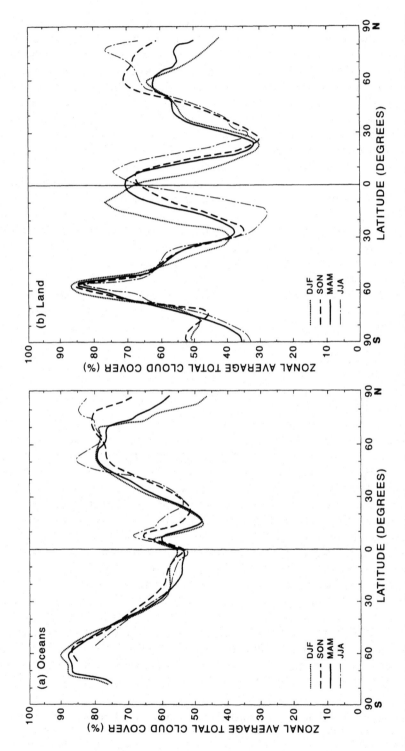

Figure 9.11 Meridional profiles of seasonal cloud cover for (a) oceans, and (b) land. Note the large ranges in cloud amount between DJF and JJA over land surfaces compared to ocean surfaces in the northern hemisphere low latitudes; a product of the monsoons. Reproduced by permission of NCAR/UCAR, USA

incursions of the ITCZ into the subtropics (Kuhnel, 1989). It is on the upper westerlies that the large amounts of heat and moisture are transported from tropical latitudes to mid-latitudes.

Climatologically, tropical–extratropical cloud bands are important as they are major rain producers in the subtropics and mid-latitudes. For example, over southern Africa, where they are called tropical–temperate troughs, around 60% of the summer rainfall may be attributed to such features (Harrison, 1984). Rainfall from these cloud systems may also generate major floods in this region (Lindesay and Jury, 1991). Tropical–extratropical cloud bands are also climatologically important for north-western and north-eastern Australia where such systems may contribute over 80% and 50–60% of the cool season rainfall respectively. During ENSO years, rainfall from these systems is much reduced (Wright, 1993).

Globally, 14 areas of tropical–extratropical cloud bands have been identified. Generally, the northern hemisphere cloud bands coincide with major areas of cyclogenesis while those of the southern hemisphere show the same position and orientation as cyclone trajectories. Of the 14 cloud band areas, the most active can be found in the Pacific Ocean, while the most inactive are in the Indian Ocean. The seasonal distribution of cloud band activity also varies (Kuhnel, 1989).

SATELLITE CLOUD CLIMATOLOGY

Increasingly, the development of cloud climatologies for the low latitudes and the understanding of the complex interaction between clouds and radiation is becoming dependent on satellite technology.

Three basic steps are involved in the analysis of satellite observations of clouds. These are detection, radiative modelling and statistical characterization (Rossow, 1989). Cloud detection methods are based on comparing radiance values for clear sky with those measured. The basis of this method is that a critical threshold value can be identified for clear sky conditions and any value above or below this threshold must indicate the presence of cloud. A variety of radiance-based detection methods exist and may be categorized according to whether radiance variations in wavelength, space or time are used. Radiative modelling, the second step, involves developing models for the removal of radiance effects due to other factors such as atmospheric moisture and temperature in clear sky conditions and surface effects. This modelling is performed in order to isolate the cloud contribution to the measured radiance and thus allow for the inference of the cloud physical properties. The final step of statistical characterization involves determining the nature of cloud variations on different time and space scales. Such statistical studies provide the basic information for developing satellite-based cloud climatologies (Rossow, 1989).

Potentially, satellite cloud climatologies will differ from purely surface-based climatologies. Reasons for this include the fact that radiance statistics are used to define cloud types in satellite analyses which contrasts to the ground-based visual identification based on pattern; the clouds are projected in a different manner so that neither earth cover nor sky cover is obtained, and satellite imagery may not detect clouds with dimensions smaller than the pixel or area sensed (Warren et al., 1986). Despite these potential problems, satellite-based cloud climatologies offer the advantage of global coverage at small time scales. For example, data from the International Satellite Cloud Climatology Project (ISCCP) have allowed the development of an eight-year global cloud climatology which provides information every three hours at a spatial scale of 280 km (Rossow, Walker and Garder, 1993). Comparison of the eight-year ISCCP climatology with a surface-based climatology has shown that for the tropical land areas cloud coverage has been estimated at 47.1% and 49.2% respectively, while that for the tropical oceans has been estimated at 72.2% and 67.3% respectively. The ISCCP climatology has confirmed that the ITCZ, as observed from satellites, extends to higher latitudes than observed from the surface; the largest latitudinal contrasts in cloud amount occurs between the tropics and the subtropics; globally longitudinal contrasts in cloud amount are greatest in the subtropics between the deserts and the marine stratocumulus regions near the west coasts of the continents (Rossow, Walker and Garder, 1993). Despite the marine stratocumulus regions being some of the cloudiest on earth, very little precipitation falls here. Therefore the spatial distribution of cloudiness is not necessarily a good guide to the spatial distribution of rainfall, one of the topics to be addressed in the next chapter.

SUMMARY

The importance of water in the atmosphere relates to its control on precipitation, the surface radiation balance and the fact that changes of state result in the release or absorption of energy. The main source of atmospheric water is the oceans, which explains why on an annual basis atmospheric humidities are highest over the tropical oceanic areas. However, during summer months, due to plentiful rainfall and high evapotranspiration rates, especially over tropical forest areas, atmospheric moisture levels over continental areas may exceed those of adjacent oceanic areas.

Greatest seasonal contrasts in atmospheric moisture occur over the monsoon regions, with maximum levels being attained at the height of the summer monsoon. This is because large amounts of water are advected over continental areas from warm oceans, the consequence of which is plentiful rainfall.

Clouds which are the physical manifestation of the condensation processes in the atmosphere vary in their coverage and type throughout the tropics. Typically over the eastern parts of the Atlantic and Pacific Basins cloud coverage is high and characterized by low level stratus, a product of cool ocean surface temperatures and an intense inversion. In contrast, in the west of these same basins, clouds are typically cumuliform in nature, a consequence of a deep convective layer and warm sea surface temperatures. Seasonally the movement of the zones of maximum cloudiness are the greatest over land. The use of satellites has revealed much about the climatic role of clouds in the tropics and the importance of tropical–extratropical cloud bands as ducts for moving heat and moisture from the tropics to the mid-latitudes.

CHAPTER 10

Tropical Precipitation

Tropical rainfall is of great importance as it can be life-giving on the one hand, but life-taking on the other, if excess rainfall produces floods or insufficient rainfall results in drought. In addition to its importance for the sustenance of life, tropical rainfall is also of importance for global climate and weather. Over two-thirds of global precipitation falls in the tropics. As a result, a large amount of energy in the latent heat form is released in the low latitudes. This energy not only balances radiation heat losses but is used to power the global atmospheric circulation. An understanding of the geographical distribution of rainfall therefore sheds light on the global distribution of heat sources that drive the globe's atmospheric heat engine.

Rainfall is probably the most variable element of tropical climates. Almost everywhere in the tropics the most important quantitative indicator, the annual total, differs from year to year and from place to place. Other rainfall characteristics, such as its seasonal and diurnal distribution, intensity, duration and frequency of rain-days also demonstrate spatial and temporal variations.

In this chapter an outline of the origin of tropical rainfall and the factors affecting its formation will be followed by an overview of the main characteristics of tropical rainfall in terms of annual amounts, inter-annual variability, seasonal distribution, diurnal distribution and frequency, and intensity characteristics.

THE ORIGIN OF TROPICAL RAINFALL

Three types of rainfall occur in the tropics: convectional, cyclonic and orographic. As the names suggest, these types are related to the origin of the rainfall. The three different types also occur over different geographical scales.

Convectional rainfall is the result of free convection due to heating alone, dynamic processes such as convergence, or physical forcing over mountain ranges. Convectional rainfall generally occurs over a limited spatial scale of between 10–20 km^2 and 200–300 km^2. This type of rainfall is therefore characterized by considerable spatial variability. The spatial scale of convectional rainfall usually depends on whether convectional cells or thunderstorms form individually and remain in the place of origin, or become organized into weather systems such as sea breeze fronts (Chapter 6) and linear disturbances or squall lines (Chapter 8). Convectional precipitation, because it is formed by rapid uplift, often to great heights in the atmosphere, is usually intense; occasionally it may be in the solid form of hail.

Cyclonic rainfall is produced by horizontal convergence of moist air in a circular area of low pressure where the vorticity maximum exists. Its most impressive expression is in tropical cyclones (Chapter 8) where the combined processes of cyclonic inflow and convection produce intense rainfall. Cyclonic storms typically last between one and five days, which contrasts with the short life span of individual convection cells. The area affected by cyclonic precipitation may be large, as throughout their lifetimes, tropical cyclonic storms can move several hundred kilometres.

Orographic rainfall is the result of condensation and cloud formation in moist air that has been physically forced over topographic barriers. Orographic rainfall formation may be aided by convectional processes in the tropics. It finds its best expression on windward slopes that face into a sustained moist flow of air such as the trade winds. For locations that experience a seasonal reversal of winds the geographic distribution of orographic rainfall can change markedly as a windward side of a mountain barrier becomes a leeward side. Orographic precipitation, unlike cyclonic precipitation, is not mobile and is limited to the mountain barrier to which it owes its origins.

Regardless of the type of rainfall or climate, all rainfall is the result of upward movements of moist air. For uplift to occur, the atmosphere needs to be in a state of conditional, potential or convective instability. These stability states depend on the relationship of the actual environmental lapse rate to the dry and moist adiabatic lapse rates. These relationships are summarized in Table 10.1.

An important atmospheric condition for convectional storms is the release of instability. Conditional instability occurs in the situation when the rising air is moist and the environmental lapse rate lies between the dry and saturated adiabatic lapse rates (Table 10.1). The stability of the air is therefore conditioned by its moisture status. As soon as it becomes saturated, the air mass will become unstable and cloud development will ensue. Orographic effects can help achieve the release of conditional instability. In the case of potential or convective instability, the depth of displaced air is important. Additionally, lower altitude air should be moister than higher altitude air, such as in the regions either side of the ITCZ where the trade wind inversion separates these two types of air. Potential instability is released when orographic uplift or strong convection forces the moist air through the trade inversion. When this happens, the upper parts of the penetrative air cool rapidly at the dry adiabatic lapse rate, whereas the lower very moist layers continue to cool at the lower saturated lapse rate as they remain humid (Sumner, 1988). Consequently, the environmental lapse rate will be steepened, eventually becoming greater than the saturated lapse rate. Spectacular convective cloud development follows, with cloud tops often reaching the top of the troposphere.

It is this style of cloud and rainfall development by convection alone or in combination with other factors (convergence, cyclonic inflow and orography) which makes the origin of rainfall in the tropics, in many cases, different to that in the extratropics. In mid- and high latitude environments rainfall is often associated with cyclonic activity

Table 10.1   Temperature lapse rates and atmospheric stability
(McIlveen, 1992)

Lapse rate relationship	Stability in relation to air parcel moisture state	
	Dry	Moist
ELR > DALR	U	U
ELR = DALR	N	U
	Conditionally	
ELR < DALR	S	U

ELR is environmental lapse rate (measured), DALR is dry adiabatic lapse rate
($9.8\,°C\,km^{-1}$), SALR is saturated adiabatic lapse rate ($4.8\,°C\,km^{-1}$), U is unstable,
S is stable and N is neutral.

or frontal surfaces. Convection does, however, occur in extratropical environments
and when it occurs it often produces rainfall with characteristics of the tropics, i.e.
short duration, high intensity rainfall.

A further tropical–extratropical difference concerns the properties of the air masses
which undergo uplift. In the tropics these are generally warmer and more humid than
outside the tropics. Warm and humid air masses reach the condensation level at
relatively high temperatures and consequently form cloud which in their low to middle
levels consists mainly of water droplets and rarely of ice crystals. Tropical clouds are
therefore often referred to as warm clouds, as the bulk of their mass has a temperature
greater than $-10\,°C$, with only the upper parts of the cloud towers having temperatures less than this (McIlveen, 1992). This contrasts with the extratropical atmosphere
where the majority of the atmosphere, and thus cloud temperatures, are below
$-10\,°C$.

In warm clouds, the majority of rainfall formation is by a two-stage process. The first
involves diffusion and condensation which results in a population of cloud droplets
being formed. The second stage involves droplet growth, which is achieved by collision
and coalescence of droplets with their neighbours. This two-stage process of diffusion
and condensation followed by collision and coalescence is believed to explain precipitation from warm clouds such as occurs in the lower to middle parts of the tropical
troposphere (McIlveen, 1992). Warm tropical cloud rainfall is large in amount and
often intense.

Precipitation formation in cold extratropical clouds is different to warm clouds.
However, cold cloud precipitation formation processes most likely occur in the cold
tops of very deep tropical clouds – the "hot towers". Here it is the Bergeron-Findeisen
process that dominates. This process involves the growth of ice crystals at the expense
of surrounding supercooled water droplets because the air surrounding adjacent water
droplets and ice crystals possesses different saturation states relative to ice and water.
Simply due to huge saturation vapour pressure differences between water and ice,
water droplets tend to lose mass while ice crystals gain mass. This process is most
effective at temperatures between $-10\,°C$ and $-30\,°C$ as in these temperatures,
supercooled water droplets and ice crystals co-exist in large numbers.

The fact that many tropical cloud tops are at temperatures below $-10\,°C$ means that the Bergeron-Findeisen process plays an important role in tropical rainfall production. This means that tropical rainfall with a partly cold cloud origin has started its life as an ice crystal. The fate of such a falling ice crystal will depend on the form of the moisture in the air through which the crystal falls. If falling through air containing smaller ice crystals, the ice crystal will grow by aggregation, whereby, smaller crystals will collide with and stick to it. This process eventually results in the development of snow crystals and snowflakes which, if air temperatures remain below about 2 °C, will be deposited on the ground as snow, as occurs in some high mountain tropical environments. Otherwise the ice crystals melt and fall as rain.

If a falling ice crystal that has grown in cold cloud tops falls through air containing supercooled water droplets, it will grow by accretion, whereby water becomes frozen onto the ice crystal. In this way the freezing nucleus of a hailstone is formed. Such a hailstone may not fall immediately to the ground as strong updraughts in the cloud may lead to recirculation of the hailstone back up into the upper parts of the cloud where the accretion process again occurs. In such a way, very large hailstones may develop. Because of their great fall velocities, hailstones are able to reach the ground still in their frozen form despite falling through warm atmospheric layers. As a consequence of their large impact forces, hailstone destruction of tropical agricultural crops often occurs.

Because convection and a mixture of cold and warm cloud processes dominate rainfall production in the tropics, rainfall characteristics can be quite different to those for mid-latitude environments. This is especially true in the case of annual amounts, intensity, duration, frequency, and its spatial and temporal distribution at a variety of scales.

TOTAL RAINFALL

The annual mean, usually computed based on a minimum of 30 years of rainfall records, is generally accepted as an indicator of rainfall conditions. Although its meaning is easily understood, its limitations are not. The annual mean has some serious disadvantages when it is used to estimate future rainfall for water balance studies and assessment of agricultural development possibilities. At most rainfall stations in the tropics the negative departures from the annual mean (occasions when annual rainfall is below the mean) are more numerous than the positive ones, resulting in a positive skewness of the frequency distribution of rainfall totals for individual years. The annual mean is therefore inflated by a very few high annual totals. This *skewness* in the frequency distribution is usually strongest where rainfall totals are low.

For the purpose of agricultural and general environmental planning, *percentiles* and other data based on measures of *variability* and *probability* are much better. Percentiles indicate the rainfall level below which a given percentage of occasions (years for annual rainfall) occur; these include quartiles, quintiles and deciles. For example, the 75th percentile (upper quartile) might be 1500 mm, meaning that 75% of years over the record length had rainfall less than this amount. Rainfall analysed in terms of its probability characteristics can also be modelled using probability distributions (Stern, Dennett and Garbutt, 1981; Jolliffe and Hope, 1996). For the calculation of percentile,

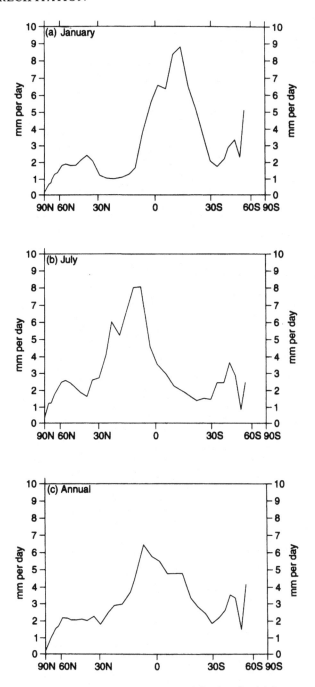

Figure 10.1   Mean meridional profiles of daily precipitation for (a) January, (b) July, and (c) annual. Note the latitudinal shifts in zones of maximum precipitation with onset of summer in the respective hemisphere. Adapted from Legates (1995)

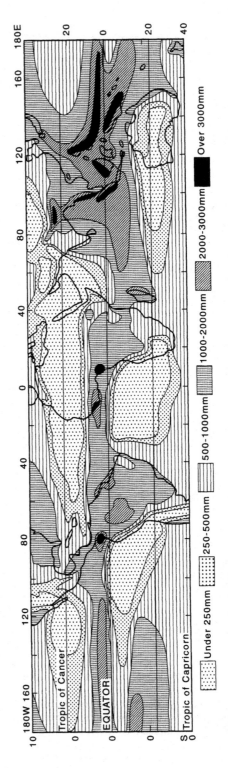

Figure 10.2   Mean annual rainfall

Under 250mm   250-500mm   500-1000mm   1000-2000mm   2000-3000mm   Over 3000mm

probability and variability statistics, a standard period of 30 years is now internationally used. Climate statistics, calculated based on this time period, are called *normals*. It is, however, important to ensure that standard 30-year periods are used in calculating rainfall normals as rainfall climatologies (mean, variability and extremes characteristics) developed for different 30-year time periods are of little use when absolute comparisons between locations are to be made. For this reason the World Meteorological Organisation has specified two standard periods. These are 1931–60 and 1961–90. The availability of standard climatic normals is also required for checking the validity of rainfall predictions from general circulation models (Chapter 13).

The measurement of rainfall in the tropics is difficult due to its great spatial variability and the fact that a large proportion of tropical rainfall occurs over inaccessible areas such as tropical oceans, tropical forests and mountainous ranges. Accordingly, there has been much interest in satellite-based estimation of tropical rainfall and the use of these estimates for development of tropical rainfall climatologies at a variety of temporal and spatial scales.

Rainfall estimates from satellite observations first began in the 1970s (Barrett, 1970). Since then, much effort has been invested in developing techniques for estimating tropical rainfall. Generally, two techniques can be identified: indirect, using the visible and thermal infrared characteristics of clouds (especially outgoing longwave radiation) as indicators of rainfall amounts; and direct, based on observations of the radiative effects of precipitation hygrometeors (Arkin and Ardanuy, 1989).

In terms of annual rainfall the tropics are of interest as some of the globe's highest and lowest rainfall totals can be found here. Overall it receives two-thirds of the global annual total. When annual means are compared to latitude they indicate a clear maximum in the tropics (Figure 10.1).

The distribution map of annual mean rainfall illustrates the latitudinal variation of rainfall very well, but it also indicates many non-latitudinal differences in the tropics (Figure 10.2). High amounts of rainfall are caused by a number of factors usually in combination. The most important of these is the ITCZ and the length of its stay over a certain area. This is demonstrated by the high totals over the Pacific Ocean, just north of the equator, where the ITCZ is nearby almost throughout the year. Over the Atlantic Ocean a similar zone exists, but it is of lower intensity, as here the ITCZ is more variable in its position (Chapter 5).

A second factor causing high rainfall totals is relief (Figure 10.3). Orographic lifting is particularly efficient where monsoonal winds are forced to rise, as illustrated by the western coasts of India, Burma, Sumatra and Borneo, but also over the Ganges plain of northern India. In West Africa, this effect is demonstrated by the coastal areas of Liberia and Sierra Leone and near Mount Cameroon. The trade winds can also yield large amounts of rainfall when uplifted by steep mountain ranges as shown along the eastern coasts of Madagascar and the northern parts of South America. Where orographic effects and onshore winds combine, very high rainfall results: over New Guinea, near the west coast of Central America and the western parts of the Amazon Basin.

A third factor causing high rainfall totals are tropical cyclones (Chapter 8). In these weather systems the zone of high rainfall follows the recurving paths of the storms.

A fourth and minor factor is related to the convergence and directional change of the

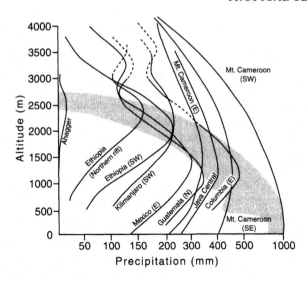

Figure 10.3   Elevation precipitation relationships. From Barry (1988). Reproduced by permission of John Wiley, Chichester

trade winds when they approach the equator. The geographically confined areas of high rainfall in the Pacific and the Indian Ocean south of the equator, which correspond to the average position of the southern branch of the ITCZ, are evidence of this factor.

Areas where rainfall totals are low are mostly caused by the effects of the subtropical high pressure cells (Chapter 5). These influences are reinforced by cold ocean currents and zones of equatorial upwelling in the eastern parts of the large ocean basins. Continentality is another major factor causing low rainfall, as demonstrated in southwest and central Asia.

THE EFFECTS OF ELEVATION

As shown by the global rainfall distribution map (Figure 10.2), mountains and highlands receive more rainfall than nearby lowlands, at least on their windward sides because of orographic effects. Generally, it is assumed that rainfall increases with elevation. While this is true for most temperate climate locations, this commonly assumed relationship breaks down in the tropics. In the tropics, elevation–rainfall relations are complex. Generally rainfall increases up to about 1000–1500 m but beyond this, it generally decreases with height. The height of the rainfall maximum for a location generally coincides with the mean cloud base height at that location. In West Africa on the north-eastern side of Mount Cameroon (4070 m), rainfall increases up to 1800 m but decreases beyond this, while on the south-western slopes, maximum rainfall is at sea level due to monsoonal influences (Figure 10.3). This contrasts with East Africa, where rainfall only increases above the plateau surface; maximum rainfall is recorded on the volcanic peaks rising above the plateau surface. Rainfall here is only produced when the air has experienced continual uplift over the plateau edges with cloud bases positioned near the top of the volcanic peaks.

For some mountainous tropical regions such as Papua New Guinea, the altitude-rainfall relationship appears to be seasonally dependent, with marked relationships apparent in the wet season but not for the dry season (McAlpine, Keig and Falls, 1983). For Sri Lanka, a mixture of rainfall–elevation relationships exist dependent on the climatic regime and the season (Puvaneswaran and Smithson, 1991).

Despite the complexity of the situation described above, it is clear that the rainfall–elevation relationship is different to that of extratropical latitudes. This characteristic is due to two conditions, which frequently prevail in the tropics but are relatively rare in the mid-latitudes. The first is a strong difference in the water vapour content between the lower and upper layers of the troposphere. Tropical air masses are often very humid up to an elevation of about 800–1500 m, but above this level they are usually rather dry. This is mainly due to the trade wind inversion but also to the large water vapour production at the earth's surface in the tropics. The steep lapse rates, frequently present in tropical air masses, also tend to reduce the capacity of the air to retain water vapour in the higher parts of the troposphere.

The second factor is the predominance of vertical air movements in the tropics, where horizontal advection of moisture is limited. Most precipitation therefore originates from the atmosphere directly above the slopes. In the mid-latitudes water vapour is transported over large horizontal distances, making the potential supply of moisture for rainfall production greater, assuming a mechanism is available to aid the release of conditional or potential instability. Furthermore, in the tropics, condensation levels for maritime air masses lie at about 500–800 m, at which elevation one would expect maximum precipitation. This contrasts with mid-latitude climates where there may be condensation and cloud levels at a variety of heights.

INTER-ANNUAL VARIABILITY OF RAINFALL

Inspection of annual rainfall totals for many tropical regions will immediately reveal that there are clear year to year variations of rainfall (Figure 10.4). This is an important feature of tropical climates, a fact not appreciated by most extratropical dwellers, who perceive the tropics as being invariably wet. Lack of regard for this climate characteristic by agricultural and environmental planners has led to some unwise economic development decisions being taken in developing tropical countries. Seasonal and non-seasonal variations in the atmosphere–ocean system, which most likely play a role in determining the nature of the inter-annual variability of rainfall, have been discussed in Chapters 6 and 7. Inter-annual variability characteristics for some of the major tropical regions will be discussed in Chapter 11.

For agricultural planning purposes relationships between mean annual rainfall and variability have been assessed. The measure of variability used for this type of exercise is the *coefficient of variation*. It is a measure of *relative variability* and allows the variability characteristics of different regions to be compared. It is simply calculated by dividing the standard deviation of rainfall by the mean rainfall, for the period of interest, and multiplying this by 100. Although this measure has been criticized for

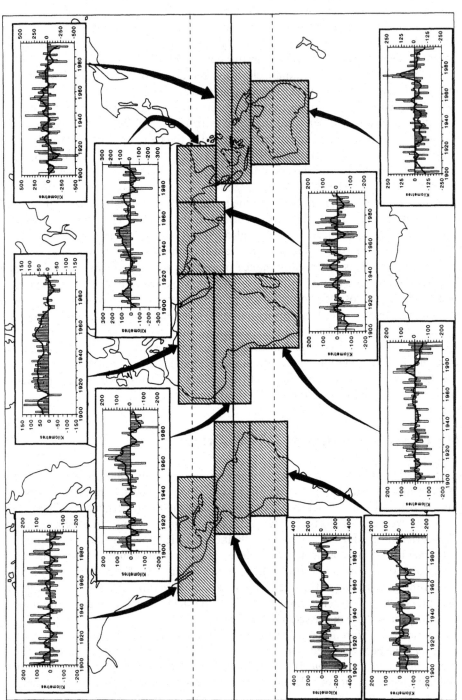

Figure 10.4   Rainfall anomaly series for various tropical regions. From IPCC (1996a)

giving a false impression of high variability in dry climates and low variability in wet climates (Jackson, 1989), it has been used widely in rainfall inter-annual variability studies (Hulme, 1992).

Generally, variability decreases with increasing rainfall, but there are many exceptions to this rule in the tropics (Jackson, 1989; Nicholls and Wong, 1990). For example, in East Africa, the inverse relationship between annual rainfall and variability varies between strong and very weak (Jackson, 1989). Variability characteristics in the tropics also appear to be a function of a location's sensitivity to the southern oscillation (Chapter 6). In the tropics, locations with strong rainfall southern oscillation (SO) associations tend to be more variable, but this variability decreases polewards (Nicholls and Wong, 1990). In the global tropics, areas of greatest variability tend to be those desert or desert-margin regions that are SO-sensitive.

A variety of *absolute measures* of variability exist. These include the *maximum* and *minimum, standard deviation* and the *mean variability*. However, these, like the measures of variability discussed above, do suffer from misapplication. For temporal analyses, the maximum and minimum can only be applied to data with long records. For spatial comparisons, maximum and minimum data should be from the same period. The standard deviation is really only applicable to rainfall data that are non-skewed and therefore normally distributed (an even distribution of rainfall values either side of the mean). However, many tropical locations are characterized by their strongly skewed (non-normal) rainfall distributions. To avoid problems associated with skewed data, the quartile deviation or the interquartile range is best applied. The latter, which is half the difference between the 75th and 25th percentile, can be converted to a relative measure by dividing the interquartile difference by the median (Jackson, 1989). A further way of treating skewed data is to transform the original data by using one of a number of transformations that often result in the "normalization" of data. These include logarithm, square root and z-score transformations, to name but a few.

In order to evaluate the temporal behaviour of rainfall, time series plots are used. These are plots of annual (or any time period) rainfall organized in chronological order. Most climate time series are plotted as normalized or standardized departures, or in statistical terms, z-scores. Normalized departures of rainfall have been used to create regional rainfall indices (Figure 10.4). These are simply compiled by averaging the normalized departures for all stations within a region. Although not as valuable as measures of rainfall reliability and probability, which are required for agricultural and water resource planning (Jackson, 1989), normalized departures of rainfall and other climate indices are of climatological interest and importance. This is because they allow an evaluation of the magnitude of the departure of an individual year from the long-term mean and thus the frequency of extreme events; persistence in departure patterns; trends in climate series; spatial patterns of variability; and comparisons with departure series for other climate indices (e.g. the SOI, Chapter 6). Because the standardized departures are in fact z-scores, they can also be used to calculate rainfall probabilities and the likely occurrence of periods of rainfall shortages often referred to as droughts.

DROUGHT

There have been many devastating droughts in the low latitudes throughout human history. However, in many cases, extratropical dwellers have only become aware of the impacts of drought, such as mass starvation, migration, famine, and economic and social decline, since the early 1980s when the occurrence of widespread droughts in Africa caught the media's attention.

Despite the clear images we readily associate with drought, much debate exists as to what constitutes a drought. Wilhite and Glantz (1987) have suggested that there are four general types of drought: meteorological, agricultural, hydrological and socio-economic. Traditionally, climatologists concern themselves with meteorological drought, as this type of drought pays attention to climate variables. It is generally defined as a period when less than expected rainfall and or greater than expected evaporation occurs. These factors also set the background for the occurrence of the other types of drought.

In most drought assessments a meteorological index of drought is used, the purpose of which is to define drought magnitude. A number of meteorological drought indices exist, such as the Palmer Drought Severity Index (PDSI) (Palmer, 1965), the Bhalme and Mooley Drought Index (BMDI) (Bhalme and Mooley, 1980) and the Rainfall Anomaly Index (RAI) (Rooy, 1965). Of the above drought indices, the BMDI has been specifically designed for use where rainfall is seasonal as in monsoon climates. However, it should be emphasized that there are a myriad of meteorological and climatological drought indices. This partly explains why it is so difficult to make a literature-based assessment of drought for the tropics. Furthermore to date, a standard drought index has not been adopted by national meteorological services in tropical countries. There is, however, a trend towards the use of normalized rainfall departures (Figure 10.4) for drought analysis at the scale of the global tropics.

Droughts are not freak events. They are a product of the same ocean and atmosphere mechanisms that give rise to the inter-annual variability of the monsoons as discussed in Chapters 6 and 7. Although these mechanisms largely explain the occurrence of drought, human-related factors are also important. Unwise agricultural practices, overgrazing, overpopulation and the stability of a country's economic and political system, especially as it relates to its ability to secure and distribute food, may well play an important role in increasing drought potential or exacerbating existing drought conditions.

As drought is largely a product of wet season failure, it is important that we understand something about the characteristics of the seasonal distribution of rainfall in the tropics; this topic is addressed in the next section. Characteristics of drought in some of the major tropical climate regions will be discussed in Chapter 11.

THE SEASONAL DISTRIBUTION OF RAINFALL

In the tropics the seasonal regime is second in importance to the total amount of rainfall. It is the major controlling factor of the calendar of agricultural activities in most tropical climates (Chapter 12). In many parts of the tropics the times of the start, duration and end of the rainy season are decisive in the struggle for sufficient food supply. Rainy seasons also bring different temperature, moisture and cloud conditions compared to the dry periods, and they influence the general weather conditions. They even have a strong effect on the way of life, because outdoor activities are much more prevalent in the tropics than in most other climates, and they are often hampered by rain.

Because other climatic elements are much more uniform, the seasonal rainfall distribution forms the basis of many classifications and subdivisions of tropical climates. Seasonal variations in the general circulation provide the basic control on the seasonal distribution of rainfall in the tropics. The elements of the general tropical circulation which are most important in this regard are the ITCZ, especially its rapidity of movement north and south of the equator, and the subtropical high pressure cells, the influences of which are carried towards the equator by the trade winds.

Based on the seasonal behaviour of these atmospheric circulation elements, it is possible to develop a simple model of rainfall regimes. This model consists of three latitudinal belts: a zone of continuous rainfall near the equator, where the ITCZ is close by throughout the year; a zone of very low rainfall around latitudes from about 15° to 25°, where the subtropical high pressure cells are situated; and an intermediate belt, where the effects of these two major controls alternate seasonally (Figure 10.5).

Over the oceans, where the ITCZ and the subtropical high pressure cells change their positions over small distances in the course of the year, the intermediate belt is rather narrow. In continental areas, however, where the seasonal displacements of the ITCZ are large, the theoretical pattern of the seasonal rainfall regimes would be more complicated (Figure 10.5). According to this model, rainfall near the equator would be continuous throughout the year, but with two maxima and two periods of less rain. The maxima would occur approximately one month after the equinoxes, when the ITCZ is over the equator.

With increasing distance from the equator, the two seasons of maximum rainfall move together in time, reducing one minimum in length, while the other dry season becomes longer and more intense. The long dry period always comes during the "winter" half of the year, when the sun is located over the opposite hemisphere (Figure 10.5, station models for 10°N and 10°S). At the same time, the total annual amount of rainfall is less than it is near the equator. At a latitude of about 15°, the short dry season disappears completely and the two maxima have merged into one.

Further away from the equator, the period of maximum rainfall becomes weaker and shorter, as the effects of the ITCZ are less noticeable and the influence of the subtropical high pressure cells predominates during most of the year (Figure 10.5, station models for 20°S and 20°N).

Over oceanic areas the actual pattern of the seasonal rainfall distribution corresponds fairly well to this model. However, a number of factors sometimes produce irregularities. Tropical cyclones and easterly waves may bring rainfall to trade wind areas which are normally rather dry. Remnants of extratropical disturbances, which occasionally travel to low latitudes, may also bring irregular rainfall. ENSO events may also result in the transposition of rainfall areas to areas that are normally dry.

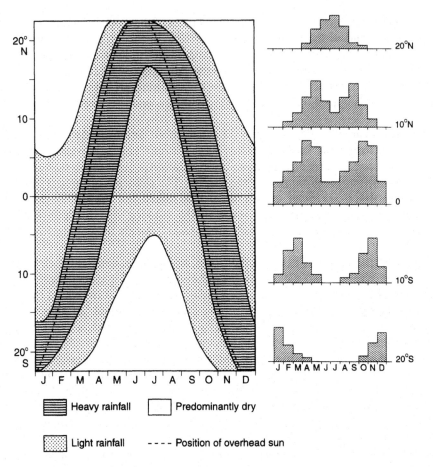

Figure 10.5 A simplified model of seasonal rainfall distribution over tropical continents. Left: latitudinal movement of precipitation zones. Right: station models at various latitudes. Partly after Miller (1971)

This is especially true for the central Pacific region where in ENSO years the dry season becomes remarkably wet because of the displacement of centres of convective activity from the western Pacific to the central Pacific (Figure 6.8).

Over the tropical continents the actual variations of the seasonal rainfall distribution are much more complicated than suggested by the model (Figure 10.5). This is because a number of factors may locally or regionally interfere with the general pattern. The most effective of these is convection, which can increase the amounts of rainfall locally. In the outer tropics convection is more prevalent during the period of overhead sun, but nearer the equator it occurs throughout the year. A second factor is orographic uplifting, which can produce enormous amounts of rainfall in the tropics (Figure 10.3). On the other hand, rain shadow effects on the leeward sides of mountain ranges are also pronounced. Where the predominant wind directions change seasonally, as in the monsoonal climates, this factor creates large regional differences in rainfall regime. These differences are often intensified by the characteristics of the air masses that are uplifted, as for instance in India and China, where the two monsoon winds bring entirely different air masses, which as a consequence, produce very different amounts of rainfall.

The combined effects of these major factors and the regional influences produce a complicated pattern of rainfall regimes over the tropical continents. There are numerous variations but it is possible to generalize these into a number of major seasonal rainfall types. Jackson (1989) has suggested that three main seasonal rainfall regimes exist in the tropics, namely humid tropical, wet and dry tropical, and dry regimes. Each of these may possess a number of subtypes. The geographical distribution of rainfall regimes according to Jackson (1989) is presented in Figure 10.6.

The humid tropics may be divided into two regimes based on annual rainfall amount. Region 1a in Figure 10.6 possesses annual rainfall in excess of 2000 mm, with all months receiving at least 100 mm (Figure 10.7(a)). Although there is a slight seasonal variation in rainfall amount, there is no marked dry season. This continuously rainy type is found mainly in equatorial regions which are almost continuously under the influence of the ITCZ and conform generally to the rainfall pattern suggested by the model in Figure 10.5. Such regimes are often referred to as equatorial rainfall regimes. One area possessing a continuous rainfall regime type, but not under the influence of the ITCZ, is Madagascar, where orographic forcing of trade flows continues almost throughout the year.

Humid Region 1b (Figure 10.6) possesses a similar seasonal rainfall pattern to that of Region 1a, but with annual rainfall totals of less than 2000 mm and some months with less than 100 mm but normally greater than 60 mm (Figure 10.7(b)). The low rainfall season is thus longer than the case of Region 1a. This is a transitional category between "humid" and "wet and dry" areas. Its wide latitudinal range on the east (windward) coasts of continents such as south-eastern South America (Figure 10.6) and the east coasts of Central America is associated with disturbances in the trade flows. Because this is a windward humid rainfall regime, orographic influences can be effective in producing marked spatial variability of rainfall amounts (Jackson, 1989). A geographic region notable for its absence from the humid rainfall group is that of East Africa. Surprisingly this area possesses dry climates, despite its equatorial location, and represents a marked variation from what would be expected given the model rainfall regimes in Figure 10.5. This climatic anomaly is a product of the relatively dry airflows over this part of Africa and the large latitudinal range of the ITCZ compared to other equatorial locations (Chapter 5).

The second major rainfall regime is that of the wet and dry tropics represented by Region 2 in Figure 10.6. This general regime is made up of a number of variations on a general theme of wet and dry seasons which may be double or single, the former being typical of locations lying immediately poleward of Region 1a (Regions 2a and 2b), while the latter is typical of tropical monsoon or semi-arid climates (Regions 2c–2e).

Regions 2a and 2b are characterized by two rainy seasons and two dry seasons. These are essentially subequatorial rainfall regimes generally conforming to the model pattern for locations at around 10°N and 10°S. In both cases the intensity of the rainy seasons usually differs and the two dry seasons are rarely of the same length. For Region 2a, the two rainy seasons are interrupted by short dry seasons or months with lower rainfall (Figure 10.7(c)). Rainfall in some dry season months may be less than

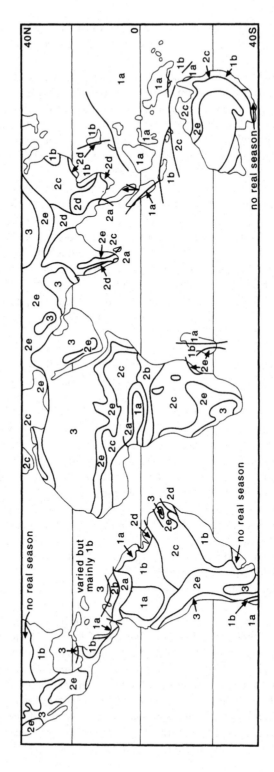

Figure 10.6 Low latitude rainfall regimes. See text for interpretation of symbols. From Jackson (1989). Reproduced by permission of Addison Wesley Longman Ltd

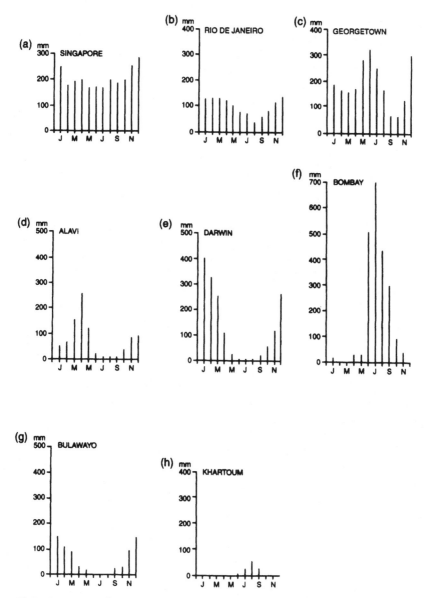

Figure 10.7   Representative seasonal distributions of rainfall for rainfall regimes in Figure 10.6. Adapted from Jackson (1989)

50 mm. Annual rainfall totals are in the range of 1000–2000 mm. Region 2b also has two rainy seasons but these are short and separated by a pronounced dry season in which monthly rainfall is below 25 mm. Annual rainfall is less than Region 2a at 650–1000 mm (Figure 10.7(d)).

The remaining type 2 rainfall regimes are all single dry and wet season regimes. However, the length and intensity of these vary. These regimes are typical of tropical rainfall regimes and represent variations on the model pattern for 20°N and 20°S in

Figure 10.5. Region 2c typically has annual rainfall amounts in the range of 650–1500 mm and is characterized by one fairly long rainy season usually of 3–5 months' duration with each month receiving more than 75 mm of rainfall. The rainy season is followed by one long dry season (Figure 10.7(e)). Region 2d possesses a rainfall regime typified by one season of exceptionally heavy rainfall followed by a single long dry season. This regime type contrasts with that of 2c in that not only is the concentration of rains greater but the annual amounts are 1500 mm and over (Figure 10.7(f)). This regime type has only limited geographical coverage in the tropical Americas as it is essentially a monsoon type found in areas lying across major summer monsoon flows (Chapter 7). The remaining dry–wet season rainfall regime 2e is characterized by one short rainy season of 3–4 months, akin to that of 2d, but with much reduced wet season monthly rainfall totals of less than 50 mm. Annual totals are therefore small at 250–650 mm. A single long dry season is typical of this type (Figure 10.7(g)). This rainfall regime type typifies the semi-arid areas of the tropics and is found extensively along desert margins. In these areas agricultural activity is marginal and desertification has a high potential (Chapter 13). These rainfall regimes are found in areas which are at the geographical limits of monsoonal flows or lie to the east of subtropical west coast deserts, as in the case of south-western Africa and South America. It is in such desert regions where the third major tropical rainfall regime is found.

Dry climates characterize the third major rainfall regime. These are low rainfall climates with annual totals less than 250 mm. Little rainfall can be expected in such climates. Rainfall, however, if it occurs, will be concentrated into just a few short weeks (Figure 10.7(h)). Such rapid and intense deliveries of rainfall can cause catastrophic events such as desert floods or wonderful sights such as desert blooms. Such climates are found beyond the geographical limits of the monsoon flows or along the coasts where cold ocean currents flow.

The classification of rainfall regimes presented above is based on monthly data. Although useful for gaining an insight into the general characteristics of the temporal distribution of rainfall, monthly based classifications give little insight into the short period dynamics of rainfall regimes. To facilitate this type of understanding, rainfall data for less than a month's duration needs to be used for classifying rainfall regimes. Studies using such short-period rainfall measures have shown clearly that locations with similar annual and seasonal rainfall characteristics often demonstrate different short-period characteristics (Jackson and Weinand, 1994).

THE DIURNAL VARIATION OF RAINFALL

While its significance is definitely less than that of the seasonal distribution, the diurnal rainfall regime is nevertheless an important feature of tropical climates.

Strong diurnal patterns of rainfall occur where the atmospheric processes at particular times of the day are conducive to the establishment of strong convection. Such processes include intense surface heating, the advance of a sea breeze front, the convergence of sea breezes with land breezes, anabatic flows in valleys and highland basins, interactions between local circulation systems and synoptic scale flows, and night-time cooling of convective cloud tops which enhances atmospheric instability.

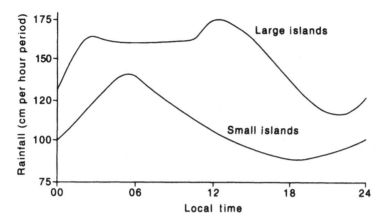

Figure 10.8   Diurnal variation of rainfall over large and small Pacific Ocean islands. From Gray and Jacobson (1977)

As the timing of these processes varies, so does the timing of the rainfall. Consequently there are a variety of diurnal rainfall patterns.

Generally, two markedly contrasting diurnal patterns exist. These are the late afternoon and early morning peak patterns. The afternoon peak pattern results from strong surface heating of the land by insolation. This induces local convective storm development in both coastal and highland areas or the advance of a sea breeze front over a coastal area (Plate 3).

Early morning rainfall maximums, common over low latitude oceanic areas, are related to night-time atmospheric radiative cooling processes in the environs of mesoscale convective cloud masses (Reynolds, 1985). At night, atmospheric cooling is more efficient in the clearer atmosphere surrounding the cloud masses. As a result, subsidence in the surrounding air occurs, and moisture convergence into the cloud mass at lower levels follows. Consequently, ascent within the cloud mass is enhanced and stronger convective cloud development results. Cooling of the rapidly developing cloud tops by longwave radiative loss leads to destabilization in the cloud mass so that the atmospheric overturning within the disturbance is enhanced (Gray and Jacobson, 1977; Reynolds, 1985). Such processes increase the opportunity for precipitation formation and thus a night-time early morning precipitation peak. During the day these processes are inhibited by strong cloud top insolation.

The effects of contrasting daytime and night-time atmospheric processes are clearly reflected in the diurnal rainfall pattern over large and small islands in the western Pacific (Figure 10.8). Both large and small islands display an early morning maximum between about 0300 and 0600 hours local time, and a minimum between 1800 and 2300 hours. The rainfall peak between 1200 and 1500 hours is a manifestation of well-developed sea breeze circulations and their associated deep convection. The role of land mass heating, as it relates to sea breeze development, is clearly evident for the case of the large western Pacific islands as small atoll islands do not possess a mid-afternoon peak.

That a variety of diurnal rainfall patterns exist is clear when the literature is

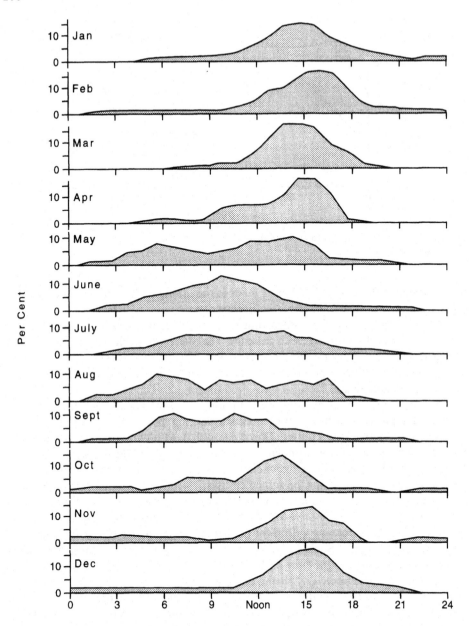

Figure 10.9   Diurnal variation of rainfall in Singapore (Nieuwolt, 1968)

surveyed. For this reason the categorization of diurnal patterns into the traditional maritime and continental types is not encouraged, although from evidence gathered to date, oceanic locations appear to be dominated by night-time early morning rainfall peaks. At the global scale, Trent and Gathman (1972) have shown that the majority of thunderstorm events (64%) occur between 0000 and 1200 hours, lending support to the contention that oceanic convection and rainfall appears to be mostly nocturnal. Over the Java and Arafura seas, in the maritime continent region, early morning rainfall and convective activity maximums occur (Murakami, 1983). This resembles in part the nocturnal maximums found for the majority of Hawaiian Islands (Schroeder, Kilonsky and Ramage, 1978). Over the tropical Atlantic, the reverse of the expected diurnal oceanic pattern has been noted (Murakami, 1983). Here there is an afternoon to early evening maximum and a nocturnal to early morning minimum. This pattern may be related to the development of squall lines (Chapter 8) over the West African coast and their subsequent movement, during the day, out over the eastern Atlantic.

Coastal locations may show a mixture of afternoon and night-time patterns, the occurrence of which may be dependent on the season. This is clearly demonstrated in the case of Singapore's diurnal rainfall regime (Figure 10.9). Here there is a tendency for rainfall to be concentrated between 0300 and 1200 hours local time in the months of May to September at the time of the south-west monsoon. During this period, mesoscale convective cloud masses developed at night-time in the vicinity of Singapore are carried in south-westerly flows over Singapore. For the remaining months of the year, a strong mid-afternoon peak dominates. This pattern, traditionally called the continental type, results from daytime convection within the north-easterly monsoon flow over Singapore. It is interesting to note that May, a monsoon season transitional month, displays bimodal rainfall peaks suggestive of night-time radiative cooling and afternoon solar heating processes, both being important at this time of the year.

Seasonal and geographical variations in the diurnal regime occur elsewhere in the tropics. In north-east Brazil the majority of coastal locations have a night-time maximum which is thought to be a product of the convergence of land and sea breezes. Inland, a sea breeze front related daytime maximum occurs. At the mouth of the Amazon there is a transition from a January–May nocturnal maximum, to a June–September afternoon maximum (Kousky, 1980). Seasonal changes are also apparent for Ibadan in Nigeria (Oguntoyinbo and Akintola, 1983). For East Africa, Griffiths (1972) has described six diurnal patterns, while for the month of August in Malaya, Ramage (1964) has described five diurnal patterns. Papua New Guinea also possesses a number of contrasting seasonal, coastal and highland diurnal rainfall patterns (McAlpine, Keig and Falls, 1983).

RAINFALL FREQUENCY

The frequency of rainfall is usually indicated by the number of raindays, which are defined as periods of 24 hours with more than a certain amount of rain. This method has many advantages, particularly in the tropics, as ordinary rain-gauges can be used, and they need only to be read once a day. However, there are many difficulties involved in accurately measuring small rainfall amounts. These include rounding errors for measurements, whether the precipitation recorded is dew fall or rainfall, the reliability of the observer, and the problem of small rainfall amounts evaporating before they are recorded.

Although the rainday concept appears a simple one, the definition of what constitutes a rainday is far from uniform. In many Commonwealth and former Common-

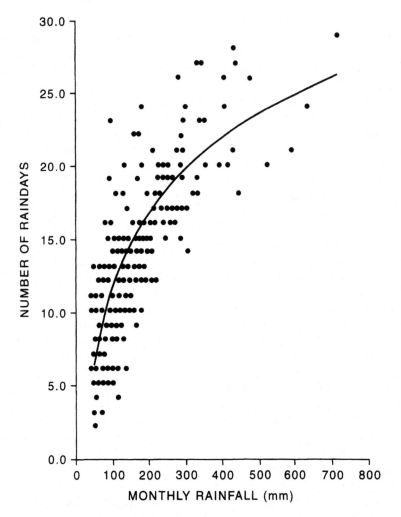

Figure 10.10   Monthly rainfall–rainday relationships. Points represent observations from various tropical locations. From Jackson (1986). Reproduced by permission of the *International Journal of Climatology*, Royal Meteorological Society

wealth countries, a lower limit of 0.25 mm (0.01 inches) is used, but other countries have used 1, 2 or 5 mm. Like many classifications, rainday definitions often vary with the application in mind; for example, meteorological as opposed to agricultural purposes. Meteorologically based definitions usually place emphasis on the occurrence of a rain event, while agricultural definitions usually emphasize the concept of effective rainfall, i.e. that amount which adequately satisfies daily plant needs. Given these problems, comparison of rainday numbers from different countries in many cases is not possible.

Another disadvantage of the rainday concept is the lower limit of 0.25 mm, used in many parts of the tropics. In warm climates, total rainfalls of less than about 2 mm are almost of no significance for agriculture or water supply since most of these small amounts will evaporate before infiltrating the soil. Since days with such little rainfall constitute a large proportion of the total number of raindays, the value of these figures for many practical usages is seriously reduced. For this reason some climatologists have encouraged the adoption of a higher threshold for the definition of a rainday (Olaniran, 1987; Stern and Coe, 1982; Nieuwolt, 1989).

Generally there is a strong association between the number of raindays and the total monthly rainfall in the tropics. The form of the relationship is logarithmic (Figure 10.10). There are, however, seasonal and geographical variations in this association (Jackson, 1972, 1986). For example, in many tropical locations at the beginning and end of the wet season, raindays are fewer than would be expected based on general monthly rainfall–rainday number relationships. The reverse is true for the middle of the wet season. This is suggestive of seasonal differences in rainfall character. Early and late wet season months may have greater daily rainfall amounts but fewer raindays than mid-wet-season months which have a greater number of raindays but lower mean daily intensities (Jackson, 1986). Variations with rainfall regime are such that humid regimes are characterized by more raindays and lower mean daily intensities than monthly totals would suggest, while the reverse is true for wet/dry and dry regimes.

Elevation is also an important determinant of rainday occurrence and produces exceptions to the above general rule. Generally raindays increase with elevation above 1000–1500 m, where the total rainfall generally decreases. With the higher number of raindays, the highlands generally possess a regular frequency of rainfall, indicated by a lower variability from year to year (Nieuwolt, 1973; Barry, 1981; McAlpine, Keig and Falls, 1983).

Raindays in the tropics often occur in groups, with wet conditions often persisting beyond one day. The same can be said for dry conditions. This characteristic, related to the prolonged lifetime of meso or synoptic scale weather systems, is referred to as *persistence*. It can be indicated by wet or dry spells which, like raindays, have a variety of definitions. Persistence coefficients and indices may also be calculated (Besson, 1924; Berger and Gossens, 1983). Application of dry and wet spell criteria demonstrate clearly that dry spells of various lengths are surprisingly frequent in many parts of the tropics, even during the rainy season (Figure 10.11). Also characteristic of wet and dry spell occurrence is marked inter-annual variability (Figure 10.11).

Because persistence is a clear characteristic of rainfall in the tropics, various attempts have been made to fit the observed distributions of wet and dry spells to mathematical probability models in order to predict the daily behaviour of rainfall. Analysis of wet and dry spell characteristics has shown that for many locations persistence of at least two days is common, with three days noted for some locations (Jackson, 1981). However, the occurrence of daily persistence and its characteristics do vary over space and time. For example, in Papua New Guinea in the rainy season, wet coastal locations can have mean wet spell periods of 4–7 days, whereas for drier coastal locations these are between 2 and 3 days (McAlpine, Keig and Falls, 1983). For Sri Lanka, the nature of daily persistence varies with seasonal atmospheric circulation changes and topography (Domroes and Ranatunge, 1993).

Where geographical contrasts in wet spell persistence exist, they are usually related to the nature of the rainfall-producing systems. For areas such as western India, where

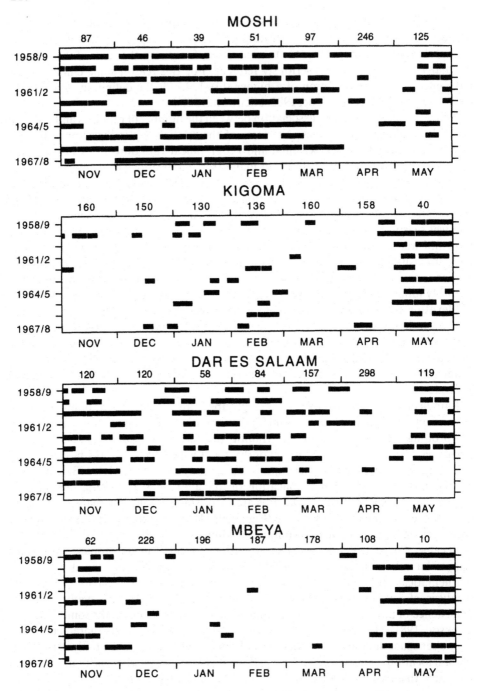

Figure 10.11   Dry spells longer than four days (each day with less than 0.25 mm of rain) at four stations in Tanzania. Figures on top of each monthly column indicate mean rainfall in millimetres for the illustrated period 1958–68. From Nieuwolt (1974)

wet spell persistence is great, rainfall-producing convective systems are organized on a larger spatial and longer temporal scale and related more to the 30–60 day tropical oscillation than other areas in South-East Asia and the equatorial Pacific. Additionally, greater wet spell persistence in western India, compared to north-eastern and central India, is due to the longer lifetimes of lows and depressions in this area (Dahale et al., 1994). Shorter wet spell persistence in Thailand compared to equatorial locations is related to the daily movement of the geographical centre of ITCZ as excursions of this system over Thailand are short-lived. For West Africa, disturbance lines that have a mean life cycle of 3.8 days (Burpee, 1974) appear to account for the 4 day rainy–dry cycle found here (Stern, Dennett and Garbutt, 1981).

Equally important as wet spells are dry spells as the climatological characteristics of these are pertinent to drought risk assessments. Generally, for highly seasonal tropical climates the persistence of dry spells is far greater than wet spells. For example, for a 1 mm dry/wet threshold for Bombay (India), Columbo (Sri Lanka), Trang (Thailand) and Penrhyn (Equatorial Pacific), dry (wet) spells have statistically significant persistence of 4 (3), > 4 (1), 2 (1) and > 4 (2) days respectively (Dahale et al., 1994).

RAINFALL INTENSITY

Intensity relates the total amount of rainfall to its duration. Rainfall intensity figures are usually presented as rainfall amount per unit time such as millimetres per hour ($mm\ h^{-1}$) or millimetres per day ($mm\ day^{-1}$). Rainfall intensity controls the probability and seriousness of local floods and is therefore a major factor to be considered in the planning and construction of dams, reservoirs, drainage canals, culverts and bridges. Moreover, it has a strong influence on the effectiveness of rainfall for agriculture, because when the intensity exceeds the maximum infiltration capacity of the soil, surface runoff results and a proportion of the rainfall is lost. Rainfall intensity also affects soil erosion through rainfall erosivity, landslides and sedimentation rates in lakes and reservoirs (Plate 8).

The mean rainfall intensity in the tropics shows large seasonal and regional variations. For example, rainfall in northern tropical Australia is generally more concentrated with fewer raindays and higher mean daily intensities compared to other tropical locations (Jackson, 1988). The occurrence of tropical cyclones and other closed circulation systems in this area may well explain these deviations. Generally mean daily rainfall intensity increases with monthly total rainfall (Figure 10.12). However, the general exception to this rule occurs in tropical highland areas at levels above 1500 m, where the number of raindays grows with higher elevations, but the total amount of rainfall diminishes, so that the rainfall per rainday decreases.

Rainfall intensity can also be computed for shorter periods than 24 hours, but this depends on the availability of rainfall records from automatic rain-gauges that provide a continuous record of rainfall. Unfortunately there is a general paucity of information on rainfall intensity in scientific journals. This, in addition to the fact that for some areas little information exists because of the lack of automatic rain-gauges, most likely explains why in tables of maximum rainfall for a variety of durations, tropical locations do not feature in the top rankings.

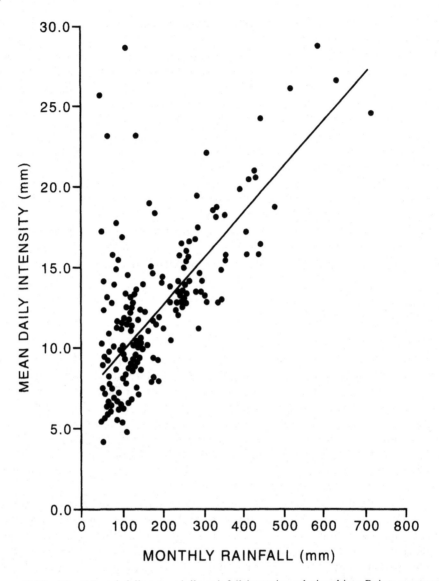

Figure 10.12 Monthly rainfall–mean daily rainfall intensity relationships. Points represent observations from various tropical locations. Note the outliers with low monthly rainfall but high daily intensities. These are semi-arid–subtropical locations which experience intense but infrequent rainstorms that deliver a large proportion of the monthly total. Tropical cyclones can produce such anomalies. From Jackson (1986). Reproduced by permission of the *International Journal of Climatology*, Royal Meteorological Society

For Cuba, rainfall intensities have been reported as 29 mm, 47 mm and 150 mm for 5, 10 and 60 minute durations respectively, and 400 mm and 624 mm for 5 and 12 hours respectively (Arenas, 1983). For Monrovia in Liberia, Griffiths (1972) has reported intensities of 18 mm, 43 mm and 63 mm for durations of 5, 15 and 30 minutes respectively. Victoria (Cameroon) and Tukuyu (Tanzania) have recorded 510 mm and

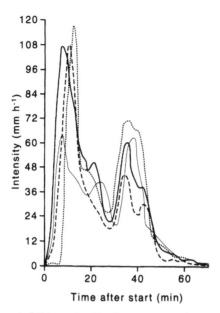

Figure 10.13   Intra-storm rainfall intensity distribution. Note the two intensity peaks separated by the intensity trough which is related to periods in thunderstorm development when down-draughts predominate. See Figure 8.2. From Sumner (1988).

432 mm in a day respectively, while Townsville (Australia) has recorded 650 mm in 21 hours (Jackson, 1989).

Throughout any one rainstorm, intensities may vary. For short duration convective storms there tends to be higher rainfall intensities at the start of the storm. This pattern is related to the stages of convectional storm development (Chapter 8). Often following the initial burst of intense rainfall, intensities fall as downdraughts in the storm cell suppress updraughts and buoyancy. Secondary intensity peaks may occur in the latter parts of the storm as intense convective processes are re-established (Figure 10.13). Rainfall intensities also vary spatially in a storm. For single cell storms highest intensities are usually found at the storm centre. For larger scale storms, involving convective complexes organized around a centre of low pressure, as in the case of a tropical cyclone, maximum rainfall intensities occur in that part of the storm where instability is greatest. For example, in tropical cyclones the rainfall intensity structure is very clear (Plate 5).

An important aspect of rainfall intensity climatologies are intensity extremes. Usually annual maximum intensities for a given duration, say 24 hours, for a number of years are compiled to form an annual series. This series is then used to calculate return periods for the 24 hour rainfall or any other duration of interest. The *return period*, or *recurrence interval*, is defined as the average period within which a specified amount of rainfall, for the given duration, can be expected to occur or be exceeded once. The 24 hour duration is often used as rainfall amounts for this duration are available for a large number of stations. This allows geographical comparisons to be made of rainfall intensity characteristics. Generally, when the rainfall intensities are

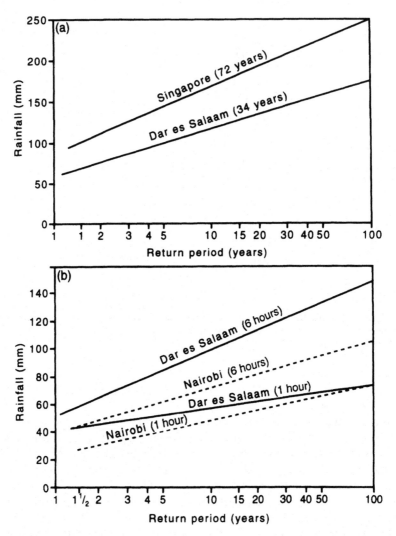

Figure 10.14   (a) Annual series of 24 hour rainfall at Singapore and Dar es Salaam. (b) Extreme rainfall during 1 and 6 hours at Nairobi (based on 27 years of records) and Dar es Salaam (21 years). Source: Taylor and Lawes (1971, p. 13.19)

plotted against their return period, on semi-logarithmic paper, they form a straight line. Such frequency–duration–intensity plots for 24 hours or shorter durations (Figure 10.14) are extremely useful in applied climatological studies.

Maximum intensities can also be assessed by computing the probable maximum precipitation (PMP) for a specific area. The PMP is generally defined as the depth of precipitation which for a given area and duration can be reached but not exceeded under known meteorological conditions. It is estimated based on empirical meteorological relationships, storm transposition or by using storm models (Jackson, 1989). Values obtained in the tropics are very high compared to the mid-latitudes, with the highest values over mountain slopes. However, it should be noted that these values are only hypothetical as PMP calculations produce a theoretical depth of precipitation for a set of extreme meteorological conditions. These in reality may never be attained. A preferred approach is to plot global values of actual maximum precipitation for a number of durations. This in effect provides information on the likely maximum precipitation for any given duration. This information can then be applied as a worst case scenario to the area and situation of interest.

RAINSTORMS

Although not a unique characteristic of the tropics, rainstorms are very common there (Plate 9). These are usually related to distinct meteorological phenomena such as tropical disturbances as discussed in Chapter 8. Rainstorms usually have distinct spatial patterns and occur over well-defined areas. Such spatial characteristics reflect the rainfall structure of the disturbance associated with the rainstorm. The usual surface conditions associated with rainstorms are periods of often intense rainfall and large storm rainfall totals. Gusty wind conditions may also be associated with such rainstorms.

Rainstorms cause soil erosion and they may render physical damage to crops and frail buildings. They are also responsible for the occurrence of flash floods which can have devastating consequences for floodplain inhabitants. For example, rainstorms over north-west India during 10–12 July 1993, which delivered daily rainfall amounts ranging from 254 mm to 482 mm for 14 locations, resulted in 140 deaths and the widespread disruption of communication and food-producing systems. Although impressive, this event was not unprecedented, with other severe storms having occurred in the past (Kulkarni, Mandal and Sangam, 1996).

Severe rainstorms appear to have a bias in their geographical occurrence which is related to the track often taken by the rainstorm-producing disturbance. For example, in India, zones of severe rainstorms are associated with the tracks of cyclonic disturbances originating in the Bay of Bengal. When crossing the east coast of the Indian subcontinent, maximum rainfall occurs in the south-western sector of these disturbances; however, when they recurve to the north or north-east, heavy rainfall occurs in their north-east sector and to the right of their track (Dhar and Nandargi, 1993). Over the Indo-China and southern China regions, heavy rainfalls are associated with tropical cyclone tracks as they are for other regions for which these disturbances occur (Chapter 8).

Rainstorms are also important in terms of their contribution to annual or seasonal rainfall totals. Often it is the case that a small number of events contribute the majority of seasonal rainfall. This is especially the case in semi-arid tropical areas. For example, in north-east Australia, maximum daily rainfalls at times can contribute 25–50% and up to 100% of the mean summer rainfall for moist coastal and inland semi-arid locations respectively. Furthermore, over the whole state of Queensland, heavy rainstorms with daily totals in excess of 50 mm account, on average, for 34% of the summer rainfall. For coastal locations, tropical cyclones are particularly important for making large contributions to seasonal totals (Lough, 1993). For the Njempa Flats in Kenya, a semi-arid location, although only 4% (1.5%) of rainstorms have totals greater than

30 mm (50 mm), these same storms account for on average 29% (11%) of the annual rainfall for this area (Rowntree, 1988).

SUMMARY

Over two-thirds of global precipitation falls in the tropics, with some of the globe's highest annual rainfalls found here, especially in those regions possessing humid tropical climates where rainfall amounts may be enhanced by orographic effects. Despite the perceived wetness of the tropics, extensive areas receive little rainfall. These regions lie at or beyond the geographical limits of the monsoonal flows or along the west coasts of continental areas where cool ocean currents predominate.

A notable feature of tropical rainfall is its inter-annual variability, which on occasions can lead to prolonged dry (drought) and wet (flood) periods. Inter-annual variability is a product of the same seasonal and non-seasonal variations in the ocean–atmosphere system which control the monsoons. Variability may be described using a number of statistical measures, of which normalized departures from 30 year rainfall normals are the most widely used.

Although the rainfall regimes of some extratropical locations may be characterized by diurnal rainfall patterns and high rainfall intensities and frequencies, in the tropics these characteristics are widespread. This is because the majority of tropical rainfall results from convection and "warm" cloud rainfall formation processes. However, "cold" cloud processes, which are almost universal in the mid-latitudes, are also important in the tropics, especially in the tops of deep convective towers. Such processes lead to the occurrence of agriculturally damaging hailstorms in tropical highland and subtropical areas.

Generally, two diurnal rainfall patterns exist, namely late afternoon and early morning. These are the result of contrasting daytime and night-time atmospheric processes; surface heating in the case of the late afternoon pattern typical of coastal and highland areas and enhanced night-time instability inducing radiative cooling of convective cloud tops in the case of the night-time pattern typical of ocean areas. However, many variants of these patterns do occur, depending on the season and the geographical setting.

In the tropics there is a strong positive association between the number of raindays and the total monthly rainfall. This association does, however, vary geographically, seasonally and with elevation. A further characteristic of tropical rainfall frequency is persistence, with runs of wet and dry days a notable feature for many locations. Like the total number of raindays, the mean daily rainfall intensity increases with monthly rainfall in the tropics. The exception to this general rule occurs at levels above 1500 mm or in locations where tropical cyclones are important climatologically.

CHAPTER 11

Tropical Climates

In this chapter a general survey of the various tropical climates of the world will be given. The geographical frame of reference will be regions which possess a set of relatively homogeneous conditions. Conditions in each climatic region will be illustrated in temperature–rainfall diagrams for representative stations. The scales for these diagrams are, with a few exceptions, uniform, so that a comparison between the various climatic types is easy. All diagrams are based on published data (Clino, 1971; *World Survey of Climatology*, Vol. 9 (1981); Vol. 10 (1971); Vol. 12 (1976); Vol. 15 (1984)).

As temperatures and other climatic elements in the tropics show little seasonal variation, our system will be based mainly on rainfall conditions, in particular the seasonal distribution and the annual average. However, the various categories will be defined differently in each major part of the tropics. The main factors of origin of the rainfall will be considered and the determination of regions will be aimed at a reasonable number of regions, each representing a clear type of tropical climate. This pragmatic approach reflects the opinion that a meaningful subdivision of climates in one part of the tropics is not necessarily also the most efficient way in another major tropical region.

The tropics, as defined and delimited in Chapter 1, occupy four main areas:

—tropical Asia, comprising most of India, South-East Asia and northern Australia,
—tropical Africa, including Madagascar,
—tropical America, including the Caribbean area and a large part of South America,
—the tropical oceans and oceanic islands (Figure 1.1).

TROPICAL ASIA

The very large area of tropical Asia is home to around 800 million people, or approximately 20% of the total world population. Most of these people live directly or indirectly from what the land produces, so the climate is a very important factor in their lives.

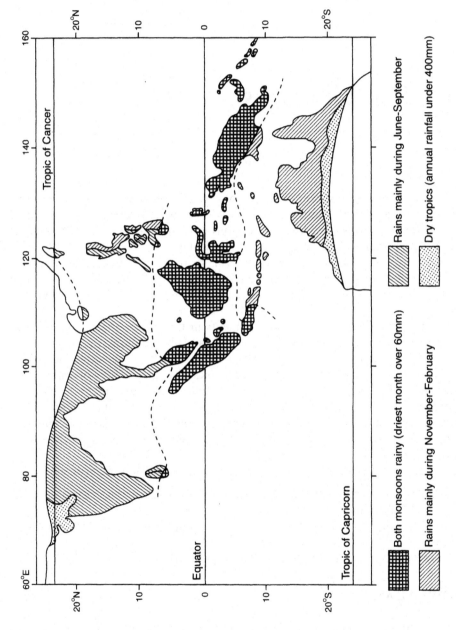

Figure 11.1 The main climatic regions of tropical monsoon Asia

Both monsoons rainy (driest month over 60mm)

Rains mainly during November-February

Rains mainly during June-September

Dry tropics (annual rainfall under 400mm)

The climates of this region are controlled by two major factors: the fact that most of the region consists of islands and peninsulas, so that marine influences are very strong; and the monsoons (see Chapter 5). The Asian monsoons also affect large areas outside the tropics, especially in eastern Asia, where the summer monsoon reaches as far north as Hokkaido and Sachalin, but these non-tropical areas are not considered here.

Tropical Asia is best subdivided according to a decisive characteristic of the monsoons, namely whether they bring precipitation or not. This factor is of prime importance for the agricultural production of the area. Most other climatic elements, such as cloudiness, humidity, wind systems and temperature, are strongly influenced by this same factor. Accordingly, tropical Asia can be subdivided into three major climatic regions, which show a relatively simple distribution pattern:

1. *Equatorial monsoon climates,* where both monsoons bring rainfall and no dry season occurs. A conventional limit is that the mean rainfall of the driest month is over 60 mm.
2. On both sides of the first type occur the *dry and wet monsoon* climates, where one monsoon brings most of the rainfall, while the other one is relatively dry. The mean rainfall of the driest month is under 60 mm. The rainy and dry seasons come at different times according to the location in this region.
3. The *dry tropics,* where both monsoons bring little or no precipitation, in the extreme north-west and south-east of the region (Figure 11.1).

These three regions are equivalent to the humid, wet and dry, and dry rainfall regimes discussed in Chapter 10.

EQUATORIAL MONSOON CLIMATES

These climates prevail over most of the Indonesian archipelago, Malaysia, New Guinea and some islands further east. All these areas are situated between about 10°S and 8°N, so they are truly equatorial (Figure 11.1). A main characteristic of the region is the mixture of land and sea surfaces, which makes it a truly "maritime continent" (Ramage, 1968). This mixture of sea and land, and the mountainous character of most islands, creates a large variety of local climates, mainly depending on exposure to the monsoons and elevation. Nevertheless, a certain climatic homogeneity prevails over the whole region, because the monsoons are remarkably similar.

The north-east monsoon, which dominates the circulation over this region from about December to March, gradually changes into a north-westerly wind near the equator (Figure 7.6). South of the equator, and especially in Java, it is often called the "west" monsoon, or, as it brings a large part of the total rainfall, the "wet" monsoon.

The south-west monsoon, which prevails from about June to September, is a continuation of a south-easterly wind in the southern hemisphere (Figure 7.4). In Java, it is usually called the "east" monsoon and because it brings rather dry air masses to eastern Java, also the "dry" monsoon, a characteristic which does not apply to western Java.

In equatorial Asia the two monsoons are very similar. They both bring predominantly warm and humid air masses. These are related to the high sea surface temperatures in the region itself and in the oceans around it, which are almost everywhere well above 25 °C throughout the year. The dense equatorial rainforests on the islands also produce large amounts of water vapour. The air over this region contains more water vapour than over any other equatorial area (Lockwood, 1974).

During the two inter-monsoon periods of March to May and September to November, winds are variable and generally quite weak. Local factors control the distribution of rainfall, which can be quite high because the same warm and humid air masses prevail over the region. These warm and humid air masses need little uplifting to produce rainfall.

Convection, convergence, orographic uplift and local circulation systems (Chapter 10) occur either alone or in combination to produce rainfall in this region, which is the rainiest in the world, with annual totals everywhere over 2000 mm, but in many areas over 3000 mm (Figure 10.2). However, large local differences occur in the region, because of its mountainous character. Not all these differences are well known because there are few rainfall recording stations in the region, especially on the smaller islands.

The seasonal rainfall distribution also shows large variations within the region (Figures 11.2 and 11.3). Local factors, particularly orographic lifting and local circulation systems, which change in importance during the course of the year, are the main reasons for these differences in rainfall distribution and total amount.

Over the region as a whole, the inter-monsoon periods are the wettest, as strong convection, widespread convergence of airstreams and active local winds prevail over most areas. Differences in the seasonal distribution are of little practical importance because dry periods are rarely prolonged, except during exceptionally dry years (Nieuwolt, 1966a) which appear to be related to ENSO events (Chapter 6).

Related to the prevalence of convection, rainfall intensity is generally high: the percentage of total rainfall which is received during cloudbursts (showers of at least 5 mm of rainfall during 5 minutes) varies in Indonesia between 8% and 37%, with an average of 22%. For comparison, the highest value in Bavaria, Germany, is 3.7%. At Bogor, over 80% of the total annual rainfall is normally received during showers which bring at least 20 mm (Mohr, van Baren and van Schuylenborgh, 1972).

The diurnal distribution of rainfall follows the usual pattern of afternoon maxima over land and night maxima over the sea. Coastal stations show large variations, due to both local and regional influences (Chapter 10).

Because of the similarity of prevailing air masses and the limited seasonal variation in the duration of day and night, the other elements of climate all show a seasonal uniformity. This is especially true for temperatures: over the whole region the annual range is below 2° (Figure 4.1). Temperature differences with place are also very small, with the exception of the effect of elevation. Many hill and mountain stations owe their existence to the relief they can bring from the high physiological temperatures in the lowlands and the related agricultural possibilities.

Where seasonal variations are small, diurnal differences assume greater importance. The diurnal range of temperature exceeds the annual range everywhere in the region. It increases rapidly with distance from the sea and with elevation, because daytime temperatures decrease much less with height than those during the night. Day to day variations in the weather conditions are more important than is generally assumed in descriptions of these climates. They are related to cloudiness and affect temperatures, humidity and local winds (Nieuwolt, 1968).

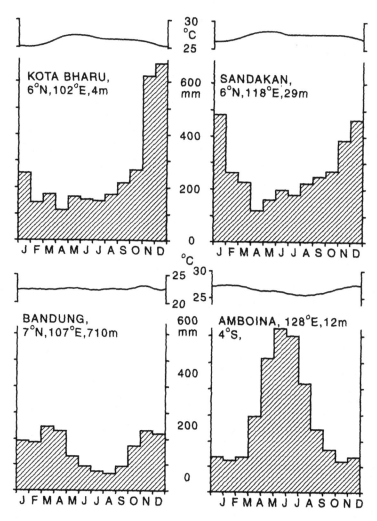

Figure 11.2 Climatic diagrams for four stations in the equatorial monsoon climates of Southeast Asia

While it might be assumed that a region without cold and dry seasons is ideal for agriculture, the seasonal uniformity has its drawbacks: many crops, particularly rice, the main subsistence crop, need a dry period for ripening. These do occur, of course, but rarely on a regular basis. Although there is generally more than sufficient rain for two or even three crops of rice per year, the rice is usually grown in irrigated fields, so that the control of the water level is easier. Other important plantation crops, such as rubber, oil palm, tea and coffee, can do without dry periods, but production is often delayed by rains.

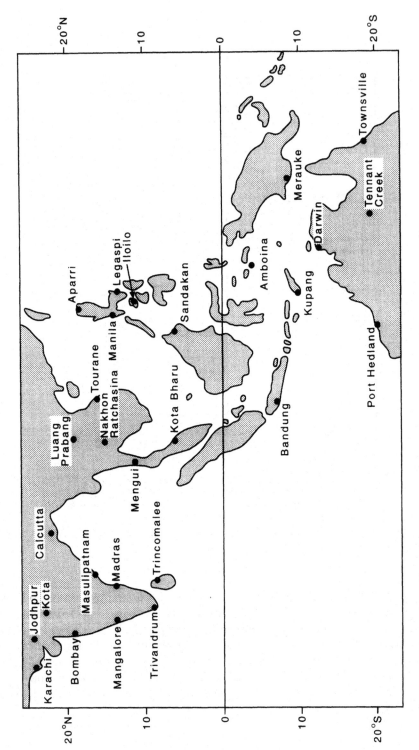

Figure 11.3   Location of stations in tropical monsoon Asia for which diagrams are given

DRY AND WET MONSOON CLIMATES

Dry and wet monsoon climates prevail in four main regions on either side of the equatorial zone:

1. most of the Indian subcontinent, including the coastal areas of Bangladesh and the northern parts of Sri Lanka;
2. Burma, Thailand and Indo-China, including the southern parts of the island of Hainan and the southern tip of Taiwan;
3. the Philippine archipelago;
4. south-eastern Indonesia, southern new Guinea and northern Australia (Figure 11.1).

The common feature of these climates is that one of the monsoons is dry, while the other one brings abundant rain (Chapter 10). It is usually the summer monsoon which brings the rain, but on a few exposed areas, such as south-eastern India, north-eastern Sri Lanka and the eastern parts of the Philippine Islands and a few smaller areas like the east coast of Vietnam, the situation is reversed: here the winter monsoon from the north-east brings rain, while the summer monsoon season is relatively dry (Figure 11.1).

The Indian Subcontinent and Northern Sri Lanka

In this region, four seasons can be recognized: the cool winter, the hot and dry pre-monsoon season, the period of general rains associated with the monsoon, and the season of the retreating monsoon.

The *cool season* is in January and February in the south, and lasts approximately one month longer in the north of the region. During this time the north-east or winter monsoon prevails over most of tropical India. It is generally dry, but brings some rains to exposed coastal regions in the south and in Sri Lanka (Figures 11.4 and 11.5, Madras and Trincomalee). In the northern parts of tropical India, westerly winds frequently prevail; they may bring some rain, as depressions, moving with these winds, are reactivated over the Ganges Lowlands, but amounts of rain are very modest (Figure 11.4, Calcutta). In this season mean temperatures range from about 25 °C in the south to below 20 °C in the north of the region (Figures 11.4 and 11.5).

The *hot and dry season* lasts from about March to May. During this time, temperatures rapidly increase over the whole region, as the sun moves into its zenithal position and the days get longer. Very high temperatures are reached in the north, where there is little cloudiness to reduce the impact of the sun and daily maxima are frequently over 40 °C (Figure 11.4, Kota). Over the rest of the region the highest temperatures of the year are reached during this season, when there is little rainfall (Figures 11.4 and 11.5). Only at coastal stations, where sea breezes bring cooler air during the hottest hours of the day and some precipitation might be caused by convectional thunderstorms, are the temperatures somewhat lower. The extreme south of the region receives some rainfall during this time, both from convectional disturbances and from the first beginnings of the monsoon (Figure 11.5, Trivandrum and Trincomalee). The north-east of the region also receives some modest amounts of rainfall, caused by local

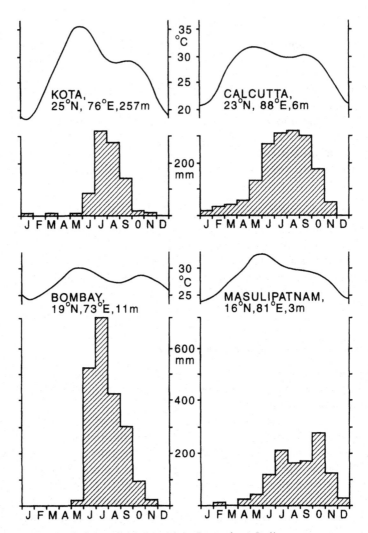

Figurue 11.4   Climatic diagrams for four stations in northern India

disturbances and by tropical cyclones, which originate during this season over the Bay of Bengal and often bring heavy rainfall, causing much damage to the coastal areas of Bangladesh (Chapter 8).

The *monsoon season* begins in May in the south, and in June in the northern parts of the region. The onset of the monsoon is frequently accompanied by violent thunderstorms and is therefore often called the "burst" of the monsoon. It progresses from south to north-west (Figure 11.6).

With the increased cloudiness, temperatures are generally lower than during the pre-monsoon period, creating the "Ganges" type of temperature curve, shown by most stations in the region (Figures 11.4 and 11.5).

Over most of the region the monsoon brings over 80% of the total annual rainfall.

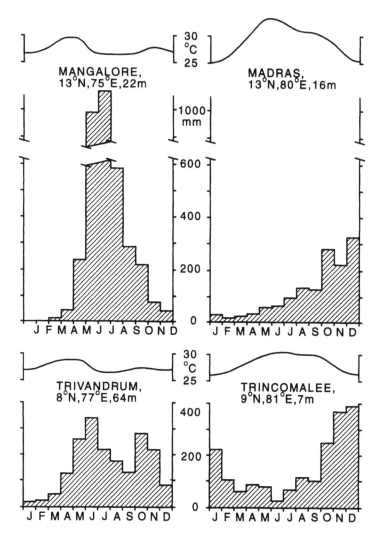

Figure 11.5   Climatic diagrams for four stations in southern India and Sri Lanka

For all India the seasonal normal monsoon rainfall is 890 mm, with a coefficient of variation of 9.8% (Shukla, 1987). However there is a fair degree of regional and inter-annual variability of monsoon rainfall (Figure 10.4). The heaviest falls and lowest variabilities are along the west coast, where orographic lifting takes place, and in the Ganges delta, where widespread convergence of the monsoon current occurs frequently (Figure 11.4, Bombay and Calcutta; Figure 11.5, Mangalore). Lowest rainfall totals during the monsoon are found in the central and eastern parts of the Deccan plateau, on the leeward sites of the Ghats. After the burst of the monsoon, rainfall decreases markedly (Figure 11.4, Masulipatnam). The south-east also receives relatively little rainfall (Figure 11.5, Madras and Trincomalee). Another dry area is in the extreme

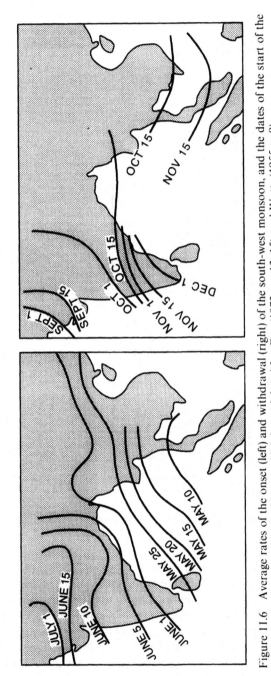

Figure 11.6   Average rates of the onset (left) and withdrawal (right) of the south-west monsoon, and the dates of the start of the north-east monsoon over Indo-China and Malaya (right). After Das (1972, pp. 12–15), and Watts (1955, p. 9)

north-west of the region, where there is a gradual transition to the dry tropics (Figure 11.1). It is here where the largest inter-annual rainfall variations occur.

Generally there is a tendency for rainfall anomalies over north-west India to oppose those over north-east and southern India (Kiladis and Sinha, 1991). For all India there appears to have been significant changes of the rainfall climatology, notably an increase in the variance and a decrease of the mean; a sign of increasing aridity from the late 1950s to the 1980s (Rasmusson and Arkin, 1993). Variance increases also appear to lead to changes in the mean by about a decade. These changes, however, may be part of long-term climatic variability, with rainfall peaks and troughs on the multi-decade scale, as India has experienced four major rainfall periods between 1871 and 1978 (Mooley and Parthasarathy, 1984). Notably, a period of marked variability similar to that of the 1950s to 1980s occurred in the 1890s to 1910.

As agriculture is largely dependent on the monsoon rains, failure of the monsoons can have disastrous consequences for the main subsistence crops; rice in the south (cowpeas are also important), and wheat in the north. Tree crops, especially coconut palms in the southern regions, are also affected by monsoon failure, as well as the production of important export crops such as tea and coffee in highlands plantations. In major drought years socio-economic decline and famine are widespread.

The Indian Meteorological Service classifies drought in terms of south-west monsoon (June to September) rainfall deficiency such that a moderate drought is equated with rainfall deficiency between 26% to 50% while severe drought is associated with deficiencies in excess of 50% of normal for a region. Areas may also be classified as "drought prone" or "chronic drought prone" based on a 20–40% and over 40% probability of a drought year respectively. With respect to rainfall deficiency, the five worst droughts occurred in (deficiencies in brackets): 1877 (51%), 1897 (27%), 1918 (26%), 1972 (25%) and 1987 (19%). With respect to the area of the country receiving deficient rainfall (the geographical impact), the worst droughts were (percentage total area in brackets): 1918 (69%), 1899 (51%), 1877 (32%), 1987 (29%) and 1972 (28%). Historically, a greater number of droughts occurred in the period 1900–20, a period of generally below normal rainfall, and the period 1965–87, when rainfall was highly variable and biased towards below normal. The majority of Indian droughts appear to have an ENSO teleconnection. For example, of the major famine-related droughts of 1877, 1899, 1918, 1972, 1979 and 1987, only 1979 was not an ENSO year. Although teleconnections most likely play a role in Indian drought occurrence, regionally, weaker meridional pressure gradients, larger seasonal shifts of the monsoon trough, larger numbers of days of breaks in the monsoon, lower frequencies of depressions and shorter westward excursions of depression tracks appear to be the major factors associated with large-scale Indian droughts. Individually or in combination, some of these factors caused the dramatic failure of the monsoon over central and north-western India in 1987 which led to large-scale drought in that year, which in socio-economic terms has been one of the worst on record. Large-scale floods are associated with meteorological patterns opposite to those of drought (Bhalme and Mooley, 1980).

The *post-monsoon season*, or period of the retreating monsoon, starts in September in the north and continues until December in the south of the region (Figure 11.6). During this period temperatures show a slight increase, due to decreased cloudiness, but they remain well below those of the pre-monsoon period (Figure 11.4, Kota, Calcutta, Bombay; Figure 11.5, Mangalore, Trivandrum).

Temperature differences within the region are very small; the October mean is everywhere around 26–28 °C. Rainfall during this period is generally light, except in the extreme south and south-east of the region. Tropical cyclones frequently develop over the very warm waters of the southern Bay of Bengal. These cyclones often move westwards and bring highly variable amounts of rainfall to the coastal areas and the central and eastern parts of the Deccan plateau (Figure 11.4, Masulipatnam). In this season the winds gradually change to westerly directions in the northern parts of the region, while in the south, north-easterlies develop as the precursors of the north-east winter monsoon.

Burma, Thailand and Indo-China

In comparison with India, climatic conditions in this area are poorly documented. The number of meteorological stations is small, particularly in mountainous areas, and there exists no uniformity between the different states of this region regarding methods of observation, recording and publication. Interruptions during the Second World War and the Vietnam War make it necessary to use rather old sources, which are often of doubtful accuracy. The following description of the climate of this region is therefore necessarily somewhat general (Nieuwolt, 1981).

Although there is a great seasonal similarity between this region and the Indian subcontinent, there are four important differences in each of the seasons (Kripalani, Singh and Nalini, 1995). This is due to the contrast in the nature of the Indian and East Asian monsoons discussed in Chapter 7.

1. During the *cool season*, the north-east monsoon is much stronger and more continuous over Indo-China than over India. It also brings colder and more continental air masses and the outer limits of the tropics, based on the sea level temperature of the coolest month, remains well south of the Tropic of Cancer (Figure 11.1). Cold surges are also a feature of this season (Chapter 8). The north-east monsoon is also relatively dry; however, it brings rain to the coastal areas of Vietnam, after it has travelled for some distance over the relatively warm South China Sea. Some rainfall during this season is also related to the upper Polar Front, which is often situated over this region, sometimes as far south as 10°N (Figure 7.6). Its associated disturbances bring small amounts of rain (Figure 11.7). Conditions of heavily overcast skies and drizzle in northern Vietnam, known as "crachin", are related to these disturbances when they have their centre over the Gulf of Tonkin.
2. The *pre-monsoon season* is generally cooler than over India. This is mainly due to heavier cloudiness, caused by more frequent thunderstorms related to the eastward movement of a trough at the 500 mb level from India (Ramage, 1955). The rains can be considerable and are called the "mango-rains" in Burma and Thailand (Figure 11.7).
3. The *rainy season* starts about two to three weeks earlier than in India, but the onset of the south-west monsoon is more gradual and not so often accompanied by

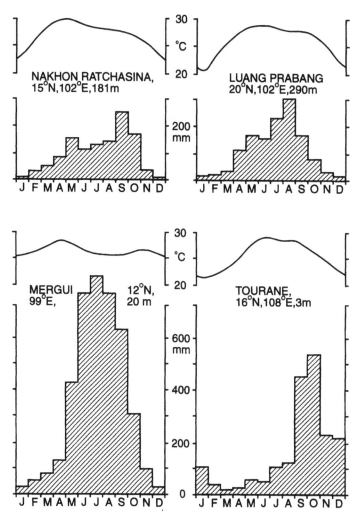

Figure 11.7   Climatic diagrams for four stations in Myanmar, Thailand and Indo-China

violent thunderstorms. The winds of the south-west monsoon are generally weaker than those of the north-east monsoon, but the air masses are much deeper, reaching up to 9000 m along the coast of Burma. The proportion of annual rainfall that is brought by the south-west monsoon varies from about 80% along the Burmese–Thai west coast to only 40% in the interior basins and on the leeward side of the mountain ranges (Figure 11.7).

4. The *post-monsoon season* is very similar to that of India, but it arrives earlier and by 15 October most of the region is under the influence of the advancing north-east monsoon (Figure 11.6). During this season typhoons develop frequently over the warm South China Sea and they produce a rainfall maximum during this period along the coast of Annam (Figure 11.7, Tourane).

Climatic differences within the region are mainly due to relief. The mountain ranges are predominantly in a north–south direction, creating large differences in exposure to both monsoons. West-facing slopes receive most rainfall during the south-west monsoon; east-facing slopes, particularly in the eastern parts of the region, get most of their rainfall during the cool season (Figure 11.7, Mergui and Tourane). The interior basins are rather dry and often show a double rainfall maximum at the beginning and the end of the south-west monsoon season, when wind velocities are low and local convection therefore more effective (Figure 11.7, Nakhon Ratchasina and Luang Prabang).

In a region which stretches over more than 15° of latitude, some climatic differences are related to this factor. The cool season temperatures show a clear decrease with increasing latitude, but summer temperatures are more uniform and largely controlled by local factors and cloudiness. Rainfall variability increases with latitude as a larger proportion of total rainfall is caused by travelling disturbances in the north of the region. Diurnal temperature ranges in this region vary greatly, depending on local relief conditions. They are generally largest during the dry season and show a clear increase with distance from the sea, producing a good indication of the degree of continentality (Nieuwolt, 1981).

The Philippines

The Philippine archipelago consists of more than 7000 islands of many different sizes, located between 5°N and 21°N. The islands are surrounded by large and warm seas. Most islands are mountainous, with a general north–south orientation of the mountains, some of which, on the bigger islands, reach well above 1000 m. These cause numerous local variations of climate but certain common features can be recognized, because the general climatic conditions are largely controlled by three major airstreams, which prevail during different periods. Accordingly, three main seasons can be recognized in this region.

1. The *north-east monsoon season*, which lasts from about October to March, attaining its maximum strength during January. This wind brings relatively warm and humid air masses to the Philippines and sea level temperatures are around 25 °C (Figure 11.8). The high temperatures of this air, and its high moisture content, are the result of its journey over the warm South China Sea; but this effect is limited to a surface layer, reaching not higher than about 1500 m. Above this layer a weak inversion persists most of the time, and the whole monsoon current is normally not deeper than about 2500 m. The north-east monsoon therefore brings much rainfall when it is uplifted orographically, as indicated by stations with an easterly or northerly exposure (Figure 11.8, Aparri, Legaspi). Areas not directly exposed to the north-east monsoon receive only small amounts of rainfall during this season.

   Aloft, the north-east monsoon is overlain by North Pacific trade winds, which cause middle and high level clouds, or in the north of the region, by temperate zone westerlies, which cause few high clouds. The convergence with the surface North Pacific trades is just to the east of the region in this season (Figure 7.6). In March, the north-east monsoon begins to weaken and it finally disappears from the Philippines by April.

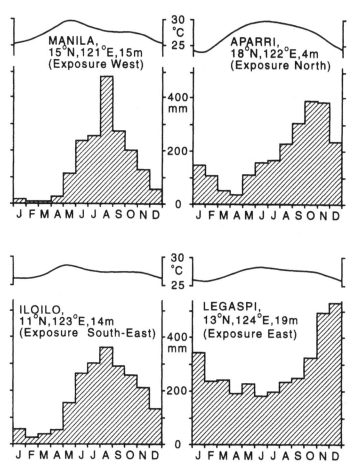

Figure 11.8   Climatic diagrams for four stations in the Philippines

2. The *North Pacific trade season*. In March, the convergence zone between the
   north-east monsoon and the North Pacific trades slowly moves westwards over the
   region as the north-east monsoon weakens and the islands come under the influence
   of the trades, easterlies which bring the warmest air that affects the Philippines. It
   brings temperatures of about 27°C at sea level (Figure 11.8). However, while its
   humidity is high at lowest levels, the trade wind inversion at about 1500 m prevents
   strong convection and limits rainfall to stations with a clear easterly exposure
   (Figure 11.8, Legaspi). In other areas, typical trade wind clouds prevail.
3. The *south-west monsoon season* starts in May. The south-west monsoon reaches its
   maximum intensity in August and disappears from the region in October. It is a
   deep current, up to 10 000 m high, of warm and very humid air masses. Almost all
   parts of the Philippines receive large amounts of rainfall during this season, with the
   maximum at coasts and slopes directly exposed to the monsoon (Figure 11.8). Much
   of this rain is caused by disturbances which travel with the monsoon. In October,
   when the south-west monsoon is in retreat and the north-east monsoon not yet well

established, the southern parts of the region may again come under the temporary influence of the north-eastern trades (Flores and Balagot, 1969).

Typhoons reach the Philippines from about June to December, reaching their maximum frequency in September–October, when sea surface water temperatures in the seas around the region are at their annual maximum. They bring widespread rainfall, particularly on the island of Luzon, where they may cause an autumn maximum of rainfall (Figure 11.8, Aparri). Typhoons are also responsible for widespread fatalities and damage to building and crops.

According to the seasonal distribution of rainfall, three main climatic regions may be recognized:

—areas with exposure to the south, south-east and west have their rainfall maximum during the south-west monsoon season and are relatively dry during the rest of the year (Figure 11.8, Manila and Iloilo);

—areas with exposure to the north and east have most of their rainfall between October and March, but generally also receive some rainfall during the south-west monsoon season (Figure 11.8, Aparri and Legaspi);

—in the southern parts of the Philippines there is a gradual transition to the equatorial monsoon type of climate, which may prevail up to a latitude of about 8°N (Figure 11.1) (Coronas, 1920).

As can be expected, the annual range of temperature shows a clear increase with latitude, and the maximum temperatures of the year occur generally during April–May, rather than in mid-summer, when cloudiness is heavier due to the effects of the south-west monsoon.

Northern Australia, South-eastern Indonesia and Southern New Guinea

In this large region there are two prevailing winds: the west monsoon, a continuation of the north-east monsoon of the northern hemisphere; and the east monsoon or South Pacific trade winds (Figures 7.4 and 7.6).

The *west monsoon season* lasts from about December to March. The monsoon is reinforced by a strong thermal low over Australia, where it is summer. The west monsoon brings heavy rainfall over the whole region (Figure 11.9); over northern Australia it is 60–90% of the annual total. This is due to the monsoon's passage over the warm seas of the Indonesian archipelago, and the convergence with south-westerlies to the west of Australia and the South Pacific trade winds to the east of that continent. Locally, orographic lifting along the coasts, and convection, particularly over the larger islands and Australia, bring a lot of rainfall.

Typhoons also bring large, though very variable amounts of precipitation during this season. They reach maximum frequency in February, which explains the high mean rainfall during that month along the coast of Queensland (Figure 11.9, Merauke and Townsville). During this season, temperature means increase with the (southern) latitude, from about 26°C in the north to 28°C along the northern Australian coast (Figure 11.9, Merauke and Darwin). Even higher means are reached away from the coast.

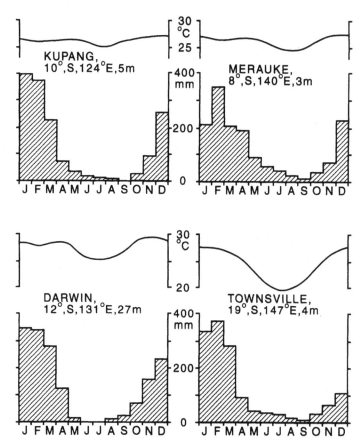

Figure 11.9   Climatic diagrams for stations in South-east Indonesia, southern New Guinea and northern Australia

The *east or dry monsoon season* lasts from about May to September. The South Pacific trade winds bring hot but stable and dry air masses to the region. These originate from the relatively dry area over the southern Pacific east of Australia and a quite low inversion layer. The relatively short journey over the sea between Australia and Indonesia changes only a thin layer of this air mass, and over the whole region rainfall is very low during this period. Orographic lifting along the Queensland coast and the south coast of New Guinea produces only modest amounts of rainfall (Figure 11.9, Townsville and Merauke). During this season, temperatures fall from north to south, with increasing latitude as winter makes its appearance. A "Ganges"-type of seasonal temperature curve is only shown by Darwin (Figure 11.9). Annual ranges generally increase with latitude, and the homogeneity of this region is mainly one of seasonal rainfall distribution.

Because this region is one of the dipoles of the Walker Circulation, much attention has been given to the role that ENSO plays in determining the inter-annual variability of climate (Quinn et al., 1978; Lough, 1991). Severe droughts in this region over the last 20 years have all been associated with ENSO events (1972–73, 1982–83, 1987, 1991–

93). These were also matched with outbreaks of fire, especially in the 1982–83, 1991–93 and 1997 ENSO events when extremely dry spring conditions in western parts of Indonesia and eastern Australia led to widespread fire. Although not normally associated with tropical climates, frost in New Guinean highland environments is a common occurrence in drought years. This is because of the prevalence of clear skies and enhanced night-time longwave radiation loss which produces cool nocturnal temperatures. Droughts in association with widespread frost occurrence during the 1940–41 ENSO event played an important part in forcing mass migrations of subsistence farmers from the Papua New Guinean highlands (Allen, Brookfield and Bryon, 1989). This also occurred to a lesser extent in the 1982–83 and 1987 ENSO events.

THE DRY TROPICS OF ASIA AND AUSTRALIA

At the outer margins of the region of the tropical monsoon climates, in north-western India and south-east Pakistan, and in central Australia, are two tropical regions where both monsoons fail to bring much rainfall, so that the annual mean remains below about 400 mm. These regions are therefore classified as parts of the dry tropics (Figure 11.1).

North-west India and South-east Pakistan

From November to about February this region is under the influence of westerly winds. These bring very little rainfall, because they have travelled over continental areas for a long distance and they are warmed and stabilized by their descent from the highlands of Iran and western Pakistan (Figure 7.6).

The pre-monsoon season, from about March to June, brings a rapid rise in surface temperatures (Figure 11.10, Karachi and Jodhpur). These cause a thermal low to develop over the region, a consequence of which are strong sea breezes in the coastal areas. These bring little rainfall as the air masses are mostly continental in origin and reach the region from Arabia after a relatively short journey over the sea.

The summer monsoon season starts around the middle of July. The main monsoon current comes from the south-east, and though it has lost most of its moisture during its long trip over the Ganges plains, it can nevertheless bring some rainfall to this region. However, during much of this season continental air masses from the north-west prevail, and these air masses are often hotter than the monsoon air. Therefore, when they converge with the monsoon, the continental air masses are uplifted and they bring little or no rainfall (Figure 11.11). Still, the south-east monsoon brings some rainfall to the region, but the amounts decrease rapidly towards the west. The increase in cloudiness reduces surface temperatures, which are considerably lower than during the pre-monsoon season (Figure 11.10, Karachi and Jodhpur). The monsoon ends around 15 September and the post-monsoon period brings a rapid return to winter conditions, with little or no rainfall. In this region the annual temperature curves show a modified type of Ganges regime combined with large annual ranges. These are caused by the latitudinal position, continentality and the predominance of clear weather during the pre-monsoon season (Figure 11.10).

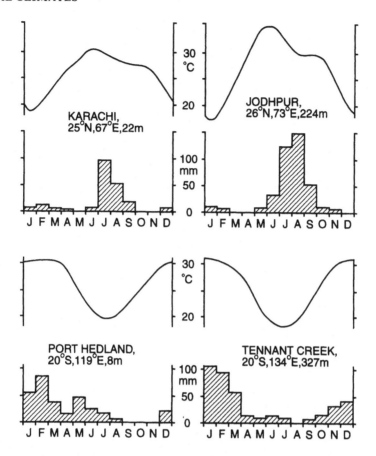

Figure 11.10 Climatic diagrams for south-east Pakistan, north-western India and tropical Australia. Note the rainfall scales on this diagram are double those for other stations

The Australian Dry Tropics

This region forms a relatively narrow transition zone between humid northern Australia and the desert of the interior continent (Figure 11.1).

There are two main reasons for the rapid decrease of rainfall towards the south. The first one is the circulation over the region. The west monsoon usually does not reach these areas because the convergence zone with the south-easterlies is situated further north (Figure 7.6). However, even when it does, it has shed most of its moisture over the coastal regions. Prevailing winds during the summer are south-westerlies from the Indian Ocean, and south-easterlies. The first bring rather stable air masses to the region because these have been cooled by the West Australia current. The south-easterlies may bring some rainfall, though amounts are generally small (Figure 11.10, Port Hedland and Tennant Creek). Most rainfall of the south-easterlies is received by the coastal areas and this is the reason why the dry tropics extend to the west coast, but are separated from the east coast by a narrow humid strip (Figure 11.1).

The second reason is that the typhoons and willy-willies usually weaken very rapidly

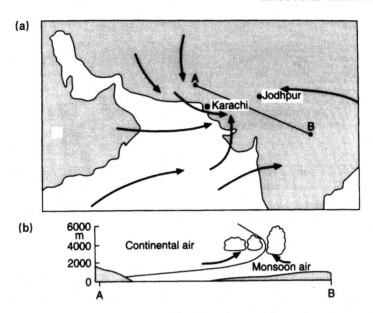

Figure 11.11 (a) Major airstreams affecting north-western India during summer. (b) Cross-section along line A–B

when they move inland. They may bring some rainfall in the western coastal areas (Figure 11.10, Port Hedland). The two reasons can be combined in the term "continentality". As in all dry climates, annual temperature ranges are large, considering that the latitude is only 20°. They increase from the coast (Figure 11.10, Port Hedland), where they are around 10–12°C, to the interior where they reach values up to about 15°C (Figure 11.10, Tennant Creek).

TROPICAL AFRICA

The largest land mass of the tropics is in tropical Africa. With a population of about 300 million, this region is second in general importance only to tropical Asia.

Climatic conditions in tropical Africa differ from those in the Asian region to a considerable extent. Firstly, there is no general system of monsoons which controls climates over the whole region. Instead, in tropical Africa there are two distinct monsoonal systems, one in the west and one in the eastern parts of the region, and they differ in a number of important characteristics (Chapter 7). For large parts of tropical Africa, like the Congo Basin, the southern regions and Madagascar, no monsoons exist at all.

A second set of climatic differences between tropical Asia and Africa is caused by surface features. Tropical Africa is continental, its coastline is generally unindented and smooth, and there are few islands. It has only a few mountain ranges, and a large part of the interior consists of extensive highland plateaux at elevations over 1000 m (Figure 11.12). Much of tropical Africa therefore has continental highland climates, a type that is almost entirely absent in tropical Asia.

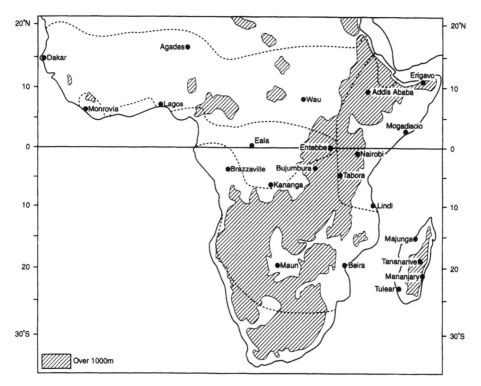

Figure 11.12   Location of stations in tropical Africa for which diagrams are given. Broken lines indicate the approximate boundaries of the main climatic regions. Shaded areas are over 1000 m above sea level

A third source of climatic differences between the two regions are ocean surface temperatures. While seas around and inside tropical monsoon Asia are all warm, tropical Africa is bordered by some rather cool seas: in the west, the Canaries and particularly the Benguela currents bring low sea surface temperatures close to the equator. In the east, upwelling of cold water creates low surface temperatures along the Somali coast, especially during the period from March to September (Chapter 5).

With the equator in the centre, and a relatively uniform surface, Africa conforms best of all continents to the pattern of climatic types as imagined on a theoretical continent (Thornthwaite, 1943; Flohn, 1950). The generalized pattern of humid climates near the equator changing into dry climates with increasing latitude is well realized in the northern hemisphere, as illustrated by annual rainfall means. However, south of the equator this pattern is less clear, as variations due to regional surface features prevail, particularly in the eastern parts of tropical Africa and in Madagascar (Figure 11.13).

In large parts of tropical Africa a zone of maximum rainfall related to the ITCZ follows the overhead position of the sun with a delay of about one month. This system is particularly well developed over East Africa, where the seasonal movements of the ITCZ are large (Figures 7.7 and 10.5). It results in two seasonal rainfall maxima near

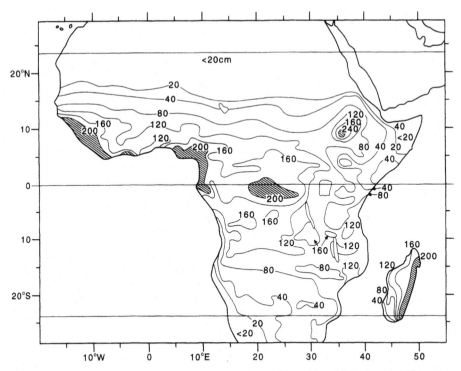

Figure 11.13   Mean annual rainfall (cm) in tropical Africa. After Nicholson (1988, p. 5) and Griffiths (1972, p. 471)

the equator, which with increasing latitude move closer together in time and at around 10° become a single maximum around the time of the highest position of the sun. Again, this general pattern is shown over most of tropical Africa, but regional departures are many and in some areas surface factors cause a trimodal distribution (Figure 11.14).

From a practical viewpoint, the total length of rainy seasons is most important: it controls the natural vegetation and the agricultural possibilities. Generally, months with a mean rainfall over 50 mm provide sufficient moisture for the cultivation of most tropical crops without irrigation. The number of months with rainfall over this minimum is used as the main basis for the climatic subdivision of Africa (Figure 11.15).

The climatic regions so determined must be considered as generalizations. They are separated by broad transition zones, which may belong to one region in one year, but to the next region during another year, according to rainfall differences from year to year. The illustrated distribution is therefore rarely realized exactly during any one year. The distribution of climatic regions, based on the duration of the rainy seasons, shows a certain symmetry about the equator (Figures 11.12 and 11.15). Five main regions can be distinguished:

1. The equatorial zone, which has at least 10 humid months. In this region there is no dry season, or it is so short that it is of little consequence. From this central zone, the rainy seasons decrease in length in three directions (2–4 below).

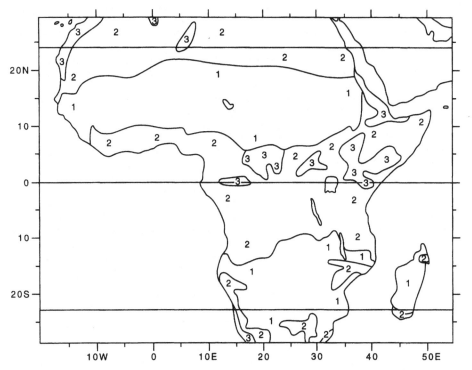

Figure 11.14   Unimodal, bimodal and trimodal seasonal rainfall distributions in tropical Africa. After Nicholson (1988, p. 9) and Griffiths (1972, pp. 484–489)

2. West Africa and the southern Sudan show a gradual decrease of the number of humid months from south to north.
3. In southern tropical Africa the duration of the rainy season diminishes from north to south, but also from east to west.
4. East Africa shows a general decrease from west to east, but in the highlands regional surface features produce a number of local climatic variations.
5. Madagascar has relief features that create a quite independent rainfall distribution pattern, which does not fit into the general scheme. It is therefore treated as a separate unit (Figure 11.15).

EQUATORIAL AFRICA

This region consists of two main areas: the Congo Basin, between about 5°N and 6°S, and extending as far east as Lake Victoria; and the southern coast of West Africa, between approximately 5°N and 9°N (Figure 11.15). The mean annual rainfall exceeds 1200 mm almost everywhere, but much higher values are reached where orographic lifting takes place (Chapter 10).

The natural vegetation in this region consists mainly of very luxuriant evergreen forests, which show little or no seasonal variations. Where a dry period occurs, it is of no practical importance as moisture is amply supplied by the soil.

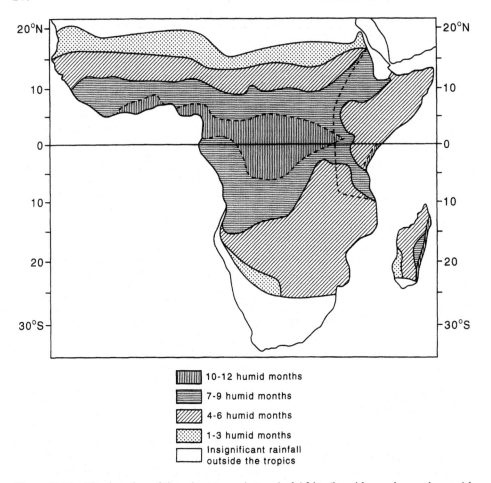

Figure 11.15 The duration of the rainy season in tropical Africa (humid months are those with a mean rainfall over 50 mm). Thick broken lines indicate approximate boundaries of the main climatic regions. After Thompson (1965, p. 54)

The large amounts of water vapour that maintain the continuous rainfall in this region originate mainly from the southern Atlantic Ocean. However, the coastal sea areas adjacent to the Congo Basin are frequently cooled by the Benguela current from the south, and this has a stabilizing influence, reducing the inflow of water vapour into the continent. Rainfall totals in equatorial Africa are therefore relatively modest compared to other equatorial regions (Figure 10.2). As over all tropical rainforests, the strong evapotranspiration from the dense vegetation provides an important local source of water vapour. However, this rapid turnover of moisture between the earth's surface and the lower troposphere is not sufficient to maintain high rainfall totals, as exhibited by the general decrease of total rainfall with distance from the Atlantic (Figure 11.13).

Although the main process causing precipitation in this region is convection, seasonal variations are mainly the result of large air mass movements. Over the Congo

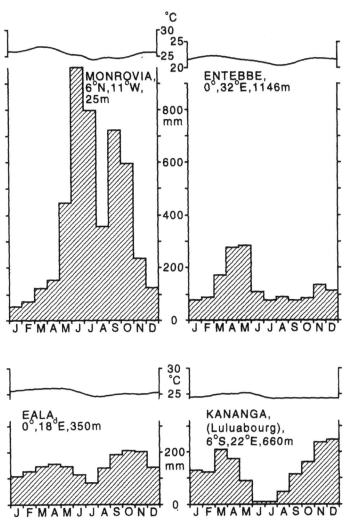

Figure 11.16   Climatic diagrams for four stations in equatorial Africa

Basin, convergences between air masses from west and east cause maxima during April/May and October/November, when the main zone of convergence is over this part of the region. Rainfall minima occur during December/January in the north and during June/July in the south (Figure 11.16, Kananga). The borderline between these two regimes is almost exactly at the equator (Griffiths, 1972, p. 288). Trimodal distributions are caused by local relief features (Figure 11.14).

Along the West African coast, the seasonal movements of the ITCZ create two periods of minimum rainfall. The major one is during December/January, when dry north-easterly winds prevail over the region (Figure 7.7). These winds are often loaded with dust and are locally called "Harmattan". The minor dry season is during July or August, when the ITCZ is located far to the north of the coast (Figure 7.7). The two

seasons of maximum rainfall occur when the south-west monsoon is well developed
and the convergence zone is near, in April/May or even June, and in September/
October (Figure 11.16, Monrovia).

The two regional anomalies of this climatic type, in western Nigeria and Ivory Coast
are caused by local factors, most likely the concave form of the coast which causes a
divergence of sea breezes (Ilesanmi, 1972). This creates a local stability in the monsoon
air.

Temperatures in this region show very small annual ranges; less than 2° in most of
the Congo Basin, under 4° in the rest of equatorial Africa (Figure 11.16). Daily ranges
are everywhere larger than the annual ones. They remain under 8° along the coast, and
increase with distance from the sea and with elevation (Figure 4.5).

WEST AFRICA AND THE SOUTHERN SUDAN

This is a very large region, stretching over about 15° of latitude, from around 5°N to
20°N, and from the Atlantic Ocean in the west to the Ethiopian highlands in the east
(Figure 11.12). The common feature of climate is the gradual decrease in rainfall, both
in total amount and in the length of the rainy season, with increasing latitude (Figures
11.13 and 11.15). This is related to the movements of the ITCZ and its associated
rainfall during the course of the year.

In December and January the convergence zone is at about 2–5°N in the west, and
farther south over the eastern parts of the continent (Figure 7.7). The region is
therefore under the influence of dry, continental and relatively stable air masses
coming from the north-east or north. Hardly any rainfall is received, except at a few
locations along the south coast of West Africa, where local wind systems of the sea
breeze type may bring some precipitation (Figure 11.17, Lagos). In many parts of the
region Harmattan-type dusty winds occur during this period. The relative humidity is
low and mean monthly temperatures are around 20°C in the north, and 25°C in the
southern parts of the region (Figure 11.17).

From about February to June the convergence zone moves in a northward direction, but this movement is not continuous. It is often interrupted or even temporarily
reversed (Griffiths, 1972, pp. 170–171). The south-west or south-east monsoon brings
warm and humid air masses from the southern Atlantic that underrun the dry
continental air, and a zone of heavy rainfall occurs about 500–1000 km south of the
surface position of the convergence, where the oceanic air masses reach a thickness of
about 4000 m (Figure 7.8). Rainfall is usually concentrated along disturbance lines
(Chapter 8), but convectional thunderstorms also frequently occur. In the southern
parts of the region rainfall starts in March or April, but in the north not until June, and
totals decrease from north to south (Figure 11.17). Temperatures are quite high in the
northern parts of the region, a condition that may prevail until September or even
October (Figure 11.17, Agades).

In July and August the ITCZ is at its most northerly position, at about 15–20°N
(Figure 7.7). Rainfall is received over most of the region, except in the coastal areas
south of about 8°, where a short dry season prevails (Figure 11.17, Lagos). However, at
many stations mean rainfall figures do not show this interruption of the rains because
it does not occur every year in the same month. Over the southern Sudan the

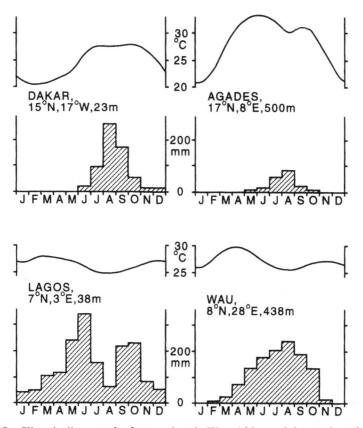

Figure 11.17   Climatic diagrams for four stations in West Africa and the southern Sahara

convergence of air masses from the south-west and the south-east causes much rainfall (Figure 11.17, Wau).

From September to November the ITCZ moves in a southward direction and its progress is generally faster than during the northward journey. However, the movements are again irregular and very different from year to year. The end of the rainy season often comes quite suddenly, in September in the north and in October in the south of the region.

Temperatures in the region show a gradual increase in the annual range with latitude and with distance from the coast (Figures 4.1 and 11.17). The hottest months usually come before the start of the rains, and a slight increase in temperature after the rainy season is common, illustrating the effects of cloudiness. At Dakar and other stations along the west coast, temperatures are relatively low throughout the year because the Canaries current keeps the sea surface cool.

In the northern parts of the region, rainfall is not only low, but also notoriously variable from year to year. A series of dry years has long-lasting effects on the vegetation and may initiate a southward movement of desert conditions. Overgrazing and more intensive use of the vegetation for the supply of wood are human factors that often increase this process of desertification (Chapter 13).

In the southern parts of this region, with 7–9 humid months, crop agriculture without irrigation is generally possible, though production figures vary strongly from year to year according to the rainfall. Further north, crop agriculture is only possible with irrigation, and moving northward animal husbandry takes over as the main source of food and income.

SOUTHERN TROPICAL AFRICA

This region covers a large north–south distance: from the equator in the north-west to 25°S in the south-east (Figure 11.15). Therefore latitudinal differences in climate are considerable. In some ways this region presents a mirror image of the last region: a decrease in rainfall with distance from the equator and a transition from bimodal to unimodal seasonal distributions (Figures 11.13 and 11.14). Again, both phenomena are closely related to seasonal movements of the ITCZ.

However, the narrowing of the African continent, the cool Benguela current to the west and the warm Mozambique Channel to the east create strong longitudinal differences in this region. In the west, sea surface temperatures are around 15–20°C throughout most of the year and the stabilizing cooling effect reaches almost as far north as the equator. This keeps the ITCZ from moving southward during the southern hemisphere summer. In the eastern parts of the region the convergence zone makes large seasonal movements, reaching as far as 15°S in January (Figure 7.7).

The difference in water surface temperatures strongly affects the air masses which move into the continent. In the west, these air masses originate mainly from the subtropical high pressure cell over the southern Atlantic. They are relatively stable with an inversion layer at levels between 1000 and 2500 m. These air masses are further stabilized by the cool ocean surface and bring very little rainfall to the western coastal areas. This effect is particularly strong during the southern hemisphere winter, when the Benguela current penetrates almost to the equator. Brazzaville, although more than 350 km from the coast, experiences a sharp drop in temperature and receives practically no rain in this season (Figure 11.18). Another effect of the cool ocean surface is a strong sea breeze, which brings no rain, but frequent fogs in the coastal region.

The situation is quite different in the east. In December–January the convergence zone is at 10–20°S and three different air masses meet here (Figure 11.19). Those from the south-east originate from the subtropical high pressure cell over the Indian Ocean and adjacent southern Africa, and are therefore relatively dry. They have dropped most of the moisture in their lowest layers along the steep eastern slopes of Madagascar. The air masses from the north-east can be quite humid and unstable, depending on the path they have followed further north. The air from the north-west is the so-called "Congo-air", it is often humid as it has moved over the Congo Basin with its luxuriant vegetation, particularly along the rivers. Therefore most rainfall is received along the Congo Air Boundary. In the area of mixed air masses itself, local convection and orographic lifting also cause much rainfall, with strong regional differences and variation from year to year (Figure 11.18; Maun, Beira).

During March and April the whole convergence zone moves northwards and these months bring rainfall to the northern parts of the region (Figure 11.18, Bujumbura). The convergence zone leaves the eastern parts of the region by the end of May. The

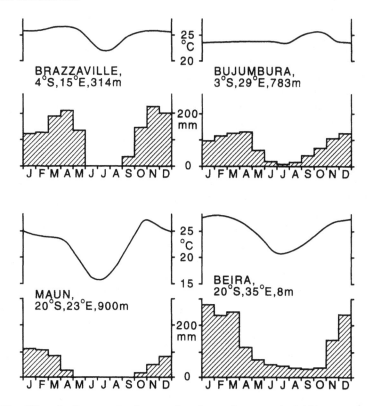

Figure 11.18   Climatic diagrams for four stations in southern tropical Africa

following months, from about June to September, are relatively dry. During this period the whole region is under the influence of south-easterly trade winds. These are often of continental origin, coming from a strong high pressure cell over the southern parts of Africa. But even when the south-easterlies come from the Indian Ocean they bring little rainfall, as they generally bring quite stable air masses, which have deposited most of the moisture in their lowest layers in Madagascar or the coastal mountain ranges (Figure 11.18, Beira). As temperatures are relatively low during these months, convection is rare and little rainfall is received.

This region illustrates clearly the increase in the annual rainfall from the south-west to east and north (Figures 11.13 and 11.15). Accordingly, agricultural possibilities are limited to animal husbandry and irrigated crops along rivers in the arid coastal areas of the west, but are quite good in the east and north of the region, where unirrigated crops can be grown.

EAST AFRICA

This region reaches from about 11°S to 14°N. The common climatic feature is a relatively low rainfall with a complicated distribution pattern showing a general increase from north to south (Figure 11.13). Low rainfall in an equatorial region is exceptional, and particularly surprising here because the region is adjacent to the

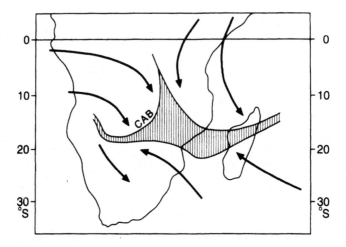

Figure 11.19   Approximate surface position of the ITCZ (shaded) during December–January. CAB is Congo Air Boundary

warm Indian Ocean. The main reason for the aridity are the monsoonal systems, which prevail over East Africa during large parts of the year (Chapter 7).

From December to about March the north-east monsoon brings air masses which are mainly of continental origin and relatively dry (Figure 7.7). The monsoon has passed over the cool Somali current, which has a stabilizing effect. It also runs mainly parallel to the coast and the edges of the highlands, so little orographic lifting takes place. A further stabilizing effect is the quasi-permanent low over Lake Victoria, which introduces a divergence in the monsoon stream. Only in the south of East Africa, where the monsoon converges with air currents from the west and east, does it bring considerable rainfall (Figure 11.20, Tabora, Lindi).

From about June to September the south-west monsoon prevails over East Africa (Figure 7.7). This is a rather shallow air current. It is partly of continental origin, coming from a high pressure cell over eastern South Africa, but most of the monsoon air masses come from the Indian Ocean. These air masses have dropped most of the moisture in their lowest layers along the steep eastern mountain ranges of Madagascar and the fetch over the Mozambique Channel is too short to pick up much humidity. The monsoon does not bring much rainfall in the southern coastal regions (Figure 11.20, Lindi). However, further north, where the fetch over the Indian Ocean is longer, the south monsoon brings some rainfall in coastal areas (Figure 11.20, Mogadiscio). Inland, there is again the effect of the low over Lake Victoria, causing a divergence in the monsoon stream and it brings little rainfall, except on some well-exposed mountain ranges. A secondary low pressure over the strongly heated Ethiopean highlands also creates a divergence in the south-west monsoon, but brings regional rainfall in those areas (Figure 11.20, Addis Ababa).

During the monsoon seasons convection is generally not very active as frequently an inversion layer is present. However, orographic lifting over exposed slopes of the mountains in the western parts of the region causes rainfall during both monsoons.

As both monsoons are relatively dry, the main source of rainfall in East Africa is the

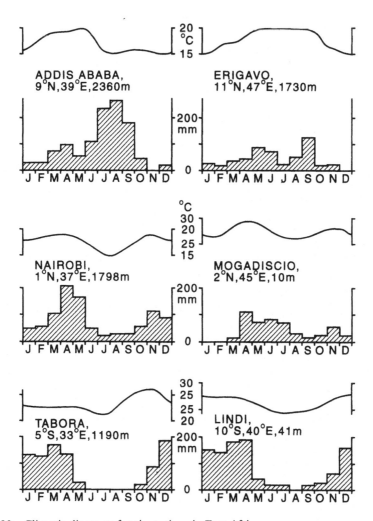

Figure 11.20   Climatic diagrams for six stations in East Africa

ITCZ and the seasonal distribution resembles the theoretical pattern (Figure 10.5). The zone of maximum rainfall follows the latitudinal position of the overhead sun with a time lag of about 4–6 weeks. Accordingly, over most of East Africa there are two rainy seasons: the "long rains" around April–May and the "short rains" around October–November. These local names are related to the fact that during the first rainy period the convergence zone tends to be very broad and move only slowly northward, so that there is extensive heavy rainfall. During October–November the ITCZ tends to move southward relatively fast, particularly over the eastern parts of East Africa and less rain is received over many areas (Figure 11.20, Nairobi). In the south, the two rainy seasons come close together, but there is often a clear interruption in February or March, which is not well shown by monthly mean rainfall figures because it does not occur during the same month every year (Figure 11.20, Lindi).

At latitudes over about $10°$ the two rainy seasons coalesce, but the mountainous nature of much of East Africa causes many departures from this general pattern and over a large part of the region a trimodal rainfall distribution prevails (Figure 11.14). Spring rains in April–May over Ethiopia and northern Somalia are related to mid-latitude upper air troughs, usually from the Mediterranean region. The warm surface air forms an upper cold front, with the cold upper air from the higher latitudes resulting in numerous thunderstorms (Figure 11.20, Addis Ababa and Erigavo).

Rainfall variability in East Africa is generally high and the arrival of the rainy seasons differs from year to year, often by several weeks (Nicholson, 1988, p. 7). This is a serious drawback in a region where in some areas rainfall is marginal for unirrigated agriculture. Nevertheless, agriculture is favoured by the two rainfall seasons, allowing double cropping in some places. In this area, plantations are common: sisal in the lowlands, and coffee and tea in the wetter highlands. Peasant agriculture is based on maize but in the highlands much wheat is produced by commercial farmers.

Temperatures show small annual ranges. Near the equator, the coolest time of the year is related to cloudiness, with the highest temperatures before the rains (Figure 11.20, Nairobi). In Mogadiscio, the low temperatures during July and August are caused by the cool Somali current (Figure 11.20). As almost everywhere in the tropics, the daily ranges of temperature exceed the annual ones and increase with distance from the coast and with elevation (Figure 4.5).

For the major African regions discussed above there appear to have been significant changes in the rainfall climatology over the two standard WMO reference periods of 1931–60 and 1961–90 (Hulme, 1992). For tropical north Africa, 1961–90 rainfall was 30% less than that for the 1931–60 period, especially for the summer monsoon months, while for the tropical margins of southern Africa there were significant rainfall reductions of around 5%. Interestingly, significant increases of 15% for East Africa and 10% for the southern coastal region of West Africa have also occurred. Increases in variability have been a feature of the rainfall climatology of Tunisia, Algeria, the Nile Basin and the extreme south of the African continent. Africa overall, however, has not experienced significant changes in rainfall variability. Regarding seasonality there have been increases for the southern coastal region of West Africa, whereas the opposite is true for East Africa (Hulme, 1992).

At times drought may affect large areas covering western, eastern and southern Africa, but at other times drought occurrence in these major regions does not always coincide. For example, southern Africa has experienced widespread droughts in the years 1946–47, 1965–66, 1972–73, 1982–83 and 1991–92, while eastern Africa suffered drought conditions in 1933–34, 1938–39, 1949–50, 1973–74, 1983–84 and 1991–92. Further to contrasts in timing, it appears that the intensity of drought, as related to rainfall fluctuation, is less severe in southern and eastern Africa than in western Africa. Certainly the persistent conditions of dryness in the Sahel region of western Africa throughout the 1970s to 1980s were not matched elsewhere in the African continent (Figure 10.4). This event, widely referred to as the Sahelian drought and notable for its duration, is not unique as similar dry spells occurred in this region in the 1820s, 1830s, 1910s and the 1940s. Records of lake levels also provide a clear picture concerning the occurrence of dry and wet periods for this region over the past 1100 years (Nicholson, 1979).

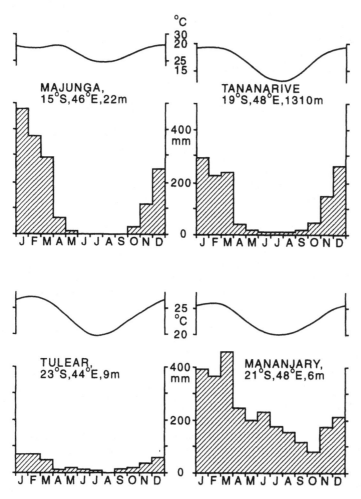

Figure 11.21   Climatic diagrams for four stations in Madagascar

MADAGASCAR

This island deserves separate treatment because it shows strong climatic differences over relatively short distances, a rare phenomenon in Africa (Figures 11.13 and 11.15). These differences are related to the main relief feature: a continuous dorsal mountain range with a height of well over 1200 m, in some places reaching over 2000 m with very steep slopes on the eastern side. This range causes four distinct climatic regions: the eastern coastal regions, the central parts of the island, the west and the extreme south.

The *east coast* is by far the wettest part of Madagascar (Figures 11.13 and 11.15). This area is under the influence of south-easterly trade winds from the Indian Ocean throughout the year. These are uplifted along the steep eastern slopes of the main mountain range and yield much rainfall (Figure 11.21, Mananjary). Most rainfall is received during the summer, when the ITCZ with its associated depressions and cyclones are over northern Madagascar (Figure 7.7). These cyclones are usually formed to the north-east of the island and often recurve east of Madagascar, but many bring large amounts of rainfall to the east coast. In winter, the trades are strongest, with the subtropical high pressure cell over the Indian Ocean reaching surface pressures around 1026 mb (Figure 5.2). The relatively low rainfall during September–October is caused by the weakening of the trade winds, before the establishment of the ITCZ in November–December.

The *central areas* of Madagascar consist of highland plateaux at elevations between 700 and 1400 m. These areas are also under the influence of the south-easterly trade winds, but these do not bring much rainfall, because little orographic lifting takes place and the humid lower layer of these winds is rather shallow. Most rainfall is received during the summer, when the ITCZ is nearby (Figure 11.21, Tananarive). However, large regional and local differences occur in this region according to exposure and elevation, and annual means may vary between 2000 and about 1000 mm (Figure 11.13).

The *western parts* of the island show strong differences between dry and wet seasons. During the winter these areas are under the influence of the trade winds, which are mainly dry, having deposited most of their humidity in the east. From about December to March the region is under the influence of north-westerlies and the ITCZ. Over the Mozambique Channel depressions develop frequently and particularly the northern areas receive much rainfall (Figure 11.21, Majunga). However, the effects of these depressions and the convergence itself is largely limited to the northern parts of this region, and the south-western areas receive very little rainfall (Figure 11.21, Tulear). Annual mean rainfall decreases from over 1600 mm in the north to 400 mm in the south (Figure 11.13). A special feature of the coastal areas are the strong land and sea breezes, which develop particularly during the summer.

The *south* of Madagascar does not belong to the tropics, because sea level temperatures during the winter are below 18 °C. It is a semi-arid region with annual rainfall totals below 400 mm.

Annual temperature ranges in Madagascar are affected by latitude. They are below 4 °C in the north but over 6 °C in the south (Figures 3.1 and 11.21). The mean daily range shows a complicated pattern, due to the mountainous nature of the island, but there is a general increase from east to west as the effects of the trade winds become weaker (Figure 4.5). Some central areas show mean diurnal ranges as high as 16 °C (Griffiths, 1972, p. 472). Exposure creates another difference between east and west: the west is generally about 2–3 °C warmer and this difference is especially clear during summer (Figure 11.21, Tulear and Mananjary). The difference is related to cloudiness.

TROPICAL AMERICA

Surface features divide tropical America into three distinctive parts:

1. the Caribbean region, an area of many islands and a predominant ocean surface;
2. Central America, the landbridge between North and South America, which is

dominated by extensive highlands in the north, where it is rather broad, and a long and narrow peninsula in the south;
3. tropical South America, a continental region, divided by the high mountain range of the Andes into a narrow western coastal zone, a broader zone of highlands and mountains, and a large lowland region in the east.

The seasonal distribution of rainfall and the annual totals indicate the many climatic differences related to these surface features in tropical America (Figure 11.22).

THE CARIBBEAN REGION

This region between latitudes 12°N and 27°N, is dominated by the ocean surface; the islands are generally too small to create regional climatic differences, they have only local effects. The most important climatic feature of this region is that it is almost never under the influence of the ITCZ; the North American continent is simply too far north to attract it during the northern hemisphere summer. Occasionally the convergence comes close to the most southern islands. This happens mainly in July, but the zone is usually diffuse and difficult to locate on these occurrences and it brings no precipitation (Figure 5.2).

The Caribbean is under the dominance of the trade winds throughout the year. These winds make only a small angle with the isobars and are generally almost exactly easterly in direction. They are strongest during the northern hemisphere winter and in December they may have a slightly more north-easterly direction. When the trades arrive in the Caribbean region, they have passed over about 6000 km of the warm North Atlantic Ocean, and carry large amounts of water vapour in their lowest layers, below the inversion. The trade wind inversion is relatively low: at about 1000–1500 m in winter when it occurs during about 80% of all days, and at around 2000 m in summer when it occurs only during 30–40% of all days (Figure 8.6). The capacity to bring precipitation is therefore strongest during the northern hemisphere summer.

Another important climatic element are the hurricanes, which visit the region from about June to November, with the maximum frequency in September. A large part of the North Atlantic Ocean between about 10°N and 15°N has water temperatures exceeding 27°C during the summer and the Caribbean is one of the most frequented regions for hurricanes (Chapter 8). The rainfall regime in most of the islands therefore shows a summer maximum. In winter, the subtropical high pressure cells often extend as far as the Venezuelan coast. The summer maximum may last up to December in the south; in the north it ends usually in October. Local differences are strong, due to exposure to the winds, the possibility of convection and orographic lifting, and the occurrence of hurricanes (Figures 11.23 and 11.24). Differences from year to year are large, due to the irregular occurrence of the hurricanes.

Temperatures in the Caribbean generally show small annual ranges, which amount to only about 3°C in the south but may reach 8°C in the north of the region (Figure 11.23). Annual and diurnal ranges of temperature generally increase with distance from the coasts. In the northern islands and the southern tip of Florida, cold air masses from the North American continent bring occasional cold spells in winter. The southern islands are free from this phenomenon.

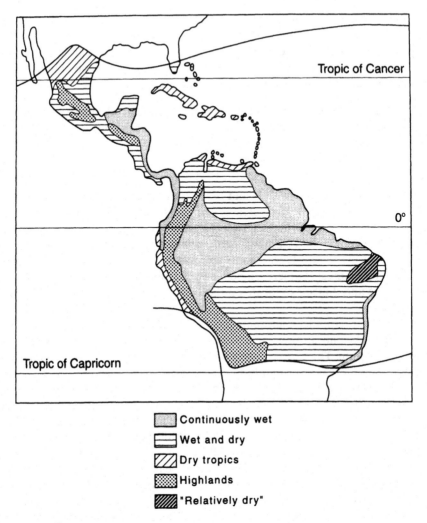

Figure 11.22   Climatic regions in tropical America

CENTRAL AMERICA

In its central parts, the American continent consists of a long and rather narrow landmass, which stretches over a distance of over 4000 km in a general NW–SE direction. Near the Panama canal it is only 65 km wide and at two other locations its width is not more than 300 km. This land mass has a high and almost continuous mountain backbone, which broadens in the northern parts (Figure 11.24). Except in the north, this mountain range constitutes a clear climatic divide, causing strong contrasts between the Atlantic and the Pacific sides.

The *extreme north* of the peninsula has a dry climate, as the easterly winds from the Caribbean rarely reach this region (Figure 11.22). In winter its climate is often

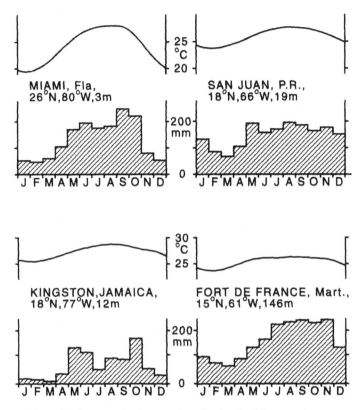

Figure 11.23   Climatic diagrams for four stations in the Caribbean region

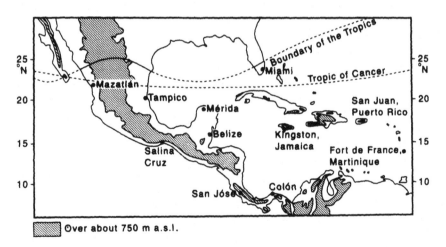

Figure 11.24   Location of stations in the Caribbean region and in Central America for which diagrams are given

dominated by continental air masses and winds from the Pacific side which are usually dry as they originate from the subtropical high pressure cell. The main rainfall season is the autumn, when hurricanes and strong disturbances of various types may visit the region.

The *Atlantic coastal areas* have climates which are rather similar to those of the Caribbean Islands, because these regions are also under the influence of the North Atlantic trade winds during most of the year. However, the trade winds bring more rainfall here than on most of the islands, because the base level of the inversion is generally higher, as it has been weakened by turbulence over some of the islands, and on arrival at the coast by local convection and orographic lifting. Also, the lowest layers of the trade winds have picked up much water vapour over the warm Caribbean Sea. As a result, rainfall on the Atlantic side of Central America has an annual mean over 2000 mm in the southern parts, with some rainfall also during the winter (Figure 11.25, Colon and Belize). Further north, the total rainfall is less and the winter is generally dry (Figure 11.26, Merida and Tampico).

There are some drier regions on the Atlantic side, as in the Yucatan peninsula, where the main cause is the leeward position in relation to the easterlies; and the coast of Honduras, which runs parallel to the prevailing winds, so that a divergence is created because of differences in friction.

The *Pacific side* of Central America is generally drier than the Atlantic side, and this is particularly true in the northern parts. Here the prevailing winds are from the north-west and they bring rather stable air masses from the eastern side of the subtropical high pressure cell over the North Pacific. These air masses are further stabilized by the rather cool California current. Only during late summer and autumn does this region receive more rainfall, as disturbances and hurricanes occasionally cross over from the eastern side of the continent, and these are reactivated by the high surface temperatures over the Pacific coastal lowlands (Figure 11.26, Mazatlan). The number of hurricanes reaching this region varies widely from year to year, the long-term average being about 10 per year.

The southern parts of the Pacific coast have generally more summer rainfall (Figure 11.25, Salina Cruz). The main reasons for this increase are the diminished influence of the California current, as the coast turns more eastward; and the increased frequency of winds from the south-west. During summer and autumn the mountain ranges of the southern parts of Central America often separate Atlantic easterlies from south-westerlies of the Pacific. These south-westerlies bring humid equatorial air masses to the landbridge, and convection and orographic lifting cause large amounts of rainfall. As sea surface temperatures over the Pacific adjacent to the Central American coast remain rather high during the autumn, the rainfall season may well run until October or even November.

The *mountain highlands* of Central America show a wide variety of climatic conditions, depending on local landforms, exposure and elevation. Most of the centrally located stations are rather dry, not only because of the sheltering effects of mountain ranges on both sides, but also because they frequently reach into the inversion layer of the Atlantic trades (Figure 11.26, Mexico City). In the south, where the landbridge is narrower, there is also more rainfall during the summer (Figure 11.25, San Jose).

In Central America, annual temperature ranges increase from south to north, and

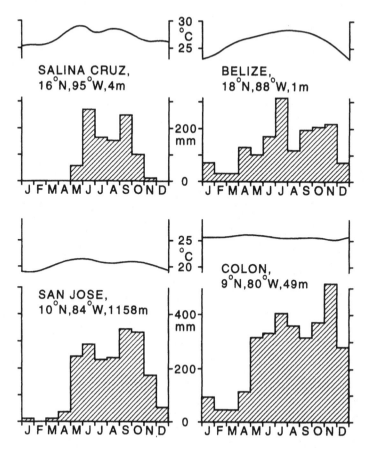

Figure 11.25   Climatic diagrams for four stations in the southern parts of Central America

also from the coast to the interior (Figures 11.25 and 11.26). Local factors, especially in the mountainous regions, create many departures from this general rule, however. Diurnal ranges of temperature increase sharply with elevation and distance from the sea.

TROPICAL SOUTH AMERICA

The largest part of tropical America is occupied by the South American continent. The most important surface feature is the high and continuous mountain range of the Andes. This range divides the tropical part of the continent into three climatic regions: a narrow coastal strip to the west of the mountains, the Andes highlands themselves, and the major part of tropical South America, to the east of the mountains (Figure 11.27). Each of these regions can be further subdivided according to climatic criteria (Figure 11.22).

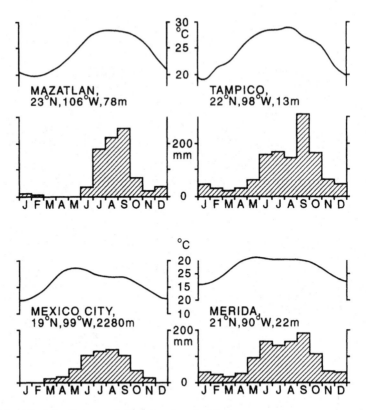

Figure 11.26   Climatic diagrams for four stations in Mexico

The Western Coastal Region

In this area, two surface features control climate: the cold Humboldt–Peru ocean current, which runs parallel and close to the coast and brings cool water as far north as the equator and the Andes range, which also runs close to the coast, leaving only a narrow strip of coastal lowlands. This strip gets wider in the northern parts of the region, and also near the Gulf of Guayaquil. The combined effect of these two surface features is to keep the ITCZ from moving into the southern hemisphere: it remains, throughout the year, at around latitudes of 3–8°N (Figure 5.2). This results in three distinct climatic regions: a continuously wet region between about 8°N and the equator, but reaching a few degrees further south along the foothills of the Andes; an intermediate area with rains only during a part of the year, between the equator and about 2°S; and a completely dry area in the south (Figure 11.22).

The *continuously wet area* is under the influence of the ITCZ throughout most of the year. Westerly winds are frequent, and they bring warm and humid air masses to the coastal regions because the equatorial counter current brings relatively warm water to the adjacent parts of the Pacific Ocean. Convergence, orographic lifting and convection result in large amounts of rainfall throughout the year, with annual averages around 5000 mm (Figure 11.28, Andagoya). In the northern parts of this region the

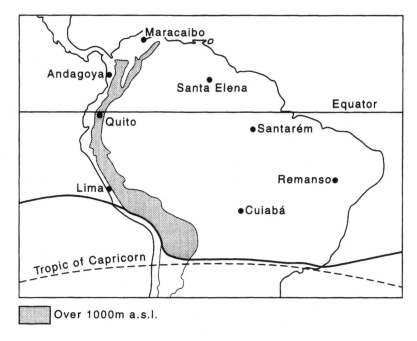

Figure 11.27   Locations of stations in tropical South America for which climatic diagrams are given

season of lower rainfall is around December–January, when the effects of the subtropical high pressure cell of the northern hemisphere are sometimes felt; in the southern parts the relatively dry season is around July–August, when the subtropical high over the southern Pacific is at its seasonal maximum strength. However, these seasons still bring substantial amounts of rain.

Directly along the coast most of the rain falls at night, due to night-time convection over the warm sea, but further inland a clear afternoon maximum prevails. Temperatures in this region show almost no seasonal variations, due to the stability of sea surface temperatures.

The *transition zone* is centred around the Gulf of Guayaquil and the surrounding lowlands. Here, the dry season caused by the strong high pressure cell over the South Pacific and the relatively cool ocean waters, starts around July, and may continue until December. In this area large contrasts in rainfall occur: north of the Gulf of Guayaquil annual means are around 1000 mm, while directly to the south they amount to no more than 250 mm.

While the coastal lowlands have a long dry season, the western mountain slopes of the Andes, about 150 km inland, receive rainfall throughout the year because of strong orographic lifting. When "El Nino" occurs, the ITCZ moves southward along the coast and may reach positions as far as 5–7°S. In this case, rain may fall in normally very arid parts of the coast, usually during January or February (Figure 6.8).

The *coastal desert* begins at about 2°S. It is a narrow strip reaching well beyond the tropics. It is one of the driest areas in the world, due to two major factors. The main one

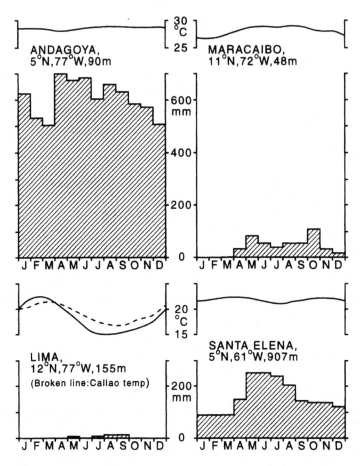

Figure 11.28   Climatic diagrams for four stations in the northern and western parts of tropical South America

is the strong stability of the air masses from the well-developed subtropical high pressure cell over the adjacent South Pacific Ocean, which are further stabilized by the cool ocean waters of the Humboldt–Peru current. The other factor is the deflection of the originally south-south-east winds by the high wall of the Andes mountains. These winds run almost parallel to the coast, but slighty offshore, causing upwelling of cold water directly near the coast. Frictional differences between land and sea surfaces may also cause some divergence in the winds, resulting in subsidence and a further stabiliz-ation of the air masses. In this area a low stratus cloud often develops where the relatively warm air masses from the continent come into contact with the cold ocean surface. This stratus cloud and its associated inversion layer often move inland with the strong sea breezes which occur almost every day in this area. These clouds occasionally produce a light drizzle, locally called "garua".

The desert extends over the coastal mountain ranges, which are too low to cause orographic rainfall from the very stable air masses. Leeward of the coastal ranges, climates are extremely dry as even the drizzles are rare. In the coastal desert, temperatures are controlled by distance from the sea. Directly along the coast, they are continuously low, with small annual and daily ranges. But even a few miles inland temperature variations are larger, as illustrated by the temperature curves for Callao, the harbour of Lima, and Lima airport, which is only 8 km from the coast (Figure 11.28).

The Mountain Highlands

In the Andes, as in all mountainous areas, climates are largely controlled by two main factors: elevation and exposure, with local landforms creating many small-scale variations.

Elevation has not only a strong effect on temperatures, as discussed in Chapter 4, but also on rainfall. In most of the Andes, rainfall increases with elevation up to around 1200–1500 m a.s.l., and decreases slowly at higher levels as remnants of the subtropical trade wind inversion frequently reduce the rain-producing capacity of the air above these elevations. Interior plateaux receive usually less rainfall than the surrounding mountains. The diurnal distribution of rainfall is also related to elevation: at higher levels, where convection is the main cause, an afternoon maximum prevails, but the lower valleys have often a night maximum of rainfall, as mountain winds converge.

Exposure is important because easterly winds prevail over the region. Therefore east-facing slopes generally receive more precipitation than those facing westerly directions. This is also reflected in the height of the snowline: it is generally lower on easterly slopes because more snow is received than on slopes with other exposures. Near the equator the snowline is at around 4500 m above sea level. Exposure to the sun is also important at higher elevations, where temperature differences between sun and shade are usually quite large. West-facing slopes, which have more sun during the afternoon, are generally warmer than other areas at the same level.

Seasonal rainfall distribution shows generally a maximum about one to two months after the overhead position of the sun. Of the dry periods, the one during June–August is usually the driest, especially south of the equator, because the south-easterly trade winds are rather weak during this time (Figure 11.29, Quito).

The Andes climates are otherwise typical mountain climates: large diurnal temperature ranges, with frequently clear and calm nights, cloudless mornings, and afternoons with rapid cloud development and strong winds over the mountains, and a rapid cooling after sunset. The Andes mountains have the highest permanent settlements in the world.

Tropical South America East of the Mountains

This is by far the largest land area of tropical America. It is a continental region, with a coastline that is not very indented. It is predominantly flat, with a shallow depression in the centre, the Amazon Basin, and mountain ranges and highlands in the north and south-east, which remain generally well below 1000 m.

The general atmospheric circulation over this area is controlled by the seasonal movements of the ITCZ and its associated wind systems. In some areas this results in a

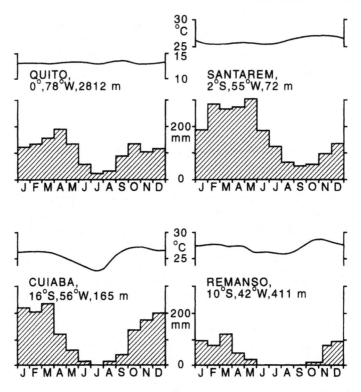

Figure 11.29   Climatic diagrams for four stations in the southern parts of tropical South America

seasonal reversal of the prevailing wind directions but the term monsoon is not used in this area, probably because the character of the winds does not change very much with the change in direction. This is because both the north-easterly and south-easterly winds from the Atlantic Ocean, which prevail, rapidly change into real equatorial winds: convection and orographic lifting near the coast destroy the inversion layer, and over the continent these winds pick up large amounts of water vapour from the luxuriant vegetation over the Amazon Basin. Changing direction near the equator causes convergence, which further destabilizes the air masses these winds carry.

From about *April to October* the ITCZ forms a broad, diffuse zone of convergence between about 1°N and 9°N. South-easterlies, turning into pure southerly winds near the equator, prevail over most of tropical South America south of the equator, but further north winds are of variable direction and heavy rainfall is produced.

From *December to March* the ITCZ at first moves to a position around 15°S. A secondary area of convergence develops around 10°S. Towards the end of this period, the ITCZ weakens and then moves northward. North of the convergences and along the east coast of Brazil north-easterlies prevail, these winds often being rather dry (Figure 5.2).

The results of these seasonal movements are two types of climate: a continuously wet one near the equator, where equatorial air masses prevail throughout the year, and a type with a dry season, at some distance from the equator, as dry air masses prevail when either the south-easterlies or the north-easterlies have not changed character enough to produce much rainfall (Figure 11.22).

The *continuously wet climate* prevails over the Amazon Basin, the Guyanas and the north-east coast of Brazil. In this region the mean annual rainfall is everywhere over 1500 mm (Figure 11.29, Santarem). The reasons for the continuously high rainfall are surface conditions. The many rivers and swamps with their luxuriant vegetation cover, produce enormous amounts of water vapour. Local convection rapidly destroys the inversion layers of the winds coming from the Atlantic Ocean, and orographic lifting, along the coast and near the Andean foothills, produces much precipitation. Inter-annual variability of the location and intensity of convection centres can produce marked climatic variability in this region (Marengo, 1995).

The season of maximum precipitation generally corresponds to the overhead position of the sun, usually with a delay of one to two months. However, the period from January to May is at most locations wetter than the other period of overhead sun, August to November, because during the latter season the ITCZ moves rather rapidly over the region.

The *climates with a dry season* occupy two areas, to the north and to the south of the central zone (Figure 11.22). In the northern areas the dry season is from about October to March (Figure 11.28, Santa Elena). Because Santa Elena is close to the central zone, the dry season is one of reduced rainfall only; further away from this zone it is really dry. This dry period is related to the dry coastal regions, where a stabilization of the north-easterlies takes place. During the rest of the year, the prevailing southerly winds bring humid, unstable air masses from the Amazon Basin.

The southern area has a dry season from about March to October, when south-easterlies prevail (Figure 11.29, Cuiaba). These south-easterlies have lost most of their rain-producing capacity along the south-east coast of Brazil, and they have not yet passed over the Amazon Basin. During the rest of the year, northerly winds prevail which have moved over the equatorial areas.

There are three small areas which deviate from the above general scheme:

1. a small dry strip along the coast of Venezuela and including the nearby islands, e.g. Curacao;
2. a relatively dry area in north-eastern Brazil;
3. a wet coastal strip along the south-east coast of Brazil, between 13°S and 23°S (Figure 11.22).

The *Venezuelan coast* is under the control of easterlies almost throughout the year. These winds blow almost parallel to the coast, and a divergence is caused by differential friction. This divergence is reinforced by acceleration, as the winds are attracted by the low pressure over the continent. The resulting subsidence stabilizes the air masses involved up to elevations of about 2000 m above sea level, in most cases high enough to prevent rainfall. There are two periods when rainfall can be expected: during October–December, when the easterlies are relatively weak, and in May–July, when the ITCZ is relatively close to the region (Figure 11.28, Maracaibo).

North-eastern Brazil, often referred to as the Nordeste, is under the influence of south-easterlies during most of the year. These winds bring relatively dry and stable air masses from the nearby subtropical high pressure cell over the southern Atlantic. These have to cross a mountain barrier of moderate elevation before reaching the area, so that most humidity in the lower layers of these air masses has been removed. Rainfall occurs only from November to March, when the ITCZ is just south of the region. Temperatures show only a small seasonal variation (Figure 11.29, Remanso).

The *south-east coast of Brazil* experiences rainfall throughout the year, caused by orographic lifting of the south-easterly trade winds from the Atlantic Ocean. The season of maximum rainfall is from about April to September, when the south-easterlies are strongest. During this period, mid-latitude cold fronts also cause widespread rainfall, not only along the south-east coast, but also over the adjacent regions of southern Brazil, up to about 10°S latitude. These cold fronts can do extensive damage to tropical crops, particularly coffee.

The humid and wet and dry tropical climate regions east of the Andes are all characterized by their inter-annual variability of rainfall (Figure 10.4) and especially the occurrence of economically damaging droughts. Concerning inter-annual variability, there is considerable regional variation. For example, for the dry Nordeste region, below-average rainfall was a feature for the years 1941–44, 1951–56, 1958–60 and 1981–83, whereas in the wet Amazonian region, the years 1960–75 and 1981–88, apart from 1983, were generally wet years (Marengo, 1992; Marengo and Hastenrath, 1993). For dry southern Brazil, the pattern of inter-annual and intra-annual variability is different to that of the Nordeste and Amazon regions as in some years wet conditions in this region are matched with dry conditions in the Nordeste and the Amazon Basin (Rao and Hada, 1990). This suggests that while the three regions may generally respond to the same rainfall controls (South Atlantic SST patterns, position of the inter-tropical discontinuity, behaviour of the St Helena anticyclone and strength of the subtropical jet), the nature of the rainfall response may be quite different. Generally, during ENSO events tropical South America receives less than normal rainfall (Moron, Bigot and Roucou, 1995).

One of the most drought suffering regions of tropical South America is the Nordeste region of north-east Brazil. Here severe droughts occurred in 1942, 1951, 1953, 1970 and 1983. Droughts in 1992 and 1993, however, exceeded these in intensity, mainly due to a succession of dry years since the late 1980s which limited recharge of depleted soil moisture levels (Rao, Satyamurty and de Brito, 1986; Rao, Hada and Herdies, 1995). By the end of 1993, millions of people in this region were on the brink of starvation, with migration seriously considered as the only drought-avoidance strategy available. Failure of cotton, sugar and other crops was widespread with rainfall deficiency related problems exacerbated by abnormally high temperatures and strong winds. Drought is also a problem for the South American subtropical countries as clearly demonstrated between 1988 and 1989 and between 1991 and 1993 when large areas of Uruguay, Paraguay, south-eastern Bolivia and southern Brazil received < 50% and 25–60% of their normal rainfall respectively. Consequently there was severe stock and crop losses throughout this entire region.

THE TROPICAL OCEANS

About 80% of the tropics are occupied by the Pacific, Indian and Atlantic Oceans (Figure 1.1). The climatic conditions over these enormous water surfaces are of great importance because they have a strong influence on many other climates, both inside and outside the tropics.

There is a spatial uniformity in temperatures over the tropical oceans, due to the homogeneous water surface. Moreover, this surface is thermally very conservative and both diurnal and seasonal differences in temperatures are very small compared to those over land areas (Figures 4.1 and 4.5). However, this spatial uniformity is disturbed by the cold ocean currents (Figures 5.15 and 5.16).

Seasonal temperature variations are very small in the tropical Pacific, where only the equatorial countercurrent changes in width and depth, having more strength during the northern hemisphere summer. The two main cold currents might shift a few degrees with the seasons, but their strength remains the same. El Nino, however, can produce rapid SST changes and accounts for most of the inter-annual variability of ocean temperature in the Pacific.

In the tropical Atlantic the equatorial countercurrent develops during the northern hemisphere summer but during the winter it is largely absent. In the tropical Indian Ocean a complete reversal of currents takes place with the monsoons, but these changes are largely limited to the northern hemisphere. The cold current off western Australia is strongest during the southern hemisphere summer when it is helped by the east monsoon further north. The Somali current is also reversed, but remains cold.

The cold currents bring not only lower sea surface water temperatures, but they also cause relatively large seasonal differences, bringing temperature conditions of the higher latitudes, where they come from, to the tropics (Figure 4.4). The cooler waters cause the normal situation over the tropical oceans, where the atmosphere is warmed and humidified, to be reversed: over the cold currents the atmosphere is cooled and stabilized from below. This is the main factor in some of the tropical coastal deserts. Diurnal temperature changes over the oceans are generally only a fraction of a degree Celsius.

Over the tropical Pacific, the zone of maximum rainfall frequency is between $3°N$ and $10°N$, but it shifts to about $5°N$ and $24°N$ in July and August. Its longitude is fairly constant between the coast of tropical South America, at $78-80°W$, and $170-180°W$, with the only seasonal variation being that from about January to May when the zone starts at about $110°W$ (*US Navy Marine Atlas*, Vol. 2, 5). However, it should be noted that the zone of maximum probability of rainfall is diffuse in character and shows no clear boundaries.

In the case of the tropical Atlantic, the zone of maximum rainfall probability lies between $4°N$ and $10°N$ and stretches from the African coast to about $50°W$. Seasonal differences are relatively small, except for an extension of the zone of probability along the South American coast during March to May (Hastenrath and Lamb, 1977).

Over the Indian Ocean, the pattern is unclear during most months, mainly because of the small number of observations over large areas, particularly in the southern hemisphere. A clear maximum develops between May and October in a large area south-west of Sumatra, where frequencies may reach 50% of all observations, and a smaller one along the coast of Bangladesh. During the period from January to April, there is frequent rainfall in all areas between the equator and approximately 15°S (Hastenrath and Lamb, 1979).

OCEANIC ISLANDS

It can be assumed that data obtained over the oceans are representative over large areas, because of the uniformity of the surface. However, this is not always true for data obtained at oceanic islands. They reflect the general climatic conditions, of course, and show seasonal variations, but local conditions may strongly affect diurnal differences, because even over the smallest islands small atmospheric circulations develop on most days, with sea breezes causing clouds and precipitation, especially where orographic lifting takes place (Chapter 10) (Plate 7).

A good example of the effect of aspect on monthly means is provided by Honolulu and Hilo in the Hawaiian Islands, which are only 330 km apart, but on different sides of the islands. Hilo, on the north-eastern coast of Hawaii, gets the full impact of the north-easterlies which prevail in this region throughout the year, while Honolulu is on the south-west of Oahu, on the leeward side of the mountains. The result is that Hilo gets much more rainfall, and is about 2°C cooler than Honolulu (Figure 11.30). Another example may be taken from the Galapagos Islands, where most stations at sea level report a mean annual rainfall between 300 and 400 mm, while Seymour, in a sheltered position, records only 62 mm per year (Nieuwolt, 1991). Still, island data can be used to illustrate the large differences which occur over the oceans when distances are measured in thousands of kilometres. Truk is in the western part of the tropical Pacific, while Canton Island is situated 4500 km away in the eastern, drier part (Figure 11.30). The heavy cloudiness in Truk seems to affect the mean temperatures also.

Although oceanic islands in the eastern parts of the ocean basins are characteristically dry (the archetypal desert island), humid tropical oceanic islands do experience drought on occasions (Nullet and Giambelluca, 1988). For the central South Pacific, droughts that have a 20% probability of occurrence in any one year (a five-year return period), can be expected to have a duration of 50 days in the southern parts of the region around the Cook, Tongan and Fijian islands. Shortest drought durations of around 14 days exist in the north-west of the region in the area of Tuvalu where rainfall activity in the SPCZ is pronounced. For this whole region the drought risk generally decreases along the SPCZ in a poleward direction. For high islands such as Western Samoa, drought duration is very much controlled by elevation and aspect. Mountainous interior areas and windward slopes experience short drought durations (Giambelluca, Nullet and Nullet, 1988). In the western Pacific, marked inter-annual variability of rainfall can also cause prolonged dry periods (Salinger et al., 1995) which are related to negative phases of the southern oscillation (ENSO events). For example, during the 1992–93 ENSO event, many parts of the western Pacific such as eastern Papua New Guinea, Vanuatu, the Solomon Islands, New Caledonia, Fiji and Tonga experienced drought, with only a fraction of the normal rainfall received for many months on end.

In the case of the Hawaiian Islands in the North Pacific, droughts can occur at the local to the entire regional scale. At the regional scale most droughts last for less than four months but may reach a maximum of six months. The regional drought return

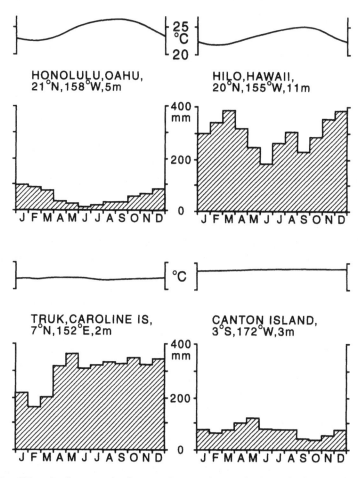

Figure 11.30   Climatic diagrams for four stations in the tropical Pacific Ocean

period is around 3.3 years. At the island scale the most drought-prone areas lie within or leeward of areas where rainfall maxima coincide with topographic peaks. At the local scale severe droughts may last between 19 and 32 months (Giambelluca, 1991). As outlined by Chu (1989), drought occurrence in the Hawaiian Islands appears to be related to contrasting phases of the southern oscillation.

SUMMARY

The tropics may be divided climatically into three geographical regions, namely tropical Asia, tropical Africa, tropical America and a fourth general environmental setting of the oceans and oceanic islands. While each of the three geographical regions, have their own climatic characteristics, the general rainfall regimes discussed in Chapter 10 can also be found in each of these regions.

In tropical Asia, the home to approximately 20% of the globe's population, the climates are controlled mainly by the monsoons which influence the entire region, the distribution of land and sea which makes marine influences important, and elevation. Because of the climatological dominance of the monsoons, distinct seasons with definite temperature and rainfall characteristics such as the Ganges type temperature regime can be recognized in this region. Geographically, equatorial and dry–wet monsoon climates dominate, with inter-annual variability of rainfall a distinct feature of the latter climatic setting as well as tropical cyclones which often wreak havoc throughout this region.

Tropical Africa contrasts in many ways with that of tropical Asia. Not only does it possess two distinct monsoon systems with their own characteristics in terms of climate, origin and structural properties but, for large parts of central and southern Africa, no monsoon exists at all. Generally, the West African monsoon is the most vigorous and wet, in contrast to the East African monsoon which is relatively dry. Because Africa has few mountain ranges and a large part of the interior consists of elevated plateau surfaces, continental highland climates are a distinct feature of Africa. Unlike tropical Asia, the seas surrounding the African continent are not universally warm as cold currents are an important climatological feature. Consequently, tropical cyclones are virtually non-existent and onshore climates in areas of cold ocean currents are dry.

The tropical Americas are made up of the Caribbean, central American and tropical South American regions. Climates of the Caribbean are dominated by the trade winds which advect moist unstable flows of air over the isolated islands making up this region. These flows are orographically forced by the high relative relief, leading to high rainfall amounts. Thermal regimes in this region are conservative with small seasonal contrasts; a product of the oceanic influence. Because of very warm summer sea surface temperatures, hurricanes are an important feature of Caribbean climate. Central America has clear Pacific coast and Atlantic coast climates because of the climate control imposed by the mountain chain that runs the length of this isthmus. Tropical South America geographically dominates the tropical Americas. Similar to the situation for Central America, a major mountain divide in the form of the Andes exerts a strong climate control in this region. To the west of the Andes, on the Pacific side, due to the competing geographical influences of the ITCZ in the north and the cold Humbolt–Peru current in the south, climates change from completely dry in the south to wet–dry between the equator and 2°S, to continuously wet in the north where the ITCZ can be found throughout the majority of the year. To the east of the high Andes mountain climates, the climates are largely controlled by seasonal movements of the ITCZ. Distinct seasonal changes are therefore a feature. Similar to the Pacific coast, areas including the Amazon Basin, the Guyanas and north-east coast of Brazil are continuously wet due to the ITCZ passing twice over these areas and the vast wet surfaces found here comprising rivers, swamps and forests from which large volumes of moisture evaporate into the atmosphere. To the north and south, climates with a dry season are found; however, variants to this general scheme are found along the Venezuelan coast (dry), in north-eastern Brazil (the Nordeste – semi-arid) and along the south-east coast of Brazil (wet). A feature of the climates to the east of the Andes is their sensitivity to ENSO events, the occurrence of which accounts for a considerable deal of the inter-annual climate variability experienced here.

Geographically, the oceans dominate the tropics. They are characterized by their spatial uniformity of temperatures, with departures from a clear zonal arrangement of thermal zones only where cold ocean currents exist. Small annual temperature ranges are also a feature. The climates of oceanic islands reflect the general climatic conditions of the islands; however, seasonal variations, aspect and local conditions may generate a range of topoclimates with some islands possessing distinct wet and dry sides. In the Pacific Basin the ENSO phenomenon exerts a very strong control on the inter-annual variability of oceanic island climates, especially those in the central Pacific and indeed the climate of the whole Pacific basin including the ocean climate.

CHAPTER 12

Tropical Climates and Agriculture

The great economic and social importance of tropical agriculture was described in Chapter 1. Further indications are the high proportion of agricultural workers, which reaches well over 75% of the economically active population, and the large part of total exports constituted by agricultural products in many tropical countries (*World Market Atlas*, 1992, p. 55).

Tropical agriculture is at least ten thousand years old, and until fairly recently, methods used were largely traditional. However, increased contacts and trade during the 20th century brought new crops and production methods to many parts of the tropics. A large part of agricultural production is now from plantations, large enterprises which generally use very progressive methods, based on local and international research. Peasant agriculture is also using new crops and improved production methods in some areas, but in the more remote parts of the tropics agriculture is still largely traditional.

As in all climates, tropical agriculture depends heavily on atmospheric conditions, which are probably the most important single factor in production. Farmers cannot change the climate, but they can adapt to it by methods like irrigation, shading, shelterbelts and mulching; and they can use climatic possibilities by selecting proper crops and careful timing of operations like seeding, fertilization, irrigation and harvesting. Yet climatic variability makes tropical agriculture often a risky business, and many famines in the tropics are related to climate (Chapter 11).

Climates affect agriculture in four different ways: by solar radiation, temperatures, precipitation and winds. Indirectly, climate influences agriculture by its effects on soil formation.

SOLAR RADIATION

All agricultural plants need solar radiation in the process of *photosynthesis*, in which plants use the visible portion of the sunlight to produce carbohydrates in the form of starch, sugar and cellulose. This process is essential to agriculture, which may be defined as the exploitation of solar energy. Obviously, photosynthesis takes place only during the day, and its rate is approximately proportional to the intensity of insolation, up to a certain maximum, the saturation light intensity. This maximum differs considerably for different plant species, as some crops prefer much sunlight, while others grow better in shady conditions. Photosynthesis generally takes place in the green parts of the plants, especially the leaves.

There are two main groups of crops regarding photosynthesis: the C3 group contains most tropical crops; the C4 group comprises maize, sugar cane, quinoa, some fodder crops and grasses. There is an intermediate group, called CAM, that consists of agave, sisal and pineapple. These groups differ in relation to the effects of temperature on the photosynthetic rate.

Not all the dry matter produced by photosynthesis is retained in the plants; part of it is used in the process of *respiration*. This is a continuous process, which takes part in many parts of the plants. It is a necessary activity, bringing nutrients from the roots to the various parts of the plant, and also cooling the leaves when they are exposed to the sun. It increases with environmental temperature, and is therefore highest during the day, but when nights are warm, it may continue at a relatively high rate, causing considerable losses of carbohydrates.

The difference between gains by photosynthesis and losses by respiration is the *net photosynthesis*: it indicates the rate of accumulation of dry matter in a crop. It is usually positive during the day but negative at night. Net photosynthesis indicates potential crop growth and is closely related to climate (Table 12.1).

Table 12.1 indicates that crop productivity in the tropics is lower than in the mid-latitudes, where the short and relatively cool nights of summer are a favourable factor. Solar radiation is almost never a limiting factor in tropical agriculture, but it explains why, for instance, rice yields in Japan and Louisiana, USA, are generally higher than in tropical south-east Asia. However, crop growing in the tropics is possible throughout the year and there is no seasonal limitation by solar radiation and related temperatures as in the mid- and high latitudes. Therefore, perennial crops and those with a long growing period (e.g. sugar cane) are the logical ones for tropical climates, and annual crops can be grown two or even three times per year, provided sufficient water is available.

Shading is widely practised in the tropics, because some crops (e.g. coffee, tobacco and cocoa) produce better with low light intensities. Reduced soil moisture losses,

Table 12.1 Potential net photosynthesis in various climates (g/m^2 day^{-1})

Type*	Humid tropics Af, Am	Wet/dry tropics Aw	Mediterranean type Cs	Mid-latitude humid Cf, Cw	High latitude Df, Dw
Year	24	25	26	24	18
Summer	24	27	32	31	33
8 months	24	26	30	29	26

* Koeppen classification.
Source: Chang (1970, p. 96).

suppression of weeds and preservation of the surface soil structure are secondary effects which explain the beneficial effects of low light intensity.

TEMPERATURE

Temperatures have a strong effect on crop growth because they affect the rate of photosynthesis. For each crop, certain temperature limits exist. For most tropical lowland crops, no growth takes place when the temperature is below 15–18 °C; fastest growth is when the air temperature is around 30–37 °C and growth stops again when temperatures reach 41–50 °C. There is a difference, however, between C3 and C4 species. C3 crops reach a maximum rate of photosynthesis around the middle of the optimum temperature range, and the rate rapidly decreases with both lower and higher temperature. C4 crops reach the maximum rate at higher temperatures, and the rate of photosynthesis remains high for quite a large interval. Therefore C4 crops have a productive advantage over C3 crops at temperatures above 25 °C. On the other hand, C4 plants cannot stand temperatures below about 10 °C.

Due to its spatial uniformity, temperature does not play an important role in crop selection and location in the tropical lowlands, which are generally at levels below 500 m a.s.l. Diurnal ranges are an advantageous factor in inland locations, because cool nights reduce respiration losses and increase net photosynthesis. Crops like sugar cane and pepper therefore produce higher yields away from the coasts. Cool nights may also act as a stimulus for the flowering of fruit trees, like papaya, durian and citrus. On the other hand, coconut, oil palm and cocoa prefer coastal areas. The same factor explains why, for instance in peninsular Malaysia, liberica coffee varieties prevail in coastal areas, while inland locations have more robusta coffee with its higher yields.

Unusually high temperatures can have deleterious effects on some crops. Cocoa, for instance, is rather sensitive to temperatures over about 30 °C, and is therefore often shaded. Root diseases, for example in papaya trees, are often caused by high soil temperatures. When high temperatures are accompanied by high air humidity, insect pests and diseases caused by fungi can become serious threats to many crops. Weeds and parasites also thrive under these conditions and these may cause losses to crops in the field, but also in storage and during transportation. The absence of cold nights and a cold season make eradication and control of pests a continuous process.

Elevation is the major factor causing differences of temperature with location (Chapter 4). The general decrease of about 0.65 °C per 100 m of elevation interrupts the general uniformity of temperatures and creates its own conditions in the tropical highlands. There is no exact boundary between lowland and highland agriculture, as some lowland crops are grown at higher levels than the traditional limit of about 500 m a.s.l. because of social or economic factors. Rubber, for instance, is an easy crop to cultivate, as it requires little care and gives good returns, even if conditions are not entirely favourable. It is therefore frequently produced well above its optimum area, which is at levels below 400 m, despite the fact that it takes the trees about six months longer to reach the productive stage for every 100 m of altitude (Wycherley, 1963). Similarly, oil palms take about one year longer to come into bearing when planted at levels over 300 m, but they may still produce sufficiently for low-cost small producers.

Table 12.2   Maximum levels of optimum conditions for tropical lowland crops

Crop	Maximum level (m a.s.l.)	Crop	Maximum level (m a.s.l.)
Bananas	900	Mango	1400
Cashew nuts	2000	Oil palm	300
Cassava (manioc)	1000	Papaya	900
Citrus	1800	Pepper	300
Cocoa	500	Pineapple	1700
Coconut	750	Rice	1000
Coffee (liberica)	600	Rubber	400
Coffee (robusta)	1000	Sisal	1700
Cotton	1400	Sugar cane	1600
Groundnuts	1400	Sweet potatoes	2100
Maize	2500	Sunflowers	2600

Sources: Acland (1971), Nieuwolt (1982, pp. 121–129).

Table 12.3   Best elevations for cultivation of tropical highland crops

Crop	Lowest elevation (m a.s.l.)	Highest elevation (m a.s.l.)
Apples	1800	2500
Barley	2100	3200
Coffee (arabica)	1200	1800
European vegetables*	800	2000
Potatoes (solanum)	1500	2800
Pyrethrum	1800	3500
Sorghum (highl. var.)	900	1500
Tea	1200	2000
Tobacco	900	1500
Wheat	1800	2900

* Beans, cabbage, cauliflower, lettuce, spinach, etc.
Sources: Acland (1971), Nieuwolt (1982).

Generally, lowland crops cultivated at higher levels grow slower and produce less, but this effect can be compensated by a better quality of the final product: citrus fruits produced at higher levels generally have a better taste and colour than those from the lowlands. The resistance to lower temperature conditions varies strongly between tropical lowland crops, and even between varieties of the same crop. Some general values are given in Table 12.2.

In tropical highland agriculture, temperature and therefore elevation is the main deciding factor in crop selection, because most highland crops have quite definite requirements (Table 12.3). Of course, the indicated ranges for best production are approximate, as local landforms may strongly affect temperature conditions in high-

lands (Chapter 4). Agro-technical methods, such as planting density, timing of operations, height and density of undergrowth, shading and mulching, may also influence
the altitude limits of highland crops. Finally, some cultivars have been developed for
specific temperature conditions.

Continuously lower temperatures in the highlands reduce respiration losses and
thereby the water needs of plants, in comparison to the lowlands. In most tropical
highlands, rainfall increases with elevation and therefore it is not an important factor
in crop selection; temperature is the main climatic consideration. The actual choice of
crops is often decided by economic factors such as transportation costs, market prices
and availability of labour. Lower temperatures generally mean slower growth. Wheat,
for instance, needs about two weeks longer to mature for every 300 m rise in elevation.

A good example of a crop that can be grown at many elevations is tea. Because the
leaves are the product, it shows wide tolerances of temperature: the optimum is around
18–22°C, but assimilation increases with temperature up to about 35°C. Tea can
therefore be grown in tropical lowlands, and it produces high yields, but the quality of
the product is poor. Quality increases with slower growth and therefore with elevation.
Many crops show this better quality of the final product in higher elevations; examples
are cotton, maize, pyrethrum, sugar cane and most European vegetables. In all these
crops, slower growth and lower production are more than compensated by higher
market prices for the final product.

ALTITUDINAL ZONES OF AGRICULTURE

In mountainous areas where settlement is continuous over various elevations, as in the
South American Andes or the New Guinean Highlands, zones of agricultural activities
can be related to altitude. It must be emphasized that the boundary levels between
these zones are only general indications of a gradual transition. They also vary locally,
especially in relation to the seasonal distribution of precipitation and local topography
(Troll, 1959). These limits move downwards with increasing distance from the equator,
until the lowest zone disappears completely at the outer latitudinal boundary of the
tropics. Near the equator the following zonal belts can be recognized:

1. *The lowlands*, up to about 500 m above sea level. In this zone, in Latin America
   called *"tierra caliente"* or *"warm zone"*, annual mean temperatures are around
   24–27°C and the normal tropical lowland crops are cultivated. Their choice and
   distribution are largely governed by rainfall conditions rather than temperature. As
   was indicated above, some lowland crops reach to elevations higher than 500 m, but
   in the upper levels of this zone some of the crops of the next zone, notably coffee, are
   cultivated where rainfall conditions are favourable.
2. *The temperate zone*, or *"tierra templada"* (Latin America), reaches from about 500 m
   to approximately 2000 m above sea level. Here the annual mean temperatures vary
   between about 16 and 24°C. Many different crops can be grown in this zone: some
   lowland crops such as sugar cane, rice and maize do very well, though their growth
   is slower than at lower levels. Cotton, bananas and pineapples are mainly limited to
   the lower parts of this zone. Some highland crops, such as tea and coffee, are widely

distributed, and mid-latitude crops can be grown quite successfully: beans, citrus fruits and vegetables in the wetter parts; grain crops, such as wheat and barley, in the drier areas.

3. *The cold zone,* or *"tierra fria"* (Latin America), lies between 2000 and 3000 m above sea level, where annual mean temperatures are between 12 and 18 °C. Here, typical highland crops prevail: where rainfall is adequate, tea and pyrethrum; in the drier areas wheat, barley and mid-latitude fruits. This zone is above the upper level where rainfall increases with elevation, and rainfall therefore is a critical factor in crop selection. Night frosts are also important; they are mainly controlled by local topography and may prevent the cultivation of tea.

4. *The paramo belt* is named after a low grass and bush formation which prevails at heights between 3000 and 4000 m in the Andes mountains. With annual mean temperatures between 6 and 12 °C, agriculture is limited to cattle grazing and some mid-latitude crops such as barley and potatoes, which can withstand low temperatures. However, large parts of this zone are not cultivated and where forests form the natural vegetation cover, they have largely remained uncleared. Most forms of agriculture are limited to the lower half of this belt.

5. *The frost zone* is found at elevations over about 4000 m, where snow and ice prevail and no agriculture takes place.

While a similar zonation could be observed in mid-latitude mountains, there are two important differences: in the mid-latitudes all agricultural activities are limited to the summer, and therefore summer conditions would be decisive. They would also cause seasonal movements, such as the transhumance of the Mediterranean regions, in search of better seasonal agricultural possibilities. The second difference is that the zonal belts outside the tropics would be much narrower. Though the lower zone of the tropical lowlands would, of course, be absent, the other belts are telescoped into a much smaller range of elevation, because the zone of snow and ice starts at a much lower level compared to tropical mountains.

PRECIPITATION

Rainfall is the main controlling factor in tropical lowland agriculture. The amount of rainfall that is normally received decides which types of agriculture can be used and which crops can be cultivated in a region; the seasonal distribution of rainfall regulates the agricultural calendar and the rainfall variability from year to year is the main factor responsible for fluctuations in yields and the related risks of production (Nieuwolt, 1986).

Water is an essential element in plant growth. Its role in photosynthesis has been mentioned, but it also acts as the solvent and transporting agent for plant nutrients, which are moved from the roots to all parts of the plant and it provides turgidity in stem and leaves. Water use in plants takes place in the process of *transpiration*, by which water, absorbed by the roots, is transformed into water vapour exhaled by the stomata of the leaves. This process is necessary for the transportation of nutrients and photosynthetic products to all parts of the plant, but also for the cooling of the leaves when these are exposed to the sun. It is clear that transpiration increases with temperature and exposure to sunlight.

Lack of water, or moisture stress, therefore reduces the growth of plants. Up to a certain maximum, agricultural yields generally show a strong positive correlation with the amount of water available for transpiration. The correlation varies with plant species, stage in the growth cycle and some environmental conditions, but remains valid for all crops. Rainfall is therefore the main limiting factor in tropical lowland agriculture, and even in the humid tropics irrigation is often used to reduce the effects of rainfall variability.

The relations between rainfall and crop water needs are often illustrated by the *water balance*, which can be based on actual rainfall data for days, weeks or months, or on monthly means (Thornthwaite, 1948). The general equation of the water balance is:

rainfall = evapotranspiration + differences in soil moisture + runoff + percolation into the subsoil

When actual rainfall data are used, the formula will indicate the approximate amount of water that might be needed for irrigation; when means are used, the general agricultural climatic possibilities are shown.

Evapotranspiration is the combined loss of water by evaporation from the soil surface and by transpiration from plants. Often the potential evapotranspiration (Eo) is used. This is the amount that would evaporate from an open water surface under prevailing conditions. Eo is obtained from evaporation pans, or it is estimated from meteorological data (Chapter 9). However, the actual water needs of a crop (Et) may differ from Eo because the transpiring surface, the total area of the stomata of the leaves, can be larger than an open water surface. This is expressed in the Et/Eo ratio (Table 12.4).

As Table 12.4 shows, the Et/Eo ratio varies with the type of crop and with the stage in its growth cycle. Some crops have water needs that exceed Eo in some parts of their

Table 12.4   Et/Eo ratios for some tropical crops and natural vegetation types

Crop	Months after planting						
	1	2	3	4	5	6	7
Groundnuts	0.45	0.80	0.90	0.90	0.90	—	—
Bananas	0.40	0.50	0.60	0.70	0.80	0.90	1.00
Sugar cane	0.30	0.50	0.70	0.90	1.00	—	—
Maize	0.55	1.10	1.20	1.20	1.20	—	—
Sorghum	0.80	1.20	1.20	1.20	—	—	—
Alfalfa	0.55	1.00	1.00	1.00/cut	0.60	1.00	—
Tea (after pruning)	0.00	0.50	0.85	—	—	—	—
Eucalyptus	0.40	0.55	0.70	0.80	0.95	1.05	1.30

Seasonal values		Annual values	
Grasses (dry season)	0.81	Forest (evergreen)	1.00
Grasses (wet season)	0.86	Forest (bamboo)	0.90
Coffee (dry season)	0.50	Forest (pine)	0.80–0.86
Coffee (wet season)	0.80	Grasses	0.54–0.85

development because several layers of leaf canopy are present and the stomata can transpire more than an open water surface. As Eo and Et are both controlled by the same meteorological factors, their ratio is largely independent of location and values obtained in one part of the tropics can generally be used in other tropical areas without serious error. Evapotranspiration values also differ little from year to year, as temperatures are usually uniform, so that means over relatively short periods can be used. Therefore the calculation of crop water requirements for water balance studies creates few practical difficulties.

Soil moisture acts as a buffer between rainfall and water use by plants. It can make water available to plants (soil moisture usage), and it can also store surplus rainfall. Estimates of the soil moisture holding capacity are based on soil type and depth of the root zone, but local conditions may differ widely from this general value (Thornthwaite and Mather, 1957). Direct measurement of this factor is difficult, and expensive and time-consuming methods have to be used (Chang, 1968, pp. 194–195). The error introduced by this factor in the water balance equation is important in climates with pronounced dry and wet seasons, where crops depend on soil moisture during the dry period.

Runoff and *percolation into the subsoil* are always losses. They are of great importance in the tropics, where rainfall is often of high intensity, so that it exceeds the maximum infiltration speed of the topsoil. A fair proportion of the rainfall is then not available to the vegetation. Underground losses can also be considerable when rainfall continues for longer periods and the topsoil is completely saturated with water.

When the water requirements of the vegetation cannot be met by the supply from rainfall and soil moisture, and when no irrigation water is applied, the actual evapotranspiration will be less than the potential one, and the difference between this value and the water needs of the plants is the crop water deficit. Such a water deficit usually means an interruption of growth and a loss of production, but a dry period can also have beneficial effects when it comes at the right time. Many tropical crops, such as rice, maize, tobacco, sugar cane, cotton, groundnuts and cocoa, need a period of moisture stress for ripening and the quality of the final product is improved when the last part of their growing cycle falls in a dry season. Yields may also increase as photosynthesis is particularly high during dry sunny days. Other crops, like cashew nuts, sunflowers and mangoes, need a dry period for fruiting, as it inhibits the outbreak of fungus diseases. For other tropical crops, like rubber and pyrethrum, a dry season simply means a rest period and production after the rest usually makes up for the temporary losses during the dry period.

CLIMATIC WATER BALANCES

When mean monthly values for rainfall and evapotranspiration are used, water balances illustrate the agricultural potential of the climate. The differences between the various tropical climates are caused mainly by the total rainfall and its seasonal distribution, as seasonal differences in evapotranspiration are generally smaller. Based on water balances, four agroclimatic types can be recognized in the tropics (Figure 12.1).

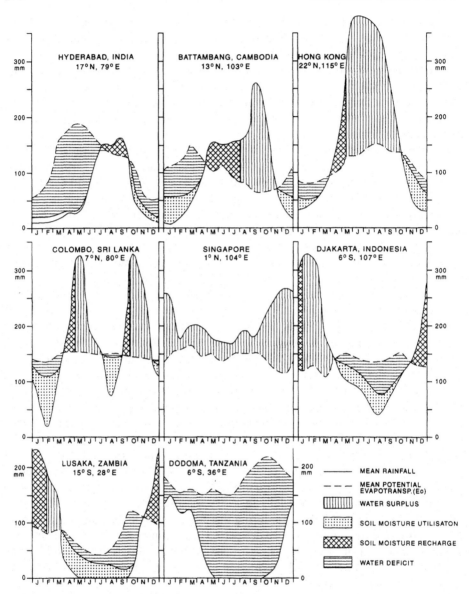

Figure 12.1 Typical water balances at selected stations in the tropics

The first is represented by Singapore, and it is typical for many areas close to the equator. Mean rainfall exceeds evapotranspiration throughout the year. Normally, there are no regular periods of water deficits and during occasional dry spells the vegetation is amply supplied by soil moisture. In these areas agriculture is not limited by rainfall conditions. However, actual rainfall may differ widely from the mean. As an illustration, the year 1963 in Malaya may be used, when water deficits of up to 8 months' duration and up to 20% of the mean annual rainfall were experienced at stations which normally had only very short droughts (Nieuwolt, 1966a). Therefore, irrigation is widely used for more sensitive crops like rice.

The second type of tropical water balance illustrates the other extreme, where the mean potential evapotranspiration exceeds rainfall throughout the year, as shown by the diagram of Dodoma, Tanzania (Figure 12.1). This category is typical for the dry tropics. Water surpluses do occur, but they are irregular and usually small. Agriculture is based on irrigation wherever possible, but some crops may be grown without it: maize and groundnuts, widely spaced to take maximum advantage of the rainfall; cashew nuts and other tree crops that can stand long dry periods. The main form of agriculture in these areas is pastoralism, based on cattle, goats or sheep, which are often moved over long distances in the search for good grazing opportunities.

The third type is shown by Colombo in Sri Lanka: two periods of water surpluses separated by short dry periods. This form of water balance is found in many monsoonal areas in South-East Asia, though the two rainy periods are usually of different intensity and duration, depending on local exposure to the rain-bringing winds. This is the water balance that prevails in areas of rice agriculture, where the rains are often supplemented by irrigation, and the dry periods are used for harvesting. In some areas, two crops per year are possible, particularly where irrigation is practised.

The fourth type is the one where one rainy period prevails and the rest of the year is quite dry. Hong Kong, Djakarta, Battambang, Lusaka and Hyderabad show this form of water balance, with decreasing water surpluses, and water deficits that grow in size and duration in the same order (Figure 12.1). Here the agriculture is completely controlled by rainfall; only one annual crop can be grown per year. In Hong Kong and Djakarta, conditions are suitable for rice, though additional irrigation is used to make the moisture supply more reliable, especially where a second annual crop is cultivated. In the other areas, only more drought-resistant crops such as maize, sorghum or peas may be grown without irrigation.

Where good and long rainfall records are available, the *Agricultural Rainfall Index* (ARI) provides a direct comparison between rainfall and crop water needs. The ARI expresses rainfall as a percentage of the potential evapotranspiration (Eo), and crop water needs are usually indicated in the same units. Seasonal and regional differences in water requirements, caused by factors like sunshine, temperature, humidity or elevation, are automatically reflected in Eo, so the ARI indicates the agricultural possibilities provided by rainfall. The ARI has the great advantage that it expresses these possibilities in one single figure. It can be improved by not basing it on mean rainfall data, which are often not representative of the amounts of rainfall that can be expected, but on probability data. However, long and reliable rainfall records are required for such data (Figure 12.2).

In Figure 12.2, rainfall amounts that can be expected with several probabilities (50–90%) are expressed in % of Eo. The water needs of maize are shown as an illustration: when planted in August the probability of sufficient rainfall is between 80 and 90%; when sown in January the chances of enough water are much lower. The method can be used to base the timing of agricultural operations on the amount of water available from rainfall and can help in crop selection. Where sufficient stations with long rainfall records are available, the spatial distribution of the ARI can be used to determine the most favourable location and season for annual crops (Figure 12.3).

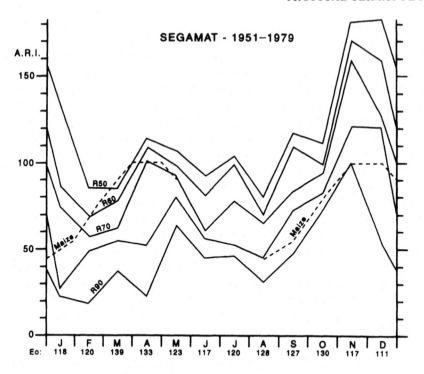

Figure 12.2   The agricultural rainfall index at Segamat, Malaysia, for different probability levels, based on rainfall records for over 30 years (full lines) and the water requirements of maize, when planted in January and August (broken lines). From Nieuwolt (1982). Reproduced by permission of MARDI, Serdang

As a risk of insufficient rainfall of about 20% is generally acceptable in most agricultural enterprises, the rainfall that may be expected with a probability of 80% is used here. The map indicates where the ARI remains below 40% of Eo, areas where normally insufficient rainfall is received for crop agriculture without irrigation.

A drawback of the ARI is that soil moisture storage is not considered. However, where a series of maps for consecutive months is available, agricultural possibilities can be estimated very well by use of the ARI (Nieuwolt, 1982, pp. 64–80). In addition to empirical approaches to agricultural resource assessment, other approaches are used. These include the use of computer-based crop–climate models which numerically model the climatic water balance and plant physiological processes in order to evaluate the best time for planting and harvest and also potential yields.

WINDS

High wind speeds can do extensive damage to tropical agriculture, particularly to tree crops. In the tropics, most destruction is done by tropical cyclones with their extremely high wind speeds. In the equatorial latitudes, less than about 5°, wind speeds are

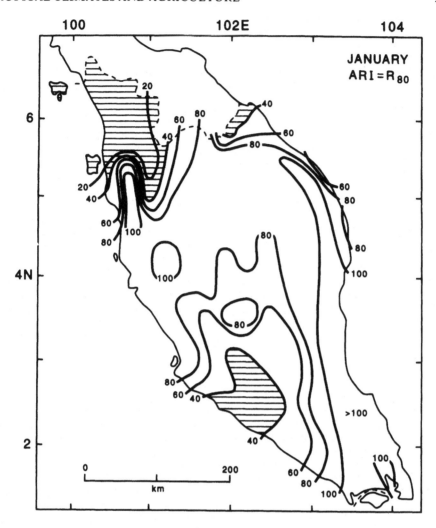

Figure 12.3   The agricultural rainfall index (ARI) over peninsular Malaysia during January for rainfall amounts that can be expected with a probability of 80%. Based on records for 41 stations over 30 years. Areas with an ARI below 40 are shaded. From Nieuwolt (1982). Reproduced by permission of MARDI, Serdang

usually low, but wind damage occurs during gusts where monsoon winds are strong or reinforced by sea breezes, or where local disturbances develop like the Sumatras of peninsular Malaysia. The cold surges associated with bursts of the East Asian winter monsoon are particularly effective at causing wind as well as chilling damage. Most wind damage is done to rubber and coconut trees and to bananas. Shelterbelts of more resistant trees may help, and for rubber the selection of cultivars that are less sensitive to wind damage has been successful.

TROPICAL SOILS

The same climatic factors that affect agriculture directly, also influence it indirectly through their role in soil formation. Where other soil-forming factors such as parent material and topography have remained fairly constant over long periods, as is the case in many parts of the tropics, a state of equilibrium between climate, vegetation and soil is established. This balance is reached particularly quickly in the tropics, because tropical climates are active in soil formation throughout the year. Soils which exhibit this type of equilibrium are called zonal soils and they show the dominating effect of climate, the most important elements being moisture and temperature conditions.

Zonal soils in the tropics are generally rather deep, because the combination of high soil temperatures and high moisture contents, which prevails frequently in many parts of the tropics, favours an accelerated chemical disintegration of parent rock material. However, the same factors limit the amount of organic material in tropical soils, because bacterial activity is intensified by high soil temperatures, causing a rapid decay of all organic matter and little formation of humus. In this respect a soil temperature of about 25 °C seems to be critical: at lower temperatures there is a general accumulation of humus in the topsoil, but at temperatures above this level the decay of organic matter proceeds more rapidly than its replacement by most types of vegetation. Therefore most tropical soils are poor in humus content and their nutrient-holding capacity is limited.

Where rainfall is plentiful, with annual means over 1500 mm, *leaching* and *eluviation* remove most nutrients from the topsoil. This effect is particularly serious where a large proportion of the rainfall comes in the form of intensive downpours, as is the case in most of the tropics (Chapter 10). In these areas, zonal soils often have bright red colours, caused by the accumulation of insoluble iron oxides and aluminium oxides. Some of these *latosols* or *laterites* are so rich in iron or aluminium that they can be mined for bauxite (aluminium ore), as in Malaysia, Guyana and Surinam.

Intensive rainfall also causes soil erosion. When the rainfall intensity exceeds the infiltration speed of the topsoil, surface runoff results and the water may not only erode the soil, but also wash away crops, particularly young crops with their relatively short root systems. Evaluations of the spatial distribution of *rainfall erosivity*, the ability of rainfall to cause erosion through its intensity, can be particularly helpful in establishing where soil erosion problems may occur (Morgan, 1974; Morgan, Hatch and Sulaiman, 1982).

Where dry and wet seasons alternate, evaporation of groundwater during dry periods may counterbalance the leaching during the rainy season to some extent. However, this process often produces a deposit of chemicals near the surface in the form of a crust, which can be hard and impenetrable. Here, soil erosion may be very serious, because in the beginning of the rainy season the vegetation cover may still be scanty providing little protection to the soil. Still, soils in the dry and wet tropics are generally more fertile than in the continuously rainy parts of the tropics.

The only areas in the tropics where deep and fertile soils are found are where either the parent material or the topography produces many nutrients. The first is over volcanic deposits, which weather and disintegrate rapidly under tropical conditions.

Favourable topography is found in many river floodplains, where frequent floods leave deposits of fertile material.

The disadvantages of most tropical soils are aggravated when the natural vegetation cover is destroyed. Under the natural equilibrium of soil, vegetation and climate, a rapid turnover of a relatively small quantity of nutrients takes place in an almost closed cycle between the top layers of the soil and the vegetation. Losses are small, as the soil is well protected against erosion by the vegetation cover, except on steep slopes. But when the vegetation cover is destroyed, all nutrients in the plants and trees are lost, temperature and moisture conditions of the soil are drastically changed, and its protection by the vegetation cover is lost. Soil surface temperatures, normally kept relatively low by the shade of the vegetation and the insulating natural mulch layer, can reach very high values in direct sunlight, resulting in a rapid decay of organic matter in the top layers of the soil. The direct impact of raindrops can start serious soil erosion and leaching, particularly of nitrogen, calcium, magnesium and potassium. In clearings, a few rainstorms can destroy a soil that took thousands of years to form (Plate 8).

These processes can be prevented by selective clearing of the natural vegetation, making sure that at all times a vegetation cover is present to protect the soil against direct sunshine and rainfall. The practice of keeping a good undergrowth under many tree crops, such as rubber, oil palm and cashew nuts, is related to this danger of rapid destruction of the soil when it is bare to the elements of a tropical climate.

AGRICULTURAL SYSTEMS

In order to minimize the effects of climatic disadvantages and to obtain maximum benefits of the opportunities offered by tropical climates, people living in the tropics have adjusted their agricultural methods and techniques over thousands of years. Popular beliefs and customs often reflect an unscientific but usually correct interpretation of climate in relation to plant growth. This has led to various systems of agricultural organization. There are many specialized agricultural systems in the tropics, but most of these can be classified into one of the three main categories.

The oldest of the agricultural systems is *shifting cultivation*, which even today is used in its various forms by millions of people, generally in remote parts of the tropical forests (Manshard, 1968). The common feature of all types of shifting cultivation is the frequent movement of the agricultural activities to different plots. To protect the natural fertility of the soil, an area of land is cleared (with large trees often left standing), then cultivated by a mixture of crops for a few years and then deserted, allowing the natural vegetation to return (Plate 10). After some years the process is repeated. Adaptation to climatic conditions is largely by the timing of operations: sowing is mainly done during wet periods; harvesting and clearing of new areas during drier seasons.

The duration of the cycle depends largely on the pressure on the land: where the population density is low, the system works well as the soil is given enough time to restore its fertility. In this condition most crops are grown for subsistence only, the food supply is improved by hunting and fishing, and contacts with the outside world are very limited. Where the population increases, the fallow periods are often reduced and this may cause a deterioration of the soil. Some more advanced systems, which have more contacts with the outside world, use fertilization and here a transition to a more permanent use of the land is in progress. Persistent shifting or swidden agriculture in some parts of the tropics can lead to an alteration of the land cover. This can result in a number of climatic effects (Chapter 13).

A second system is *peasant agriculture*, in its many different forms. It is the system used by the large majority of tropical farmers, in many different tropical climates. In this system, the emphasis is usually on subsistence crops, but some cash crops are often added. Frequently the farmers try to obtain optimal results by growing a combination of many different crops or by adding animal husbandry to their activities, thereby reducing the effects of crop failure in any part of their enterprises. The introduction of new crops or new varieties which often have been developed in entirely different parts of the tropics, have recently improved yields in such basic food crops as rice and maize.

Because this system is used in many parts of the tropics, adaptations to climatic conditions can only be given in very general terms. In many parts of the tropics, rainfall is the main limiting factor in agriculture. In areas where rainfall is usually sufficient for crop agriculture, its irregular occurrence and seasonal distribution may make it necessary to use some form of irrigation to improve the water supply to the fields. In areas where irrigation is not possible, wide spacing of plants or trees will reduce water requirements, and mulching can reduce evaporation from the soil. Where too much rainfall is received, good drainage facilities are essential; the leaching of soils can be remedied by the frequent use of fertilizers. Excessive use of these, however, may lead to a deterioration of water quality and eutrophication of nearby standing water bodies such as lakes. Soil erosion can be reduced by terracing, contour ploughing, and the protection of the undergrowth cover of vegetation. In many parts of the tropics, both droughts and floods can be expected, though they usually do not come during the same year. Efficient protection against both risks is often too costly or technically impossible.

The effects of high temperatures and high humidity in the form of insect and other pests and diseases can be controlled by modern chemicals. However, these are often too expensive for the small farmer. A good alternative is the use of disease-resistant crop varieties. Tropical diseases in animals can also be controlled by modern chemical and biological methods, yet many regions have to be avoided, such as those infected by the tsetse fly.

In some tropical areas net photosynthesis rates are low. The farmer cannot do anything about it, except allow his crops more time to grow. Where sunshine is too abundant for some crops, shading is an efficient and widely used protection.

In most forms of peasant agriculture, the main crop is for subsistence. It is usually a grain crop, maize, sorghum, millet or rice, but cowpeas or cassava also occur. Surpluses may be sold, and there are usually a few cash crops around the farm, like rubber, fruit trees, vegetables or tobacco. Animal husbandry is often added to crop agriculture, and becomes increasingly important in the drier parts of the tropics.

A special form of peasant agriculture is rice cultivation, as practised in South and East Asia. Similar systems have been introduced by Asians in Africa and South America, and it is also used in the mid-latitudes where the summers are sufficiently warm and humid. In this system, rice is first grown in small seedbeds and then transplanted to flooded fields (paddies, sawahs), which are only dried towards the end of the growing cycle. Therefore the supply of water is controlled during the growing period, the soil is well protected against erosion and leaching, weeds are easily kept under control and when the source of water is a natural stream, nutrients are supplied with it. In other cases fertilization is necessary.

This system supports over 400 million people in Asia. It is very labour-intensive, and allows dense populations. The risks are small, the main one being flooding, which can destroy the small dams around the fields, which are often terraced. The recent introduction of new rice varieties has increased yields, in some areas dramatically. Where this system prevails, rice is both subsistence and cash crop.

A third form of tropical agriculture are *plantations*. These are large enterprises of commercial agriculture, introduced in many cases by the colonial powers in their quest for tropical products. Many plantations have been nationalized by the governments of the newly independent tropical states, but in some cases the old firms were allowed to continue production. Plantations concentrate on one or two crops only and usually prepare the products for transportation and storage. Crops such as tea, rubber, oil palm, sugar cane, coffee and sisal are largely produced on plantations. The commercial attitude of the management of plantations have generated scientific interest in the problems of tropical agriculture and many research centres are maintained by contributions from plantations. The results have often helped to reduce losses and to increase yields and the quality of the final product. The research also benefits the small farmers who grow the same cash crops. Generally, plantations make optimum use of climatic possibilities by using the best techniques and planting methods and by introducing new varieties of crops adapted to the climate and soil conditions.

As emphasized in this chapter, agricultural activity and systems are very climate-sensitive (Kuhnel, 1996). Agricultural yields can vary from year to year and because of climatic variability, there may be "feast or famine". In relation to this, human-induced climate change is highly relevant, as this may result in a change in the nature, availability and distribution of the climatic resources on which tropical agriculture systems are based. Depending on the nature of the climatic change, some regions may expect increases in agricultural production while others may expect an increase in the frequency and magnitude of climatic disadvantages. This and other potential climate change related impacts on natural and human tropical systems will be discussed in the next chapter.

SUMMARY

In many ways climate sets the environmental limits for agricultural production and determines what specific type of agricultural activity may be undertaken in a region. The climatic elements of importance are solar radiation, temperature, precipitation and wind. The first two of these are especially important for the process of photosynthesis, solar radiation providing the energy to drive this process while temperature determines the rate. Precipitation is the main controlling factor in tropical lowland agriculture, especially its amount and seasonal distribution. In highland environments, elevation, through its effects on temperature, is the main control. The agricultural potential of an area is often assessed by considering the water balance, i.e. the relationship between rainfall, crop water needs, evapotranspiration, soil water storage and runoff. Calculation of water balances can help determine when soil moisture deficits are likely to occur and thus the time when irrigation is necessary. Climate also has an indirect influence on agriculture via its control on soil processes, especially leaching and salinization.

CHAPTER 13

Climate Change

TYPES OF CLIMATE CHANGE

All climates vary on many different time scales. *Long-term* climate change occurs at time scales greater than 20 000 years. The Quaternary ice ages which lasted for up to 100 000 years, and some older ice ages, which may have persisted for several millions of years, belong to this category of climate change. *Short-term* changes are generally classified as changes from 100 to 20 000 years (Goodess, Palutikof and Davies, 1992). *Climatic variability*, on the other hand, is used to refer to changes in climate over time scales of less than 100 years. Variations which occur in irregular cycles of a few years, such as those caused by the ENSO phenomena (Chapter 6), belong to this category. Changes, such as interdiurnal weather differences, seasonal and inter-annual variations are also considered to be part of climatic variability. Climatic variability is the main reason why climates are usually described in long-term averages, obtained over periods of 30 years or more.

The time scales of climate change are related, in very general terms, to different climate change mechanisms. Long-term climate change is a product of changes *external* to the climate system, mostly to do with earth–sun relationships, but *internal* causes may also operate. Short-term changes are partly related to external causes but internal mechanisms such as changes in ocean circulation and volcanic activity become important. Climatic variability is related almost entirely to internal mechanisms although variations of solar activity may also be important. A further characteristic of climatic changes is that they can come in *cycles*, but some show only a one-time trend. Since several may occur simultaneously over different time scales and with unequal intensities, the identification of past climatic changes is not always easy.

The main intention of this chapter is not to give an account of past tropical climates as good accounts of this can be found elsewhere (Hastenrath, 1985). Instead, we will introduce briefly the major causes of climate change at a variety of time scales and then focus on the possible importance of humans for bringing about climate change. With regard to this, the role played by increased greenhouse gas emissions, deforestation and desertification in forcing climate changes at the local to global scale will be considered. The chapter will be rounded off with a broad overview of the possible implications that human-induced climate change may hold for the low latitudes.

CAUSES OF CLIMATE CHANGE

The causes of climatic changes can be classified into three categories: external, internal and human. External or extra-terrestrial causes include changes in the nature of the earth's orbit around the sun (its orbital geometry), often referred to as Milankovitch variations, and variations in solar output. Internal causes relate to changes in the nature and behaviour of the major components of the climate system, namely the atmosphere, hydrosphere, biosphere, land surface and cryosphere. Human causes include global atmospheric pollution, and possibly deforestation and desertification.

EXTERNAL CAUSES

Changes in the orbit of the earth around the sun occur quite regularly: the earth's eccentricity varies in a cycle of approximately 95 800 years (Goodess, Palutikof and Davies, 1992). As a result of these cyclical variations, the irradiance varies. Over the last 5 million years changes from a more circular to a more eccentric orbit have been matched with changes in irradiance between + 0.014 and − 0.17% of current levels respectively (Henderson-Sellers and McGuffie, 1987).

Obliquity, the tilt of the earth's axis with the plane of its orbit, varies on a time scale with an average periodicity of 41 000 years. Currently the earth's axis is tilted at about 23.5°, but this has fluctuated between 22° and 24.5°. When the axis is at its maximum (minimum) tilt, the poles receive more (less) solar radiation in the summer (winter). Consequently, obliquity is an important control on seasonality. It has its maximum impact on high latitudes but the low latitudes are little affected as the strength of the effect decreases towards the equator. Obliquity does not alter the amount of radiation received, only its distribution.

Because of the gravitational pull of the sun and the moon on the equatorial bulge of the earth, the earth's axis of rotation wobbles. Such "wobbling", in combination with the elliptical nature of the earth's orbit, causes the timing of the solstices and equinoxes to change in relation to the time when the earth is at its maximum and minimum distances from the sun. This has a periodicity of around 22 000 years and is termed precession of the equinoxes. Its main effects are on the intensity of the seasons with greatest impacts at low latitudes. If the shortest sun–earth distance (termed the perihelion) occurs in mid-June then northern hemisphere summer solar radiation will increase (Goodess, Palutikof and Davies, 1992).

As the orbital characteristics described above are all varying at the same time, combinations of eccentricity, obliquity and precession can occur which are conducive to global cooling or warming. Concerning cooling and the onset of glaciation, a combination of minimum obliquity, high eccentricity and the maximum sun–earth distance (aphelion) occurring during the northern hemisphere summer is the best. This is because minimum obliquity means that there is less radiation for the high latitudes; high eccentricity produces maximum differences in solar radiation receipt between the aphelion and perihelion; and the aphelion occurring in the northern hemisphere summer produces cool summers. All these lead to low summer radiation amounts and thus cool summers, which may help sustain winter snow and ice at middle to high latitudes throughout the summer. Consequently, as global albedos increase (Chapter 3), snow and ice fields will grow through positive feedback.

Solar variability, as a cause of short-term climate change and climatic variability, has not been accepted unanimously by the climatological community. Although variations in solar output do occur, as manifest by variations in sunspot number, the evidence that such a deterministic link between sunspot activity (magnetic storms on the sun's surface) and global climate remains circumstantial. Nevertheless, many attempts have been made using sophisticated statistical techniques to relate tropical climatic variability to sunspot numbers (Currie, 1996). However, the possibility exists that any relationship between sunspot activity and climate is pure statistical coincidence. Furthermore, it seems possible that any sunspot climate link may be masked by interactions of the sunspot cycle with the QBO (Chapter 6) (Labitzke and Van Loon, 1990). Although uncertainties exist about the role of sunspots in climate change and variability, the fact remains that sunspots are highly variable in number from year to year; for example, from a low of 45 in 14 years to 190 in one year. For extended periods, sunspot activity has been extremely low and at times non-existent. Such periods would be matched with a lower value of the solar constant (Chapter 3); based on satellite observations, the solar constant decreased by about 0.1% from a high sunspot activity peak in 1981 to a minimum in 1986 (Goodess, Palutikof and Davies, 1992). Two low activity periods, called the Sporer (1450–1534) and the Maunder (1645–1715) minimums have been suggested as possibly related to the occurrence of the Little Ice Age which began sometime during the 13th–14th centuries and ended between the mid-16th and mid-19th centuries.

INTERNAL FACTORS

Internal factors may include relatively slow but long-lasting processes related to changes in oceanic circulation, atmospheric gaseous content and changes in geographic boundary conditions (Goodess, Palutikof and Davies, 1992). By their very nature, these factors may change climates on very long time scales of tens of thousands to millions of years.

Internal factors are related to non-linear feedbacks or interactions between the components of the climate system such that changes in one of these components affect in turn the other components. Such *feedbacks* may be positive and therefore increase the effect of the initial change, or they may be negative and produce a dampening of the initial change.

Changes in atmospheric gaseous composition, especially carbon dioxide (CO_2) concentrations, relate to the efficacy of the natural greenhouse effect as a "thermal blanket" (Chapter 3). Throughout geological time the concentration of atmospheric CO_2 has varied markedly. Generally, during cool (warm) periods it is low (high) in concentration. However, some debate exists as to whether on geological time scales CO_2 is a driving or feedback mechanism for climate change. General consensus seems to favour CO_2 as a feedback mechanism as CO_2 changes have been shown to follow changes in insolation and temperature (Genthon et al., 1987). Although these links seem clear, the nature of the mechanism which controls the change in CO_2 levels is less clear.

A possibility is oceanic productivity, which may be linked to changes in the intensity of equatorial upwelling and vertical mixing in the subtropical oceans (Goodess, Palutikof and Davies, 1992). Generally, ocean productivity increases during cool periods as ocean circulation strength rises along with the intensity of upwelling and mixing which brings nutrients valuable to the marine food chain to the surface. As marine phytoplankton require CO_2 for the process of photosynthesis, during periods of high marine productivity atmospheric CO_2 levels fall, which leads to greater losses of outgoing longwave radiation and thus cooling.

Atmospheric methane (CH_4), another greenhouse gas, has also varied in its concentration over geological time. On occasions, from glacial to interglacial periods, CH_4 has almost doubled. This reflects changes in the total area of wetlands, the major source of natural CH_4 emissions. Wetland area, especially in the low latitudes, is possibly related to long-term changes in the strength and humidity of the monsoons (Goodess, Palutikof and Davies, 1992).

A further atmospheric component possibly involved in climate change is atmospheric aerosols, of both marine and terrestrial origin, which have a positive feedback effect on temperatures; increases in aerosol loading lead to further falls in temperature. The main marine source is planktonic algae (Restelli and Angeletti, 1993), which produce dimethyl sulphide. On contact with air this oxidizes to an aerosol known as non-sea salt. High concentrations of non-sea salt are associated with increases in ocean productivity at times of cool global conditions. Non-sea salt is important as it acts as cloud condensation nuclei, enhancing marine cloud formation, which in turn may lead to cooler ocean surface temperatures. Important terrestrial aerosol sources are the extensive drylands, especially the great subtropical deserts, during glacial periods. Strong winds blowing over these deserts could have mobilized dust and transported it into the atmosphere, thus interfering with the transmission of radiation to the earth's surface.

The atmospheric factors of carbon dioxide, methane, non-sea salt and terrestrial dust all act at the same time and can have a cumulative effect on decreasing global temperatures in the manner of a positive feedback mechanism. Of these, the first three factors appear to be related to changes in ocean circulation and chemistry through oceanic productivity. The question remains though as to what starts these feedback mechanisms off. From studies of geological time series of CO_2, CH_4 and non-sea salt concentrations taken from ice cores, it appears that the main factor is changes in the orbital geometry of the earth; variations in the concentration of these gases and aerosols have similar periodicities to those of the orbital geometry parameters (Goodess, Palutikof and Davies, 1992).

Oceanic circulation is also considered to play a major role in climate change. The critical region in this regard appears to be the North Atlantic where North Atlantic Deep Water (NADW) is formed due to physical differences between the waters of the North Atlantic and North Pacific oceans. As the North Atlantic is warmer than the North Pacific, evaporation and therefore salinity is greater there. This, in addition to the fact that there is net water vapour transport from the Atlantic across the Panama peninsula to the Pacific, means salinity levels are further enhanced in the Atlantic. A consequence of these salinity differences is the development of a *thermohaline* circulation, one of its main components being NADW. This is formed in the North Atlantic as water becomes cooler and denser as it moves north in the Gulf Stream current. Because of these physical changes the water sinks. This sinking water is the NADW which feeds a large deep ocean circulation. It has been suggested that the formation of NADW may falter in times of global cooling and glaciation, leading to a reversal of the major ocean circulation patterns, which in the North Atlantic would lead to the enhancement of ice sheet growth and further cooling (Broecker and Denton, 1990). In such a case, oceanic circulation operates as a positive feedback mechanism. Oceanic circulation is also responsible for short-term climate change and climatic variability as this can vary on time scales of centuries (Lamb, 1979) or years, as in the case of the El Nino phenomenon (Chapter 6).

Geographic boundary changes occur as a result of tectonic plate movements, polar wanderings, mountain building and associated changes in the distribution of land and sea and general sea level. These are all believed to play a role in climate change (Frakes, 1979). Changes in the distribution of land masses can alter surface patterns of seasonal heating and cooling, which in turn will have impacts on the geographical position of the major wind belts. Mountain building can divert major airflows or generate large columns of ascending air in summer and descending air in winter (Goodess, Palutikof and Davies, 1992). Large, high snow- and ice-covered plateau surfaces in subtropical latitudes are also important because they strongly affect the albedo of the earth's surface (Ruddiman and Kutzbach, 1991).

Volcanic activity was an important factor in the evolution of the earth's atmosphere mainly through its role in increasing CO_2 concentrations which laid the basis for the development of the earth's greenhouse effect. More recently though, volcanic eruptions have been important for short-term climate change and climatic variability, through altering the global radiation balance by injection of gases and particles into the atmosphere. The climatic effect of volcanic eruptions depends on the explosivity of the volcano and the height to which the eruptive products reach. Explosivity is dependent on the rock type making up the volcano, silicate volcanoes being the most explosive. The most effective volcanoes in terms of climate change are those that eject gas and ash into the stratosphere. Here sulphur dioxide gas quickly oxidizes to form sulphate aerosols which, in combination with volcanic ash, can have a marked effect on the shortwave radiation transmissivity of the earth's atmosphere. In this regard volcanoes such as Tambora (1815), Krakatoa (1883) and Mount Pelée (1902), and more recently Mount Agung (1963), El Chichón (1982) and Mount Pinatubo (1991), have been important (Figure 13.1). The net effect of increases in planetary albedo and a reduction of atmospheric transmissivity are reductions in surface temperature (Rapino and Self, 1982; Sear et al., 1987). For example, for the five largest eruptions over the last 100 years, drops in average global temperature from 0.1 to 0.4 °C have occurred. The average cooling associated with these eruptions was 0.2 °C, with maximum effects over a two-year period starting early in the year after the eruption (Parker et al., 1996) (Figure 13.2). The climatic effects of volcanic eruptions also vary geographically; the most marked negative temperature departures have occurred over North America and the western Pacific, with widespread cooling over the equatorial zone (Kelly, Jia and

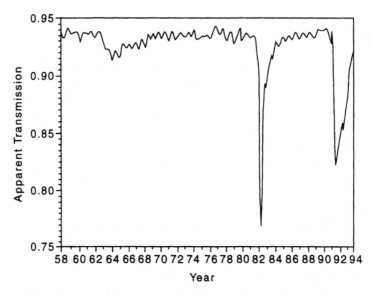

Figure 13.1   The transmission ratio of the atmosphere 1958–94. Measurements are for Mauna Loa and for clear-sky mornings. The plotted line is for smoothed monthly averages. From WMO (1995, p. 28)

Jones, 1996). The climatic effect of volcanic eruptions also depends on a number of other factors such as the phase of the QBO and ENSO (Chapter 6), the latter resulting in no marked volcanic signal in eastern Pacific temperatures (Kelly, Jia and Jones, 1996; Parker et al., 1996).

Clearly, volcanic activity has a noticeable effect on global temperatures and is an important causal factor in climatic variability. Such climatic effects from a prolonged period of greater than current volcanicity between 1250–1500 and 1550–1700, in conjunction with the climatic effects of the Sporer and Maunder minimums (reduced sunspot numbers), have been suggested as the possible cause of the Little Ice Age. Furthermore, it has been suggested that volcanic eruptions may have a moderating effect on the global warming attributed to a human-induced enhancement of the greenhouse effect. This, however, does not seem likely unless over the next century there is a marked increase in volcanic activity (Kelly, Jia and Jones, 1996).

HUMAN ACTIVITIES AND CLIMATE CHANGE

As was seen in Chapter 4, humans have the potential to inadvertently modify climates at the urban scale. Through burning fossil fuels and increasing the atmospheric concentrations of the natural greenhouse gases (GHGs) and the processes of deforestation and desertification, humans also have the capability to modify climates at the global to regional scale. Humans therefore constitute a very important component of the climate system; they can alter the nature of the other climate system components and in turn be affected by these changes.

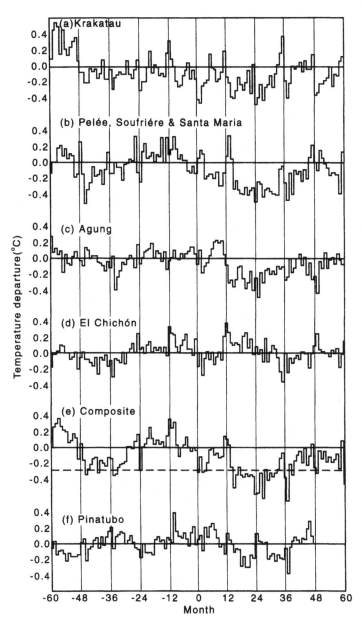

Figure 13.2   Monthly global average air temperature 12 months before and after five major volcanic eruptions. (a) Krakatoa (1883); (b) Pelée, Soufriére and Santa Maria (1902); (c) Agung (1963); (d) El Chichón; (e) average (composite) of (a)–(d); and (f) Mount Pinatubo. Temperatures are departures from the monthly averages for the 5 years preceding month zero which is the January of the eruption year. From Parker et al. (1996). Reproduced by permission of the *International Journal of Climatology*, Royal Meteorological Society

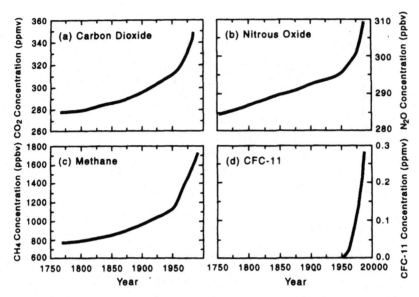

Figure 13.3   Trends in main greenhouse gases. From Glantz and Krenz (1992). Reproduced by permission of Cambridge University Press, Cambridge

GREENHOUSE GASES

Although the natural variability of climate makes it difficult to say exactly how much human activities have contributed to climate change, studies of the variations in natural GHG concentrations over the geological time scale strongly suggest that human-induced increases in GHG levels have the potential to enhance the efficacy of the greenhouse effect. There is ample evidence that concentrations of the natural GHGs (CO_2, CH_4 and N_2O) have increased markedly over the last century (Figure 13.3). Increases have come about due to the burning of fossil fuels, the expansion of wetland agriculture, the increased usage of artificial fertilizers in agriculture, and the "invention" of a group of non-natural GHGs termed halocarbons. Besides these, other activities have also contributed to increased GHG concentrations. Apart from the individual activities themselves, the main reason why GHGs continue to increase is because their rate of emission, from human-related activities, far exceeds their rate of consumption by the environment; the balance between the various sources and sinks has been altered (Table 13.1).

GHG emissions vary geographically. Generally, the largest emissions are from the industrialized nations, with little contribution from tropical countries. For example, taking 1986 energy consumption as a measure of emission potential, 50% of the globe's energy usage occurred in Canada, the United States and the former USSR. Moreover, these countries, at that time, had only around 12% of the world's population. In contrast, for tropical regions such as India, Africa and Asia, the proportion of global energy usage in 1986 was 0.3%, 0.5% and 0.9% respectively (Glantz and Krenz, 1992).

Year to year variations in the annual growth rates of the GHGs are related to the

Table 13.1   Sources and sinks of main greenhouse gases and aerosols (McMichael et al., 1996, p. 25)

Gas	Source	Sink
CO_2	Fossil fuels, deforestation, biomass burning, cement production	Ocean and land biosphere
CH_4	Rice paddies, natural wetlands, ruminant livestock, biomass burning, fossil fuels, termites, animal and domestic waste	Reaction with hydroxyl radicals in the atmosphere
N_2O	Biological sources in soils and water, fertilization, biomass burning, industry	Photolytic destruction in stratosphere
Halocarbons (CFCs)	Industrial sources: propellants, refrigerants, solvents, fire retardants, foam-blowing agents	Photolytic destruction in stratosphere
H_2O	Evaporation (ocean), contrails (air traffic), combustion, cooling towers	Cloud droplets, precipitation
Aerosols	Fossil fuel and biomass burning, soot, volcanic activity, soil dust, sea salt, plants	Washed out by precipitation

source–sink relationship. Growth rates may be increased by more fossil fuel usage and increased rates of deforestation. They may be decreased by either legislation at the national or global scale, or naturally by greater uptake by plants, soils and the oceans. For example, the fall of annual growth rates from an average of 1.5 parts per million (ppm) in the 1980s to a recent low of 0.8 ppm in 1992 and 1993 is possibly attributable to greater uptake of CO_2 in the oceans due to cooler global temperatures as a result of the Mount Pinatubo eruption (Conway et al., 1994). This is possible because during this period there was not a decrease in fossil fuel usage.

Although CO_2 is the dominant GHG by volume, the other GHGs have longer lifetimes in the atmosphere and higher warming capabilities so that they are just as effective as CO_2 in enhancing the greenhouse effect. Of particular note is the role which the halocarbons play. Of these human-made gases, the chlorofluorocarbons have received most attention in the literature. This is because not only are they a GHG, but they are destroyers of atmospheric ozone, one of the gases that makes life on earth possible through its absorption of ultraviolet radiation. Water vapour is another important GHG but it is often not treated as such when considering global warming. This is because on a global scale it is not significantly affected by anthropogenic sources and sinks. However, global atmospheric moisture levels may increase in the future due to increased evaporation rates, a product of predicted greenhouse-related increases in global temperatures.

The fact that global mean temperatures (surface land and sea) have increased by 0.3 to 0.6 °C over the same period as atmospheric concentrations of GHGs (Figure 13.4) is convincing evidence that a link appears to exist between global atmospheric pollution and an enhanced greenhouse effect. Over the last forty years the temperature increase has been very marked: 0.2–0.3 °C (Nicholls et al., 1996). Furthermore, with the exception of 1944, the nine warmest years since 1861 have occurred between 1983 and 1996,

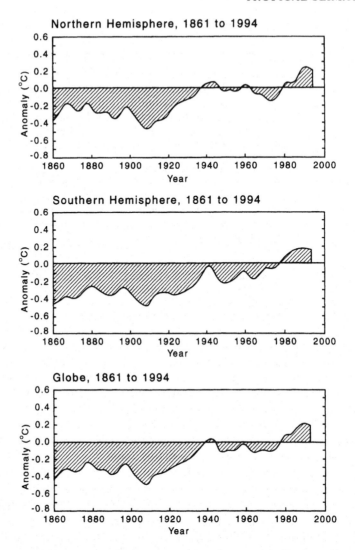

Figure 13.4   Annual combined land and sea surface temperature anomalies for 1861 to 1994 compared to the 1961–90 average. Note marked increases post-1980 when the majority of the current century's warmest years have occurred. From IPCC (1996a)

1995 being the warmest at 0.4 °C above the 1961–1990 global average. For the northern and southern subtropics, as well as the equatorial regions, not only have the last 100 years been marked by inter-annual variability of temperatures, but clear increases relative to the 1961–1990 average have occurred, especially in the subtropical regions, since the late 1970s (Figure 13.5). Regionally and locally, within the low latitudes, climate changes may be due to deforestation and desertification, the subjects of the next two sections.

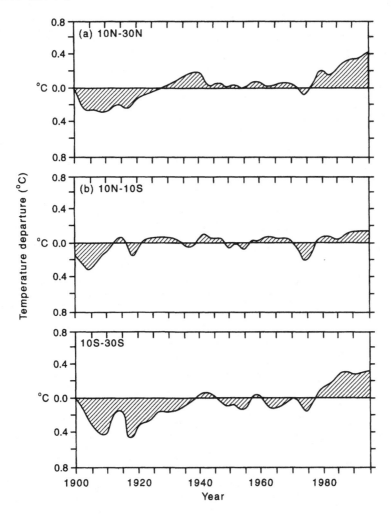

Figure 13.5   Smoothed temperature anomalies for the period 1900–90 compared to the 1961–90 mean for low latitudes: (a) 10°N–30°N; (b) 10°N–10°S; (c) 10°S – 30°S. Note that increases in the subtropics are greater relative to the equatorial regions. From IPCC (1996a)

DEFORESTATION

Deforestation is the process by which forest is removed and replaced by an alternative land cover. Tropical deforestation is of concern because not only does this have a possible climatic effect but it results in the displacement of indigenous cultures, the loss of biodiversity and soil fertility, as well as the degradation of the quality of water resources. In 1990 it was estimated that of the remaining tropical forests, 59% are found in Latin America, with the Amazon making up approximately 50% of the globe's tropical forest area; 22% are found in Asia (Indonesia, Malaysia, the Philippines, Indo-Chinese peninsula and southern China), with 10% of the world total in

Indonesia; and 19% are found in Africa, with Congo and Zaire accounting for 10% of the global total (Glantz and Krenz, 1992).

The causes of tropical deforestation are socio-economic factors, fire and drought. While the most devastating tropical forest fires in regions such as South America and South-East Asia are associated with drying or drought conditions related to ENSO events, the primary cause of tropical deforestation is through the human activities of commercial logging (Plate 11, Plate 12), ranching, large farms and plantations, mining and hydroelectric power developments (Henderson-Sellers, Zhang and Howe, 1996).

During the last fifty years or so, tropical natural vegetation has been destroyed at an increasing and unprecedented rate: recent cuttings run to about 140 000 km^2 per year, which represents about 1.8% of the forest cover in the tropics. However, in some countries the rate at which the forests disappear is much higher: in Thailand it is over 8%; in Brazil, the country with the largest forests in the tropics, it is 2.4%; in India it is 2.3% per year. Most of the clearings are done as clearcuts, and the trees that cannot be used commercially are burned. In some cases the tropical forests are usually lost for ever: replanting of trees is rare and largely limited to the few areas where tree crops are planted. Where regrowth of tropical forests is possible, it is often slowed down by soil erosion and the secondary forests are rather poor. Some cleared areas, especially in South-East Asia, turn into more or less permanent grasslands (Meyers, 1991) but this will depend on the length of time for which an alternative land cover has existed. The highlands of Papua New Guinea are a good example of where permanent grasslands now exist where formerly there was lush upland tropical forest; thousands of years of intense agricultural use of the land has brought about this situation. Although some deforested areas are replaced in the long term by grassland, this is not the situation globally, as other post-deforestation land covers can occur. In the case of northern Thailand, the land cover replacing forest is dominated by crops and areas in short and long fallow and not by pasture. In the Amazon the post-deforestation land covers, in descending order of importance, are forest, intermediate secondary succession, pasture and advanced secondary succession (Giambelluca, 1996).

The effects of deforestation on the local climate are clear (Meher-Homji, 1991; Wright et al., 1992; Bastable et al., 1993; Gash et al., 1996; Giambelluca, 1996). The removal of the canopies formed by the trees and the insulating air layers under them increase the diurnal range of temperature near the surface because direct heating of the soil during the day is followed by rapid cooling at night by terrestrial radiation loss. Deforestation results in surface albedo increases (0.12–0.13 before, 0.16–0.20 after), a consequence of which is a decrease in the net radiation available at the surface for absorption; this is estimated to be about 8% of the shortwave flux before change (Gash and Shuttleworth, 1991). Surface roughness is also altered with the removal of forest cover which has impacts on the efficiency of latent and sensible heat exchanges (Giambelluca, 1996). The water balance is also changed because infiltration of rainfall into the soil is strongly reduced, so that there is little storage of water. This means rapid runoff after rains, often combined with serious soil erosion, and no runoff in dry spells.

At the regional scale, deforestation may have an impact on climate because of increases of surface albedo, a reduction of surface roughness and decreases in the amount of rainfall intercepted by forest canopies. The main outcomes of these changes are a decrease in surface net radiation, the same as at the local scale, and a decrease in

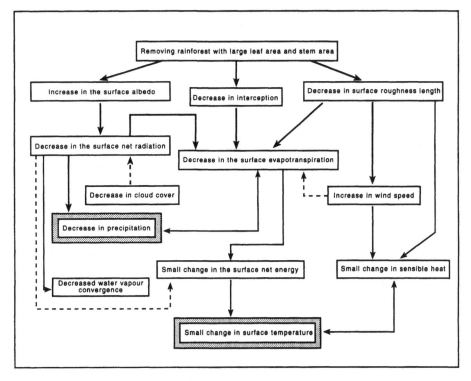

Figure 13.6   Model of tropical deforestation and regional climate interactions. From Hender-son-Sellers, Zhang and Howe (1996). Reproduced by permission of John Wiley, Chichester

evapotranspiration; both eventually lead to less convective activity and overall decreases of precipitation (Figure 13.6). Currently there is no real observational evidence at the regional-scale for decreases in precipitation as a result of deforestation. The only way in which the climatic impacts of regional scale deforestation can be assessed is through the use of computer models which describe the interactive processes between the components of the climate system. These are called general circulation models (GCMs). Their nature will be discussed in a later section of this chapter.

Many climate simulations of deforestation have been performed using GCMs (McGuffie et al., 1995). A characteristic of their output is the variability in the magnitude and type of climatic impact due to deforestation. This may be partly accounted for by the different ways in which the models represent or parameterize climate system processes. Nevertheless, GCM results do show some common agreement in terms of the climatic impacts of deforestation, especially decreases in precipitation and evaporation. Notably, evaporation decreases less markedly than precipitation, which has implications for regional water balances.

For the Amazon and South-East Asian regions, GCM simulations have been made for the extreme case of total removal of forest cover and replacement by degraded grassland (Henderson-Sellers, Zhang and Howe, 1996). The results of these "experiments" are shown in Table 13.2. For both areas there are similar impacts in terms of decreases of total, convective and large-scale precipitation as well as decreases in

Table 13.2   Climate variable over the Amazon and South-East Asia regions before and after total deforestation (modified after Henderson-Sellers, Zheng and Howe, 1996)

Climate variable	Before deforestation		Decrease after deforestation	
	Amazon	SE Asia	Amazon	SE Asia
Precipitation (mm)				
Total	1897	3169	− 420	− 251
Convective	1164	2213	− 306	− 236
Large-scale	733	956	− 96	− 15
Evapotranspiration (mm)	1243	1345	− 222	− 137
Temperature (°C)				
Average	24.8	24.5	+ 0.3	− 0.2
Maximum	31.2	28.6	+ 1.4	+ 0.8
Minimum	21.5	21.4	− 0.3	− 0.4

evapotranspiration, all attributable to the linkages outlined in Figure 13.6. As the reductions of precipitation are greater than evapotranspiration, regional moisture convergences are also reduced. The only difference between the results of the deforestation simulations are for temperature; the Amazon displays surface temperature increases while the South-East Asian region displays decreases. These differences relate to the interactions between surface albedo and surface roughness. In the case of the Amazon the opposing effects of increased surface albedo and decreased surface roughness (decreased latent heat flux) and the competing influences of greater incoming and outgoing radiation, as a result of decreased cloudiness, explain the small temperature increases. In the case of South-East Asia the change in evapotranspiration is less than the reduction in net radiation; consequently less energy is available at the surface following deforestation leading to a decrease of air and surface temperatures (Henderson-Sellers, Zhang and Howe, 1996).

The effects of deforestation on global climates are not clear. In relation to this an important aspect is the amount of carbon dioxide released into the atmosphere as a result of forest burning. Forests store about 250 tons of carbon per hectare, but in tropical forests over 80% of it is stored in the trees themselves, and less than 20% in the soil, while in mid-latitude coniferous forests about 50% of the carbon is located in the soil. When tropical trees are cut and burnt, most of the stored carbon is released into the atmosphere, while in coniferous forests around half of the carbon remains in the soil, directly available for regrowth. It is estimated that about a third of the recent increases in carbon dioxide and other GHGs in the atmosphere originate from tropical deforestation and the associated burning of large quantities of vegetation. Secondary tropical forests have a smaller storage capacity of carbon, so even if regrowth takes place, there is a net gain in atmospheric carbon. As for energy consumption related emissions of GHGs, there is geographical variation in the proportion of CO_2 emissions from forest burning; Brazil and Indonesia alone account for 32% of the total emissions from this source (Table 13.3).

Tropical deforestation may also affect the general circulation of the atmosphere, of which the low latitudes form the "heat engine" (Chapter 5). Less precipitation in the tropics might cause a reduced poleward transport of heat and moisture. This might

Table 13.3   Percentage of total CO_2 emissions from burning of
tropical forests (Glantz and Krenz, 1992, p. 35)

Country	Percentage of emissions
Brazil	20
Indonesia	12
Columbia	7
Ivory Coast	6
Thailand	6
Laos	5
Nigeria	4
Philippines	3
Burma	3
Peru	3
Other tropical countries	31

weaken the Hadley–Walker circulation and cause reduced precipitation in the mid-latitudes (Hastenrath, 1985, pp. 353–355). There is, however, no direct evidence of this effect although modelling studies indicate that large-scale deforestation has the potential to affect the general circulation in the low latitudes (Zhang, Henderson-Sellers and McGuffie, 1996a; Zhang, McGuffie and Henderson-Sellers, 1996b).

DESERTIFICATION

Desertification can be generally defined as the process by which desert-like conditions are brought about in areas where such conditions did not previously exist. Two major factors are involved in the desertification process but the relative magnitudes of these are unknown; they are drought (climatic variability) and human activities. It is in the arid, semi-arid and sub-humid areas of the globe (precipitation: evaporation ratios of 0.05 to 0.70) where desertification has the greatest potential. In such environments a run of exceptionally dry years and/or human interference with the natural ecosystem can engender a cascade of processes leading to desert-like conditions.

Drought can in many ways provide the catalyst for desertification, especially when it persists for several years, as is the case for the Sahel region. In this region it appears that an unfortunate string of events involving both natural and human aspects of the climate system may have brought about the expansion of desert-like conditions. The Sahel, like other dryland areas, is well renowned for its rainfall variability (Chapter 11). Unfortunately, humans, in opening up new areas for development in such environments, are often unaware of this climatic characteristic. This certainly seems to be the situation for the Sahel. Here during the 1950s and 1960s there were plentiful rains which encouraged the conversion of rangeland to farmland and the expansion of herd sizes and human populations. However, by the early 1970s it was quite apparent that the previous wet and prosperous period was slowly being replaced by one of increasing dryness due to a run of dry years. A consequence of this was a falling land carrying capacity; less people could be supported per square kilometre than before. An immediate response was for people to abandon the farmland that had once supported arable crops and also to graze herds over larger areas. Once the natural grasses started to become depleted, cattle and goat herds turned to shrubs and trees as a source of food; destruction of natural vegetation systems ensued. As a form of insurance in this area, large herd sizes are kept which exacerbated the overgrazing problem. The outcome of the obvious mismatch of climatic characteristics with inappropriate land uses was widespread environmental degradation typified by a loss of natural vegetation resources, a fall in soil fertility and increased soil erosion. Total abandonment of large tracts of land resulted. In this case climatic variability played a clear role in the desertification process as it essentially transformed the climate from one that could temporarily sustain agriculture and human settlement to one which could not, thus exposing bad land management practices and an ill-perception of the reliability of the climatic resource.

Not only does much debate concerning the details of the desertification process exist, but the estimated rates of desertification (calculated using a mixture of ground survey, conventional aerial photography and satellite imagery) are by no means accepted universally. For large areas such as the dryland areas of central Asia, the Sahel and northern Africa, annual desertification rates of 0.5–0.7% of the arid zone have been suggested. Taking an average rate of 0.5% per year, this translates as an annual increment of 80 000 km^2 which would mean that almost all of the world's arid and semi-arid lands will become desertified within the next century (Bullock and Le Houreou, 1996). If the robustness of arid lands is compromised any further by increasing livestock and human pressure then predicted climate change could accelerate the rate of desertification (Hulme and Kelly, 1993).

Increasing the area of desert-like land has a possible impact on climate and atmospheric circulation via the Charney mechanism (Charney, 1975). This is a positive feedback mechanism and is based around the role that surface albedo plays in the surface and atmospheric energy balances, and large-scale atmospheric stability. Briefly, overgrazing and vegetation removal result in increases in surface albedo. Consequently, less incoming shortwave radiation is absorbed at the surface, leading to a decrease of sensible heat flux to the atmosphere; cooling of the atmospheric column would eventuate. This would be compensated by a dynamic readjustment in the atmosphere in the form of adiabatic sinking and warming; a regime of subsidence would develop. Under such a regime the atmosphere would tend towards a state of stability and dryness, leading to decreased precipitation (Henderson-Sellers, Zhang and Howe, 1996). This in turn would result in vegetation stress and perhaps death; an increase in the total area of bare and higher albedo surfaces would follow. Such surfaces would exacerbate the processes of atmospheric cooling, subsidence and falling precipitation levels initially brought about by human- and livestock-induced devegetation. This mechanism has also been suggested to operate when deforestation occurs, but although the outcome of decreased precipitation is the same, the details of the mechanism may be different for the case of deforestation (Dickinson, 1992). A possible indirect result of the geographical expansion of desert-like areas is increased loadings of atmospheric dust. This, via its impact on atmospheric radiative transfer, could provide an additional positive feedback mechanism on climate with the same outcome as the Charney mechanism (Sagan, Toon and Pollack, 1979).

METHODS FOR ASSESSING THE EFFECTS OF CLIMATE CHANGE IN THE TROPICS

Before assessing the effects of climate change, the nature of the change itself must be established. This is done by developing *scenarios* which are plausible future climates. Scenarios may be developed in three broad ways (Carter et al., 1996). *Synthetic* scenarios involve adjusting the baseline climate by a given level, for example increasing mean annual temperature by 2 °C and then assessing whether human thermal comfort thresholds will be exceeded under such a future climate (McGregor, 1995a; 1995c). *Analogue* scenarios use past climate data to provide an idea of what future conditions might be. Three analogue approaches exist: analogues based on defining the conditions from instrumental records that existed for groups (ensembles) of warm or cold years and a comparison of ensemble characteristics with average conditions (Wigley, Jones and Kelly, 1980); using palaeoclimate data to reconstruct regional climate for past warm periods (Kutzbach, 1992; Chappell and Syktus, 1996); and to bring together a knowledge of atmospheric dynamics and empirical climate relationships and correlations to develop an "informed guess" about what future conditions might be like (Pittock and Salinger, 1982). The third type of scenario is that developed from *general circulation models* (GCMs). Currently, the majority of climate change impact assessments are based on GCM scenarios (IPCC, 1996a). Assessments are made by comparing existing climatic conditions, which are established using baseline climatological data or those conditions produced by "control" runs of the GCMs (the predicted current climate), with those future climatic conditions predicted by the GCMs.

General circulation models are computer models which use mathematical equations to describe the behaviour of, and the interactions between, the components of the climate system. These are the most sophisticated of a range of climate models as they are three-dimensional models (Henderson-Sellers and McGuffie, 1987; Trenberth, 1992). However, within the range of available GCMs there are varying levels of technical sophistication; for example, some are atmosphere-only models while others are fully *coupled* atmosphere and ocean models. The essential components of a coupled atmosphere–ocean GCM are sets of equations which represent the laws that describe the dynamics and physics of the atmosphere and oceans as well as the energy exchanges between the globe's many different surfaces and the atmosphere (Chapters 2 and 3). As a geographical basis for these calculations a grid is used; the globe's surface is divided into grids usually 250 km by 250 km in size while the atmosphere is divided into about 17 layers. For each of these grids, at all layers, all the equations describing the three-dimensional dynamics, physics and energy exchanges are calculated for a given time step. The outputs from these calculations are then "passed" to the adjacent grids (above and to the side) as input into the equations for the next round of calculations. For example, the sensible and latent heat fluxes for a particular ocean surface grid box will be passed up to the overlying atmospheric grid box. The final output from all these calculations is gridded climatic data, typically temperature and precipitation.

Although GCMs are the most sophisticated models for describing the complex processes of and interactions within the climate system, they do suffer from a number of problems. Because they are coarse models in a geographical sense (calculations are

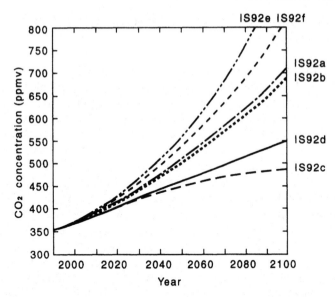

Figure 13.7   Projected levels of atmospheric CO_2 concentrations for various IPCC scenarios.
From IPCC (1996a)

made for 250 km by 250 km grids), they are unable to model many sub-grid scale
processes such as individual thunderstorms. GCMs, it must be remembered, are only
models. They therefore *represent* the actual climate processes in a simplified manner.
Furthermore, the results from different GCMs produced by different research labora-
tories are often not the same; the predicted temperature and precipitation changes may
vary between models. This is usually a product of the various research institutions
using different parametrization schemes for climate system processes.

GCMs can be used to make weather forecasts for up to about ten days in advance;
however, their main purpose is for climate prediction. Used in such a way, a "control
run" of the GCM is first made. This involves running the model in order to simulate
many decades of the globe's climate. Observations of the current climate (e.g. global
climatologies of temperature and precipitation) are then used as a basis for assessing
the faithfulness of the control simulation. If there are differences between the "control
run climate" and the actual climate, the models are calibrated and run again until they
are able to represent the current climate with a fair degree of accuracy. Once this is
achieved they can be applied for the purpose of climate prediction.

When GCMs are run to make climate predictions they are usually "forced" with
different levels of greenhouse gases (GHGs); that is, the atmospheric concentrations of
the GHGs used for the control run are changed, usually increased. These concentrations vary according to a number of GHG gas emission scenarios which are developed
based on estimates of future global population levels, economic growth rates and the
amount of fossil fuel to be burned. Currently, six such scenarios are used to estimate
the concentrations of CO_2 levels by the year 2100 (IPCC, 1996a) (Figure 13.7). For a
given emission scenario, a GCM may be run in two experimental modes: equilibrium-
response and transient-forcing. In the former case the atmospheric concentration of
GHGs (usually a doubling of CO_2) is changed abruptly with the model run until it

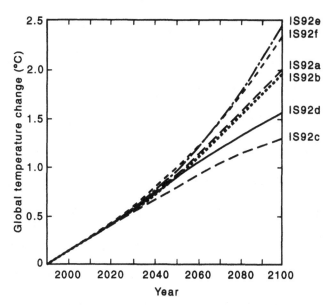

Figure 13.8   Projected global mean temperature change from 1990 to 2100 for various IPCC scenarios. From IPCC (1996a)

reaches climatic equilibrium. In the latter case GHG concentrations are changed gradually over time, which is closer to what will actually happen. Transient-forcing GCM experiments are preferred as they allow the model climate to evolve in relation to the slowly changing composition of the atmosphere. They also facilitate estimates of not only the amount of climate change but the rate. Associated with each of the emission scenarios (level of forcing) will be a different climate, as for example described by temperature (Figure 13.8).

PREDICTED CLIMATE CHANGE

Future climate predictions have been made using both equilibrium and time-dependent (transient) experiments. Although the majority of GCM experiments have to date used GHG-only forcing, some GCMs now include the effects of anthropogenic sulphate aerosols which result from the oxidation in the atmosphere of sulphur dioxide gas to sulphate aerosols. The inclusion of aerosols has revealed that these have a marked cooling effect on climate (they increase planetary albedo), with the result that GCM results from GHG and aerosol-forced GCMs demonstrate more conservative changes in temperature compared to GHG-only GCM results. Despite this, and the fact that there is some variability between models in the level of detail, there is general consensus amongst the climate modelling community that global climate is likely to change due to an anthropogenic enhancement of the greenhouse effect. For the full range of emission scenarios, and taking into account the cooling effect of anthropogenic aerosols, global temperature change is predicted to be between 1 °C and 3.5 °C by the year 2100 (IPCC 1996a, pp. 289–291). Maximum annual warming is predicted for the high northern latitudes because of the reduction of sea ice cover. Comparing land with the oceans in the tropics, the warming over tropical continental areas will be greater, especially in the dry subtropics, where low moisture levels will limit evaporation and thus make more energy available for sensible heating of the atmosphere. Although all tropical continental areas are expected to become warmer, the exact magnitude of this change is uncertain as GCM results vary in this regard. For example, taking the results from two GCMs (including aerosol effects) for the Sahel and South-East Asian regions, summer (winter) temperatures are predicted to increase between 1.5–3.0 °C (1.8–2.2 °C) and 1.6–2.8 °C(0.5–1.8 °C) respectively (IPCC, 1996a). The results from one of these GCMs is shown in Figure 13.9. The oceans will warm less than the continents because of their great thermal inertia.

Globally, annual precipitation is expected to increase due to an enhancement of the hydrological cycle. Most marked changes are predicted for the high latitudes. For the tropics there is considerable inter-GCM variability in the predictions with regards to shifts or changes in the intensity of the rainfall maxima (IPCC, 1996a, p. 309). Notably, predictions from a wide range of GHG-only forced GCMs suggest an increase in rainfall over the Indian subcontinent and South-East Asian areas (Figure 13.10), a result of an increase in the intensity of the Asian monsoon, whereas results from GHG and aerosol-forced GCMs predict a decrease in Asian monsoon precipitation. This is because the cooling effects of aerosols will decrease the intensity of the summertime heat lows developed over continental areas, a factor which is extremely important for the development of the various precipitation-bearing monsoon systems (Chapter 7). Clearly, the future situation regarding continental tropical precipitation is a lot more sensitive to the climatic effects of aerosols than it is for temperature.

For oceanic areas, precipitation patterns are closely tied to sea surface temperature (SST) patterns (Chapter 10). This is especially the case for the tropical Pacific. For this region some GCMs predict a differential mean warming of SST from the east to the west, with the eastern Pacific warming faster than the west. This resembles in many ways the SST pattern for ENSO events (Chapter 6). Associated with such a decrease in the zonal sea surface temperature gradient, models predict a slight change in precipitation patterns. These changes duplicate those experienced during ENSO events; precipitation increases over the central equatorial Pacific while in other central Pacific areas there are decreases in ITCZ precipitation. Similarly there are also precipitation decreases in the western Pacific, especially in the Indonesian–northern Australian area (IPCC, 1996a).

CLIMATE CHANGE EFFECTS IN THE LOW LATITUDES

As the situation regarding climate change related shifts in the spatial and temporal distribution of temperature and especially precipitation are far from clear, only general statements concerning the likely effects of climate change on low-latitude biophysical and socio-economic systems will be presented here. Although a full treatment of possible climate change effects is beyond the scope of this book, a comprehensive global survey of possible effects can be found in IPCC (1996a, b, c).

The climatic, ecological and cultural importance of tropical forests has been out-

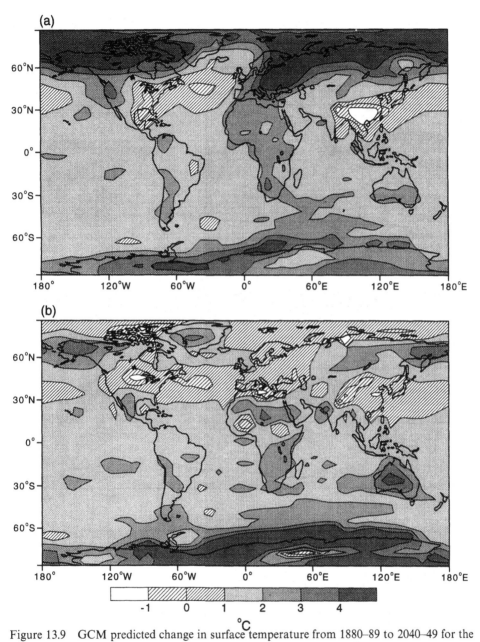

Figure 13.9   GCM predicted change in surface temperature from 1880–89 to 2040–49 for the
case of aerosol effects and greenhouse gases. (a) December to February; (b) June to August. From
IPCC (1996a)

lined above with the point made that the survival of the tropical forests is very much
dependent on human activities. This will remain a major factor in the future and
possibly will be more important than climate change itself for the survival of tropical
forests, unless indirect effects of climate change, such as socio-economic pressures, lead

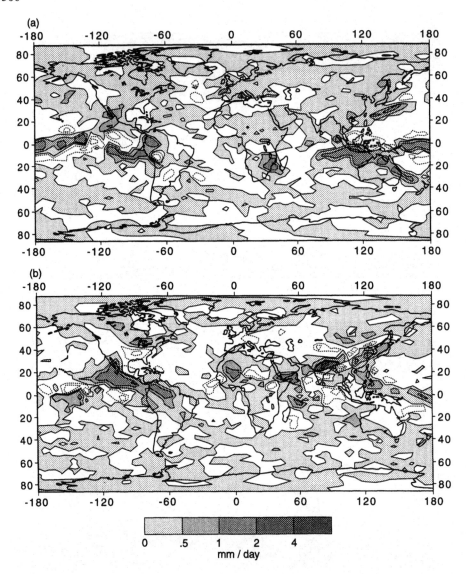

Figure 13.10   GCM predicted changes in precipitation resulting from a 1% per year increase in CO_2 to levels double that of the present. This GCM did not include aerosol effects and predicts precipitation increases in the Asian monsoon area in contrast to GCMs that include aerosol effects. From IPCC (1996a)

to greater rates of deforestation. Where soil moisture availability is currently a problem for forest growth, a climate change induced alteration of precipitation: evapotranspiration ratios may bring about some forest decline in marginal areas. Although present global vegetation models do not agree on whether the total area of tropical forests will increase or decrease, it is certain that major changes in rainfall and land use patterns will bring about changes in the distribution of tropical forests.

Indirectly, tropical forests could be affected by increased erosion rates as a result of increased rainfall amount and intensities in some areas (IPCC, 1996b).

The geographical neighbours of the tropical forests are the major grass and savanna lands of the tropical–subtropical latitudes. As this vegetation type occurs in areas where soil moisture and nutrient reserves are marginal, any adverse effects of climate change may be realized at an early stage in these environments, especially if there are decreases in rainfall. An increase in the frequency of extreme events, especially drought, is also likely to prejudice grass and savanna land stability, especially if this biome type continues to be subjected to human pressures. The effect of climate change on tropical dryland vegetation will depend on the balance between increasing aridity favouring C4 grasses and increasing CO_2 concentration favouring woody C3 vegetation (Allen-Diaz, 1996). Theoretically, with increasing CO_2 concentrations and hence enhanced water-use efficiency, C4 grasses should be favoured. There is, however, some evidence that within the time period of CO_2 increase, C3 woody species have increased in arid grassland environments. It is possible that this unexpected effect is a result of increasing aridity and grazing pressure in tropical grassland environments (Archer, Schimel and Holland, 1995).

Subtropical desert environments are likely to become more extreme in terms of temperature, with little increase in rainfall. If tropical dryland climatic variability increases as a response to climate change, a possibility given current rainfall amount variability relationships (Chapter 10), then some dryland areas will become more prone to desertification. Intensely used soils in such environments will also be affected by rainfall extremes, with leaching and acidification rates possibly increased. Allied to this will be increased salinization and possibly wind erosion in areas where evaporation rates increase (Bullock and Le Houerou, 1996).

Mountain and highland tropical environments will also experience some environmental changes with predicted climate change. The main effect will be an elevation of the mean annual isotherms. For example, in the case of the highlands of Papua New Guinea, where the altitude–temperature relationship is very strong above the 500 m level and the mean annual lapse rate is $6\,°C\,km^{-1}$, a $2\,°C$ warming could result in an altitudinal increase of isotherms by approximately 330 m (McGregor, 1988). Associated with such translocations of thermal zones will be potential shifts in the altitudinal zonation of vegetation and agriculture, a decrease in the number of frost days, an increase in highland growing season and duration, and a reduced time to harvest. Although this sounds positive, expansion of altitudinal limits for agriculture may have some negative effects in that subsistence farmers can be expected to extend their agricultural activity to higher altitudes, with associated impacts of highland devegetation and land degradation.

Wetland environments, especially in the subtropical latitudes, play an important role in sustaining water-dependent ecosystems in otherwise harsh environments. On an annual basis, because of predicted increases in rainfall, some African wetland systems could benefit from climate change, such as in eastern Niger and Chad (Oquist and Svensson, 1996). Wetland environments in the Sahel region may not be so fortunate as here decreases in rainfall and increases in temperature are predicted, leading to decreases in wetland extent. This situation could be made worse by the increasing demand for wetlands to support agriculture and growing populations.

The main effects on tropical oceans will be increases in sea surface temperatures (SST). The magnitude of these increases in tropical ocean hot spots such as in the equatorial western Pacific may be limited by the same feedback mechanisms that presently limit maximum SST (Chapter 9). An extreme weather event related to tropical SST is the tropical cyclone. Currently it is unclear whether tropical cyclones will increase in absolute numbers due to SST increases. However, the possibility exists that SST increases could cause sensible and latent heat fluxes to increase, making these systems more intense. The geographical distribution of tropical cyclones is also likely to expand with poleward shifts of the mean annual isotherm of 26.5 °C which defines the spatial extent of the tropical cyclone regions (Chapter 8). In those areas where coral reefs are an important part of the marine ecosystem the incidence of coral bleaching, as occurs during ENSO events, is likely to increase as a result of warmer ocean waters. Changes in ocean dynamics may well have an effect on the distribution and intensity of oceanic upwelling, which has implications for the fishing industries in low latitude countries (see below).

In the warmer and more humid tropics, human thermal comfort conditions will be somewhat different to those presently experienced. This is significant as current human bioclimatic conditions in many low latitude locations are marginal. Although some natural acclimatization may be expected, increased thermal stress loads are likely to elevate levels of discomfort. Changes in the frequency, duration and intensity of periods for which physioclimatically stressful conditions occur may be expected (McGregor, 1995a). This will result in increased demands for active ventilation or conditioning of indoor environments (McGregor, 1995c), especially in the large tropical cities of the future (Chapter 4). On a seasonal basis the duration of the cool comfortable dry season is likely to contract. Changes in the hygrothermal conditions also hold implications for morbidity and mortality. For some tropical locations clear temperature thresholds exist, above which mortality increases (Auliciems and Skinner, 1989; McMichael et al., 1996). Possible changes in the frequency of extreme events as manifest by heat waves may engender higher mortality rates, especially in the aged and the socio-economically disadvantaged as these sectors of the population are particularly weather-sensitive. A range of other health effects may also occur, especially an increase in the geographical range of a number of tropical diseases (Curson, 1996).

Climate change in the tropics will have a number of important consequences for agriculture (Chapter 12). The agricultural responses to climate change will probably vary from one agricultural system to another and from country to country due to differences in climate, economic characteristics, resources and the presence or absence of a strong agricultural policy (Reilly, 1996). Generally, for areas where rainfall is likely to increase, agricultural production should benefit as long as soil erosion and nutrient leaching are managed effectively. The reverse, however, may be the situation for sub-humid areas where rainfall amounts are predicted to decrease, making such areas marginal for agricultural production. A number of assessments of the effects of predicted climate change on agriculture in the low latitudes have been made. These assessments use the outputs from GCMs as part of the inputs into climate–crop models. Results of such studies for Zimbabwe, Senegal and Kenya have revealed possible decreases in yield for maize and millet, while South Africa may experience increases in the yield of maize; the Niger may also experience a reduction in the growing season (Reilly, 1996). For southern and South-East Asia, models show a possible wide variation in crop response not only across the region but between models. Taking rice for example, one model estimates drops in yield for Indonesia and Malaysia while another study for the same region suggests yield increases with climate change (Reilly, 1996). Clearly the situation is not unambiguous. For countries such as Bangladesh it has been suggested that due to the diversity of agricultural systems it is difficult to make an assessment of the possible effects of climate change on agriculture (Reilly, 1996). Estimates of effects for the small islands in the Pacific basin are few and far between. If cloud amounts increase, thus reducing surface irradiance, the potential exists for falls in crop yield. This may be exacerbated by the possible increased incidence of drought or salt water intrusion of groundwater. Increased temperatures may also reduce the life-cycle duration of maize and rice, thus leading to a decrease in yield for these crops. For Mexico, Brazil and Uruguay, declines in maize production are predicted ranging from 5 to 50% (Reilly, 1996).

For some tropical countries both the domestic and international fish trade is an important income source. This is particularly so for African countries such as Mauritania, Senegal, Morocco, Namibia and South Africa, and the South American countries of Ecuador, Chile and Peru, where cool nutrient-rich upwelling offshore waters (Chapter 5) support large pelagic fish populations. As these are sensitive to alterations in the intensity and distribution of zones of upwelling, any climate change related alterations in upwelling will affect ocean productivity. For Ecuador, Peru and Chile, an increasing frequency of ENSO events could have disastrous consequences for pelagic fishing industries (Chapter 6). The shrimp harvests in Mexico and the Arafua Sea between Australia and New Guinea, which are also ENSO sensitive, may also be affected by climate change related alterations of ENSO occurrence. In some tropical countries large lakes such as Lake Victoria support important freshwater fishing industries. Already such lakes are suffering from the pressures of human-induced environmental change. Current problems include eutrophication, over-extraction of water, mining of lake sands and over-fishing. Some of these existing problems are likely to be exacerbated by climate change, especially if pressure is put on lakes to supply large volumes of freshwater for irrigation and domestic purposes. In deep lakes, such as Malawi and Tanganyika, where there is a clear thermal structure, increased lake surface temperatures may prevent overturning of lake waters in the autumn to winter periods thus affecting lake food webs; this process is important for replenishing nutrient levels in surface waters.

In a warmer tropics there will be a greater demand for water for agricultural crops and large settlements. This will especially be a problem in the subtropics where water demand for agricultural, industrial and municipal activities is already great. Lakes and rivers will be put under increasing pressure to supply water for irrigation. Water resources in a warmer tropics are likely to become a contentious geopolitical issue.

A further effect of climate change will be sea level rise due to, in decreasing order of importance, thermal expansion of the oceans, melting of mountain glaciers and polar ice sheets as well as changes in ocean dynamics (Warrick et al., 1996). Different levels of sea level rise will be achieved depending on the future trends of GHG concentrations as modelled by the variety of GHG emission scenarios (Figure 13.11). The current best estimate from coupled GCMs is that sea level will rise by about 0.5 m by the year 2100,

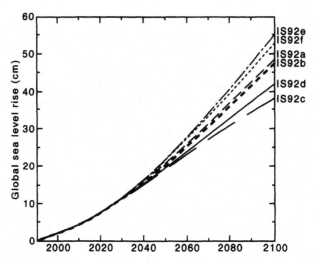

Figure 13.11   Predicted sea level rise from 1900 to 2100 for various IPCC emission scenarios.
From IPCC (1996a)

Table 13.4   Potential impacts of a 1 m sea level rise on selected low
latitude countries (modified from McMichael et al., 1996)

Country	People affected		Land at loss	
	1000's	% total	km²	% total
Bangladesh	71 000	60	25 000	17.5
Belize	70	35	1 900	8.4
Guyana	600	80	2 400	1.1
India	7 100	1	5 800	0.4
Kiribati	100	4	12.5	?
Malaysia	—	—	7 000	2.1
Marshall Is.	20	100	9	80.0
Nigeria	3 200	4	18 600	2.0
Senegal	110	> 1	6 100	3.1
Tonga	30	47	7	2.9

equivalent to two to three times the average rate of rise over the last 100 years.
Notably, sea level rise will continue into the future due to the longevity of GHGs in the
atmosphere (Wigley, 1995). This effect has often been referred to as "commitment
warming" as even if all GHG emissions were shut off immediately the atmosphere and
oceans will continue to respond to the thermal effects of current high levels of GHGs;
warming and sea level rise will only tail off when the climate system has attained
equilibrium. This could take tens of decades.

Sea level rise has a number of implications for tropical countries, especially those
countries where deltaic landforms and low islands, such as coral atolls, make up a large
proportion of the total land area (Plate 13, Plate 14). For some tropical countries,
estimates of the percentage population likely to be affected by sea level rise have been

made (Table 13.4). Of the low latitude countries, Bangladesh stands out, as 17.5% of its land could be lost, with 71 million people being affected – equivalent to 60% of the population. A logical and pertinent question arising from this likelihood is "who will take responsibility for such *environmental refugees?*" Other countries with notable population numbers likely to be affected are India, Nigeria and Guyana. Although countries with large deltaic areas stand to lose significant proportions of their total land area, it is the small island countries where the effects of projected sea level rise will be felt most drastically (Table 13.4). This is especially true for the Marshall Islands in the North Pacific, the Maldives in the Indian Ocean (Pernetta, 1992) (Plate 14) and Kiribati, Tokelau and Togatapu which are only a few of the South Pacific Island nations likely to be affected (Campbell, 1996).

In addition to loss of land due to sea level rise, the vulnerability to inundation of coastal zones due to episodic extreme weather events and storm surges is likely to increase. This will exact a heavy toll in terms of death, injury and economic losses, especially in countries such as Bangladesh, where extreme weather events already have significant socio-economic impacts. On a climatological time scale, coastal erosion will increase thus threatening agricultural land. Existing problems of saltwater contamination of coastal and small island groundwater reserves, as occurs annually in the Maldives, will be exacerbated by sea level rise. Health implications also exist related to population displacement and changes in the distribution of disease vectors (McMichael et al., 1996).

SUMMARY

Climate change can be considered at two broad temporal scales: long-term changes at greater than 20 000 years, and short-term changes between 100 and 20 000 years. Climatic variability refers mainly to decade to decade and year to year changes. Causes of climate change can be classified as either external or internal. The former includes changes of the earth's orbital geometry, often referred to as Milankovitch or astronomical variations. Internal causes result from changes to the basic nature of the components of the climate system and feedbacks between these components.

Currently there is great concern about human-induced climate change mainly due to increased global pollution levels which have, through an enhanced greenhouse effect, the potential to increase global temperatures, alter climate patterns and raise sea levels. At the local to regional scale, deforestation and desertification are also of concern. Although the magnitude of climate change in the tropics is likely to be less than at higher latitudes, climate change impacts will be considerable. A full understanding of the gravity of these changes will, however, only be achieved when the causes, nature and availability of climate resources in the tropics is fully appreciated. This will continue to be one of the main goals of tropical climatology.

References

Acland J D 1971 *East African Crops*. FAO and Longman, Rome.

Adedoyin J A 1989 Global-scale sea surface temperature anomalies and rainfall characteristics in northern Nigeria. *Int. J. Climatol.*, **9**, 133–44.

Allan R J 1988 El Nino Southern Oscillation influences in the Australasian region. *Prog. in Phys. Geog.*, **12**, 313–48.

Allan R J 1991 Australasia. In Glantz M H, Katz R W, Nicholls N (eds) *Teleconnections Linking Worldwide Climate Anomalies*. Cambridge University Press, Cambridge.

Allen B J, Brookfield H, Bryon Y 1989 Frost and drought through time and space, Part I: The climatological record. *Mount. Res. Devel.*, **9**, 252–78.

Allen-Diaz B 1996 Rangelands in a changing climate: impacts, adaptations and mitigation. In *Climate Change 1995: Impacts, Adaptations and Mitigations of Climate Change: Scientific and Technical Analyses*, Intergovernmental Panel of Climate Change. Cambridge University Press, Cambridge, pp. 131–9.

Angell J K 1981 Comparison of variations in atmospheric quantities with sea surface temperature variations in the equatorial eastern Pacific. *Mon. Wea. Rev.*, **109**, 230–43.

Anyamba E K, Weare B C 1995 Temporal variability of the 40–50 day oscillation in tropical convection. *Int. J. Climatol.*, **15**, 379–402.

Archer S, Schimel S A, Holland E A 1995 Mechanisms of shrubland expansion: land use, climate or CO_2? *Climatic Change*, **29**, 91–100.

Arenas A D 1983 Tropical storms in central America and the Caribbean: characteristics rainfall and forecasting of flash floods. In Keller R (ed.) *Hydrology of Humid Regions*. IAHS Publication No. 140, pp. 39–48.

Arkin P A, Ardanuy P E 1989 Estimating climate-scale precipitation from space: a review. *J. Clim.*, **3**, 1229–38.

Atheru Z K K 1994 Extended range prediction of monsoons over eastern and southern Africa. In *Proceedings of International Conference on Monsoon Variability and Prediction*, Trieste, Italy, 9–13 May, 1994. WCRP-84, WMO/TD No. 619, pp. 460–5.

Atkinson B W 1981 *Dynamical Meteorology: An Introductory Selection*. Methuen, London.

Auliciems A, de Dear R J 1986 Air conditioning in a tropical climate: impacts upon European residents in Darwin, Australia. *Int. J. Biometeorol.*, **30**, 259–82.

Auliciems A, Skinner J L 1989 Cardiovascular deaths and temperature in subtropical Brisbane. *Int. J. Biometeorol.*, **33**, 215–21.

Barnett T P, Dumenil L, Schlese U, Roeckner E, Latif M 1989 The effect of Eurasian snow cover on regional and global climate variations. *J. Atmos. Sci.*, **46**, 661–85.

Barrett E C 1970 Estimation of monthly rainfall from satellite data. *Mon. Wea. Rev.*, **98**, 322–7.

Barry R G 1981 *Mountain Weather and Climate*. Methuen, London.

Barry R G 1988 *Mountain Weather and Climate*. John Wiley & Sons, Chichester.

Bastable H G, Shuttleworth W J, Dallarosa R L, Fisch G, Nobre C 1993 Observations of climate, albedo and surface radiation over cleared and undisturbed Amazonian forest. *Int. J. Climatol.*, **13**, 783–96.

Bell M A, Lamb P J 1994 Temporal variations in the rainfall characteristics of disturbance lines over subsaharan West Africa 1951–1990. In *Proceedings of International Conference on Monsoon Variability and Prediction*, Trieste, Italy, 9–13 May, 1994. WCRP-84, WMO/TD

No. 619, pp. 35–41.

Berger A, Gossens C H R 1983 Persistence of wet and dry spells at Uccle (Belgium). *J. Climatol.*, **3**, 21–34.

Besson L 1924 On the probability of rain. *Mon. Wea. Rev.*, **52**, 308–14.

Bhalme H N, Mooley D A 1980 Large scale droughts/floods and monsoon circulation. *Mon. Wea. Rev.*, **108**, 1197–209.

Bigg G R 1993 Comparison of coastal wind and pressure trends over the tropical Atlantic 1946–1987. *Int. J. Climatol.*, **13**, 411–21.

Bigg G R 1995 The El Nino event of 1991–94. *Weather*, **50**, 117–23.

Bjerknes J 1969 Atmospheric teleconnections from the tropical Pacific. *Mon. Wea. Rev.*, **97**, 103–72.

Bolton D 1984 Generation and propagation of African squall lines. *Q. J. R. Met. Soc.*, **110**, 695–721.

Brand S, Belloch J W 1973 Changes in the characteristics of typhoons crossing the Philippines. *J. Appl. Met.*, **12**, 104–10.

Brand S, Belloch J W 1974 Changes in the characteristics of typhoons crossing the island of Taiwan. *Mon. Wea. Rev.*, **102**, 708–13.

Broecker W, Denton G 1990 What drives glacial cycles? *Sci. Amer.*, **262**, 43–50.

Bullock P, Le Houreou P 1996 Land degradation and desertification. In *Climate Change 1995: Impacts, Adaptations and Mitigations of Climate Change: Scientific and Technical Analyses*, Intergovernmental Panel of Climate Change. Cambridge University Press, Cambridge, pp. 171–90.

Burpee R W 1974 Characteristics of the North African easterly waves during summers of 1968 and 1969. *J. Atmos. Sci.*, **29**, 77–90.

Byers H R, Rodebush H R 1948 Causes of thunderstorms of the Florida peninsula. *J. Meteorol.*, **5**, 275–80.

Campbell C G, Von der Haar T H 1980 Climatology of radiation budget measurements from satellites. *Atmos. Sci. Paper No. 323*, Colorado State University.

Campbell J R 1996 Contextualizing the effects of climate change in Pacific Island countries. In Giambelluca T W, Henderson-Sellers A (eds) *Climate Change: Developing Southern Hemisphere Perspectives*. John Wiley, Chichester, pp. 349–76.

Canby T Y 1984 El Nino: global weather disaster. *Nat. Geogr.*, **165**, 144–83.

Carter T, Parry M, Nishioka S, Harasawa H 1996 Technical guidelines for assessing climate change impacts and adaptations. In *Climate Change 1995: Impacts, Adaptations and Mitigations of Climate Change: Scientific and Technical Analyses*, Intergovernmental Panel of Climate Change. Cambridge University Press, Cambridge, pp. 823–34.

Chan J C L, Gray W M 1982 Tropical cyclone movement and surrounding flow relationships. *Mon. Wea. Rev.*, **110**, 1354–74.

Chang C P, Erickson J E, Lau K M 1979 Northeasterly cold surges and near equatorial disturbances over the winter MONEX area during December 1974, part 1: synoptic aspects. *Mon. Wea. Rev.*, **107**, 812–29.

Chang J H 1968 *Climate and Agriculture*. Aldine, Chicago.

Chang J H 1970 Potential photosynthesis and crop productivity. *Annals Ass. of Am. Geogr.*, **60**, 92–101.

Chao Y and Philander S G H 1993 On the structure of the Southern Oscillation. *J. Clim.*, **6**, 450–69.

Chapman S, Lindzen R S 1970 *Atmospheric Tides, Thermal and Gravitational*. Gordon and Breach, New York.

Chappell J, Syktus J 1996 Palaeoclimatic modelling: a western Pacific perspective. In Giambelluca T W, Henderson-Sellers A (eds) *Climate Change: Developing Southern Hemisphere Perspectives*. John Wiley, Chichester, pp. 349–76.

Charney J G 1975 Dynamics of deserts and droughts in the Sahel. *Q. J. R. Met. Soc.*, **101**, 193–202.

Cheang B-K 1977 Synoptic features and structures of some equatorial vortices over the South China sea in the Malaysian region during the winter monsoon, December 1973. *Pure Appl.*

Geophys., **115**, 1303–34.

Cheang B-K 1991 Short and long-range monsoon prediction in Southeast Asia. In Fein J S, Stephens P L (eds) *Monsoons.* John Wiley, New York.

Chiyu T 1979 A preliminary study of low-level winds over Peninsula Malaysia during the 1967–77 north-east winter monsoon. *J. Met. Soc. Japan,* **57**, 354–7.

Chow S D, Chang C 1984 Shanghai urban influences on humidity and precipitation distribution. *Geojournal,* **8**, 201–4.

Chu P S 1984 Time and space variability of rainfall and surface circulation in the Northeast Brazil–tropical Atlantic sector. *J. Met. Soc. Japan,* **62**, 363–70.

Chu P S 1989 Hawaiian drought and the southern oscillation. *Int. J. Climatol.,* **9**, 619–31.

Clino S 1971 *Climatological Normals for Climate and Climate Ship Stations for the Period 1931–1960.* WMO, Geneva, No. 117.TP.52.

Conover J H, Lanterman W S, Schaefer V J 1969 Major cloud systems. In Rex D F (ed.) *Climate of the Free Atmosphere, Volume 4 of World Survey of Climatology.* Elsevier, New York, pp. 232–43.

Conway T J, Tans P P, Waterman L S, Thoning K W, Kitzis D R, Masari K A, Zhang N 1994 Evidence for interannual variability of the carbon cycle from the National Oceanic and Atmospheric Administration/Climate Monitoring and Diagnostics Laboratory Global Air Sampling Network. *J Geophys. Res.,* **99**(D11), 22831–56.

Coronas J 1920 *The Climate and Weather of the Philippines*, 1903 & 1918. Manila Bureau of Printing.

CPC 1996 Climate Diagnostics *Bulletin.* Climate Prediction Center, NOAA, Washington DC, USA.

Currie R G 1996 Mn and Sc signals in North Atlantic tropical cyclone occurrence. *Int. J. Climatol.,* **16**, 427–39.

Curson P 1996 Human health, climate and climate change: an Australian perspective. In Giambelluca T W, Henderson-Sellers A (eds) *Climate Change: Developing Southern Hemisphere Perspectives.* John Wiley, Chichester, pp. 310–48.

Dahale S D, Panehawagh N, Singh S V, Ranatunge E R, Brikshavana M 1994 Persistence in rainfall occurrence over tropical Southeast Asia and equatorial Pacific. *Theor. Appl. Climatol.,* **49**, 27–35.

Das P K 1972 *The Monsoons.* Edward Arnold, London.

Davidson N E, McBride J I, McAvaney B J 1983 The onset of the Australian monsoon during winter MONEX: synoptic aspects. *Mon. Wea. Rev.,* **111**, 496–516.

de Dear R J 1989 Diurnal and seasonal variations in human thermal climate of Singapore. *J. Trop. Geog.,* **10**, 13–26.

de Dear R J, Leow K G, Foo S C 1991 Thermal comfort in the humid tropics: field experiments in air conditioned and naturally ventilated buildings in Singapore. *Int. J. Biometeorol.,* **34**, 259–65.

Dey B, Bhanu Kamar O S R U 1983 Himalayan winter snow cover area and summer monsoon rainfall over India. *J. Geophys. Res.,* **88**, 5471–4.

Dhar O N, Nandargi S 1993 The zones of severe rainstorm activity over India. *Int. J. Climatol.,* **13**, 301–11.

Diaz H F, Markgraf V 1992 *El Nino: Historical and Palaeoclimatic Aspects of the Southern Oscillation.* Cambridge University Press, Cambridge.

Dickinson R 1992 Changes in land use. In Trenberth K E (ed.) *Climate System Modeling.* Cambridge University Press, Cambridge, pp. 689–704.

Dickson R R 1984 Eurasion snow cover versus Indian monsoon rainfall – an extension of the Hahn-Shukla results. *J. Clim. Appl. Meteorol.,* **23**, 171–3.

Ding Y 1990 Build-up, airmass transformation and propagation of Siberian high and its relation to cold surge in East Asia. *Meteorol. Atmos. Phys.,* **44**, 280–92.

Ding Y 1994 *Monsoons Over China.* Kluwer Academic, London.

Domroes M, Ranatunge E 1993 Daily persistence over Sri Lanka. *Mausam,* **44**, 281–6.

Domroes M, Peng G 1988 *The Climate of China.* Springer-Verlag. Berlin.

Douglas M W, Maddox R A, Howard K, Reyes S 1993 The Mexican monsoon. *J. Clim.,* **6**,

1665–77.

Drosdowsky W 1993 An analysis of Australian seasonal rainfall anomalies 1950–1987 – II: temporal variability and teleconnections patterns. *Int. J. Climatol.*, **13**, 111–49.

Fein J S, Stephens P L 1987 *Monsoons*. John Wiley, New York.

Findlater J 1972 Aerial exploration of the low-level cross-equatorial current over Eastern Africa. *Q. J. R. Met. Soc.*, **98**, 274–90.

Findlater J 1974 The low-level cross-equatorial air current of the Western Indian Ocean during the northern summer. *Weather*, **29**, 411–16.

Findlater J 1977 Observational aspects of the low-level cross-equatorial jet stream of the West Indian Ocean. *Pure Appl. Geophys.*, **115**, 1251–62.

Flohn H 1950 *Neue Anschauungen uber die Allgemeine Zirkulation der Atmosphare und ihre Klimatische Bedeutung*, Erdkunde, Vol. 4, pp. 161–75.

Flohn H 1960 Recent investigations on the mechanism of the summer monsoon of southern and eastern Asia. In *Symposium on Monsoons of the World*, New Delhi, pp. 75–88.

Flohn H 1969 Investigations of the atmospheric circulation above Africa. *Bonner Met. Abh.*, **10**, 38–9.

Flores J F, Balagot V F 1969 The climate of the Philippines. In *World Survey of Climatology 8, Northern and Eastern Asia*. Elsevier, New York, pp. 159–213.

Folland C K, Palmer T N, Parker D E 1986 Sahel rainfall and worldwide sea surface temperatures. *Nature*, **320**, 602–7.

Folland C K, Owen J A, Ward M N, Colman A W 1991 Prediction of seasonal rainfall in the Sahel region using empirical and dynamical methods. *J. Forecasting*, **10**, 21–56.

Frakes L A 1979 *Climates Throughout Geological Time*. Elsevier, Amsterdam.

Frost R 1960 Pressure variation over Malaya and the resonance theory. Scientific Paper No. 4, Air Ministry Meteorological Office, London.

Fu R, Liu W T, Dickinson R E 1996 Response of tropical clouds to the interannual variations of sea surface temperature. *J. Clim.*, **9**, 616–34.

Gamache J F, Houze R A 1982 Mesoscale air motions associated with a tropical squall line. *Mon. Wea. Rev.*, **110**, 118–35.

Gash J H, Shuttleworth W 1991 Tropical deforestation: albedo and the surface energy balance. *Climatic Change*, **19**, 123–33.

Gash J H, Nobre C A, Roberts J M, Victoria R L (eds) 1996 *Amazonian Deforestation and Climate*. John Wiley, Chichester.

Genthon C, Barnola J, Raymaud D, Lorius C, Barkov N, Korotkevich Y, Kotlykov V 1987 Vostok ice core: climatic response to CO_2 and orbital forcing changes over the last climatic cycle. *Nature*, **329**, 414–18.

George J E, Gray W M 1977 Tropical cyclone recurvature and non-recurvature as related to surrounding wind-height fields. *J. Appl. Met.*, **16**, 34–42.

Giambelluca T W 1991 *Drought in Hawaii*. State of Hawaii, Department of Land and Natural Resources, Commission on Water Resources Management, Report No. R88.

Giambelluca T W 1996 Tropical landcover change: characterizing the post-forest land surface. In Giambelluca T W, Henderson-Sellers A (eds) *Climate Change: Developing Southern Hemisphere Perspectives*. John Wiley, Chichester, pp. 293–318.

Giambelluca T W, Nullet D, Nullet M 1988 Agricultural drought on south central Pacific islands. *Prof. Geogr.*, **40**, 404–15.

Glantz M H, Katz R W, Nicholls N 1991 *Teleconnections Linking Worldwide Climate Anomalies*. Cambridge University Press, Cambridge.

Glantz M H, Krenz J 1992 Human components of the climate system. In Trenberth K E (ed.) *Climate System Modeling*. Cambridge University Press, Cambridge.

Glynn P W 1990 *Global Ecological Consequences of the 1982–83 El Nino-Southern Oscillation*. Elsevier, Amsterdam.

Goodess C M, Palutikof J, Davies T D 1992 *The Nature and Causes of Climate Change: Assessing the Long Term Future*. Belhaven Press, London.

Gourou P 1935 *Les Pays Tropicaux*. Paris, pp. 1–2.

Graham N E, White W B 1988 The El Nino cycle: a natural oscillator of the Pacific Ocean–

atmosphere system. *Science*, **240**, 1293–302.

Gray W M 1988 Seasonal frequency variations in the 40–50 day oscillation. *J. Climatol.*, **8**, 511–19.

Gray W M, Jacobson R W 1977 Diurnal variation of deep cumulus convection. *Mon. Wea. Rev.*, **105**, 1171–88.

Gray W M, Schaeffer J 1991 El Nino and QBO influences on tropical cyclone activity. In Glantz M H, Katz R W, Nicholls N (eds) *Teleconnections Linking World Wide Climate Anomalies.* John Wiley, Chichester, pp. 257–84.

Gray W M, Schaeffer J, Knaff J A 1992 Influence of the stratospheric QBO on ENSO variability. *J. Met. Soc. Japan*, **70**, 975–95.

Gray W M, Landsea C W, Mielke P W Jr, Berry K J 1994 Predicting Atlantic seasonal tropical cyclone activity by 1 June. *Weather Forecasting*, **9**, 103–15.

Gregorczuk M, Cena K 1967 Distribution of effective temperature over the globe. *Int. J. Biometeorol.*, **2**, 35–9.

Griffiths J F 1972 *Climates of Africa, Vol. 10 of World Survey of Climatology.* Amsterdam.

Gueremy J F 1990 Heat and moisture fluxes on a time scale of 20 to 60 days over the Indian monsoon area. *Meteorol. Atmos. Phys.*, **44**, 219–50.

Gueremy J F 1994 Influence of land atmosphere feedbacks on Indian monsoon intraseasonal variability. In *Proceedings of International Conference on Monsoon Variability and Prediction,* Trieste, Italy, 9–13 May, 1994. WCRP-84, WMO/TD-No. 619, pp. 653–60.

Harrison M S J 1984 A generalized classification of South African rain-bearing synoptic systems. *J. Climatol.*, **4**, 547–60.

Hastenrath S 1985 *Climate and Circulation of the Tropics.* D Reidel, Dordrecht.

Hastenrath S, Lamb P J 1977 *Climatic Atlas of the Tropical Atlantic and Eastern Pacific Ocean.* Madison, Wisconsin.

Hastenrath S, Lamb P J 1979 *Climatic Atlas of the Indian Ocean, Part 1, Surface Climate and Atmospheric Circulation.* Madison, Wisconsin.

Hastenrath S, Rosen A 1983 Patterns of India monsoon rainfall anomalies. *Tellus*, **35A**, 324–31.

Hastings P A 1990 Southern oscillation influences on tropical cyclone activity in the Australasian south-west Pacific region. *Int. J. Climatol.*, **10**, 291–8.

Henderson-Sellers A, McGuffie K 1987 *Climate Modelling: A Primer.* John Wiley, Chichester.

Henderson-Sellers A, Zhang H, Howe W 1996 Human and physical aspects of tropical deforestation. In Giambelluca T W, Henderson-Sellers A (eds) *Climate Change: Developing Southern Hemisphere Perspectives.* John Wiley, Chichester, pp. 259–92.

Hendon H H, Leibmann B 1990 A composite study of onset of the Australian monsoon. *J. Atmos. Sci.*, **47**, 2227–40.

Holland G J 1986 Interannual variability of the Australian summer monsoon at Darwin: 1952–1982. *Mon. Wea. Rev.*, **114**, 594–604.

Holland G J 1993 Ready Reckoner. In *Global Guide to Tropical Cyclone Forecasting.* World Meteorological Organization, Geneva, WMO/TC-No 560, Report No. TCP-31.

Houghten F C, Yaglou C P 1923 Determining lines of equal comfort. *ASHVE Trans.*, **29**, 163–76.

Hulme M M 1992 Rainfall changes in Africa 1931–1960 to 1961–1990. *Int. J. Climatol.*, **12**, 685–99.

Hulme M, Kelly M 1993 Exploring links between desertification and climate change. *Environ.*, **35**, 4–19.

Ilesanmi O O 1972 Aspects of the precipitation climatology of the July–August minimum of southern Nigeria. *J. Trop. Geog.*, **35**, 55–57.

Inoue M, Bigg G R 1995 Trends in wind and sea level pressure in the tropical Pacific Ocean for the period 1950–1979. *Int. J. Climatol.*, **15**, 35–52.

IPCC (Intergovernmental Panel on Climate Change) 1996a *Climate Change 1995: The Science of Climate Change.* Cambridge University Press, Cambridge.

IPCC (Intergovernmental Panel on Climate Change) 1996b *Climate Change 1995: Impacts, Adaptations and Mitigations of Climate Change: Scientific and Technical Analyses.* Cambridge University Press, Cambridge.

IPCC (Intergovernmental Panel on Climate Change) 1996c *Climate Change 1995: Economic and*

Social Dimensions of Climate Change. Cambridge University Press, Cambridge.

Iqbal M 1983 *Introduction to Solar Radiation*. Toronto.

Jackson I J 1972 Mean daily rainfall intensity and number of raindays over Tanzania. *Geograf. Annal.*, **54A**, 369–81.

Jackson I J 1981 Dependence of wet and dry days in the tropics. *Arch. Met. Geophys. Bioklimatol.*, Ser. B, **29**, 167–79.

Jackson I J 1986 Relationships between raindays, mean daily intensity and monthly rainfall in the tropics. *Int. J. Climatol.*, **6**, 117–34.

Jackson I T 1988 Daily rainfall variation over Northern Australia: deviations from the world pattern. *Int. J. Climatol.*, **8**, 463–76.

Jackson I J 1989 *Climate, Water and Agriculture in the Tropics*. Longman, London.

Jackson I T, Weinand H 1994 Towards a classification of tropical rainfall stations. *Int. J. Climatol.*, **14**, 263–86.

Janicot S 1992 Spatiotemporal variability of West African rainfall. Part II: Associated surface and airmass characteristics. *J. Clim.*, **5**, 499–511.

Janicot S 1994 The west African monsoons of 1987 and 1988: Pacific or Atlantic signal? In *Proceedings of International Conference on Monsoon Prediction and Variability*, Trieste, Italy, 9–13 May 1994. WCRP-84, WMO/TD No. 619, pp. 765–72.

Jauregui E 1986 The urban climate of Mexico City. In *Urban Climatology and its Application with Special Regard to Tropical Areas*. World Climate Programme, WMO Publication No. 652, World Meteorological Organisation, Geneva, pp. 26–45.

Jauregui E 1991 The human climate of tropical cities: an overview. *Int. J. Biometeorol.*, **35**, 151–60.

Jolliffe I T, Hope P B 1996 Representation of daily rainfall distributions using normalised rainfall curves. *Int. J. Climatol.*, **16**, 1157–64.

Jones P D 1991 Southern Hemisphere sea level pressure data: an analysis and reconstruction back to 1951 and 1911. *Int. J. Climatol.*, **11**, 585–608.

Jury M R, McQueen C, Levey K M 1994 SOI and QBO signals in the African region. *Theor. Appl. Climatol.*, **50**, 103–15.

Jury M R, Parker B A, Raholijao N, Nassor A 1995 Variability of summer rainfall over Madagascar: climatic determinants at interannual scales. *Int. J. Climatol.*, **15**, 1323–32.

Kelly P M, Jia P, Jones P D 1996 The spatial response of the climate system to explosive volcanic eruptions. *Int. J. Climatol.*, **16**, 537–50.

Kidron M, Segal R 1991 *The New State of the World Atlas*, 4th edition. New York.

Kiladis G, Sinha S K 1991 ENSO monsoon and drought in India. In Glantz M H, Katz R W, Nicholls N (eds) *Teleconnections Linking Worldwide Climate Anomalies*. Cambridge University Press, Cambridge.

Knaff J A, Gray W M 1994 Extended range prediction of Asian monsoon rainfall. In *Proceedings of International Conference on Monsoon Variability and Prediction*, Trieste, Italy, 9–13 May, 1994. WCRP-84, WMO/TD No. 619, pp. 436–43.

Kodama Y 1992 Large-scale common features of subtropical precipitation zones (the Baiu Front, the South Pacific Convergence Zone, the South Atlantic Convergence Zone). Part I: Characteristics of the Subtropical frontal zones. *J. Met. Soc. Japan*, **70**, 813–35.

Kodama Y 1993 Large-scale common features of subtropical precipitation zones (the Baiu Front, the South Pacific Convergence Zone, the South Atlantic Convergence Zone). Part II: Conditions for generating the subtropical convergence zones. *J. Met. Soc. Japan*, **71**, 581–610.

Koeppen W 1931 *Grundriss der Klimakunde*. Berlin.

Koeppen W 1936 *Das Geographische System der Klimate*. Berlin.

Koteswaram P 1958 The easterly jet in the tropics. *Tellus*, **10**, 43–57.

Kousky V E 1980 Diurnal rainfall variation in northeast Brazil. *Mon. Wea. Rev.*, **108**, 488–98.

Kraus H, Alkhalaf A 1995 Characteristic surface energy balances for different climate types. *Int. J. Climatol.*, **15**, 275–84.

Kripalani R H, Singh S V, Nalini P 1995 Variability of summer monsoon rainfall over Thailand – comparison with features over India. *Int. J. Climatol.*, **15**, 657–72.

Krishnamurti T N 1985 Summer monsoon experiment: a review. *Mon. Wea. Rev.*, **113**, 1590–

626.

Krishnamurti T N, Ardanuy P A 1980 The 10–20 day westward propagating mode and "breaks in the monsoons". *Tellus,* **32,** 15–26.

Krishnamurti T N, Bhalme H N 1976 Oscillations of a monsoon system, Part I: Observational aspects. *J. Atmos. Sci.,* **33,** 1937–54.

Krishnamurti T N, Molinair J, Pan H L, Wong V 1977 Downstrean amplification and formation of monsoon depressions. *Mon. Wea. Rev.,* **105,** 1281–97.

Kuhnel I 1989 Tropical–extratropical cloudband climatology based on satellite data. *Int. J. Climatol.,* **9,** 441–63.

Kuhnel I 1996 Relationship of Australian sugar cane yields to various climate variables. *Theor. Appl. Climatol.,* **54,** 217–28.

Kulkarni A K, Mandal B N, Sangam R B 1996 Analysis of severe rainstorms over Gujarat and Punjab during the 1993 monsoon. *Int. J. Climatol.,* **16,** 35–48.

Kutzbach J 1992 Modeling large climate changes of the past. In Trenberth K E (ed.) *Climate System Modeling.* Cambridge University Press, Cambridge, pp. 669–88.

Labitzke K, van Loon H 1990 Association between the 11 year solar cycle, the QBO and the atmosphere: a summary of recent work. *Phil. Trans. Roy. Soc. Lond.,* **330,** 577–90.

Lahiri M 1984 Indices for comfort analysis. *Mausam,* **35,** 275–6.

Lamb H H 1979 Climatic variations and changes in wind and ocean circulation: the Little Ice Age in the Northeast Atlantic. *Quatern. Res.,* **11,** 1–20.

Lamb P J, Peppler R A 1991 West Africa. In Glantz M H, Katz R W, Nicholls N (eds) *Teleconnections Linking Worldwide Climate Anomalies.* Cambridge University Press, Cambridge, pp. 121–89.

Landsea C W 1993 A climatology of intense (or major) Atlantic hurricanes. *Mon. Wea. Rev.,* **121,** 1703–13.

Lau K M, Chan P H 1986 The 40–50 day oscillation and the El Nino/Southern Oscillation: a new perspective. *Bull. Am. Met. Soc.,* **67,** 533–4.

Legates D R 1995 Global and terrestrial precipitation: a comparative assessment of existing climatologies. *Int. J. Climatol.,* **15,** 237–58.

Linacre E 1992 *Climate Data and Resources: A Reference and Guide.* Routledge, London.

Lindesay J A, Jury M R 1991 Atmospheric circulation and controls and characteristics of a flood event in central South Africa. *Int. J. Climatol.,* **11,** 609–28.

Liou K N 1992 *Radiation and Cloud Processes in the Atmosphere.* Oxford University Press, Oxford.

List R J 1958 *Smithsonian Meteorological Tables,* 6th edition. Smithsonian Institute Press, Washington, DC.

Lockwood J G 1974 *World Climatology, An Environmental Approach.* Arnold, London, p. 182.

Lough J M 1991 Rainfall variations in Queensland Australia 1891–1986. *Int. J. Climatol.,* **11,** 745–68.

Lough J M 1993 Variations in some seasonal rainfall characteristics in Queensland, Australia. *Int. J. Climatol.,* **13,** 391–409.

Madden R A, Julian P R 1971 Description of a 40–50 day oscillation in the zonal wind in the tropical Pacific. *J. Atmos. Sci.,* **28,** 701–8.

Madden R A, Julian P R 1972 Description of global scale circulation cells in the tropics with a 40–50 day period. *J. Atmos. Sci.,* **29,** 1109–23.

Madden R A, Julian P R 1994 Observations of the 40–50 day oscillation – a review. *Mon. Wea. Rev.,* **122,** 814–37.

Manshard W 1968 *Einfuehrung in die Agrargeographie der Tropen.* Mannheim, Bibl. Institut.

Marengo J A 1992 Interannual variability of surface climate in the Amazonian Basin. *Int. J. Climatol.,* **12,** 853–63.

Marengo J A 1995 Interannual variability of deep convection over tropical South America as deduced from ISCCP C2 data. *Int. J. Climatol.,* **15,** 995–1010.

Marengo J A, Hastenrath S 1993 Case studies of extreme climate events in the Amazonian Basin. *J. Clim.,* **6,** 617–27.

McAlpine J R, Keig G, Falls R 1983 *Climate of Papua New Guinea.* CSIRO, Canberra.

McBride J L 1987 The Australian summer monsoon. In Chang C P, Krishnamurti T (eds) *Monsoon Meteorology.* Oxford University Press, Oxford, pp. 203–31.

McBride J L, Nicholls N 1983 Seasonal relationships between Australian rainfall and the Southern Oscillation. *Mon. Wea. Rev.,* 111, 1998–2004.

McGregor G R 1988 Possible consequences of climatic warming in Papua New Guinea. In *Potential Impacts of Greenhouse Gas Generated Climatic Change and Projected Sea Level Rise on Pacific Island States of the South Pacific Regional Environmnetal Programme.* 4th Consultative Meeting of Research and Training Institutions and Second Intergovernmental Meeting on the South Pacific Regional Environmental Programme, Noumea, New Caledonia, June–July 1988, pp. 3–15.

McGregor G R 1989 An assessment of the annual variability of rainfall: Port Moresby, Papua New Guinea. *J. Trop. Geog.,* 10, 43–54.

McGregor G R 1992 The spatial and temporal characteristics of rainfall anomalies in Papua New Guinea and their relationship to the Southern Oscillation. *Int. J. Climatol.,* 12, 449–68.

McGregor G R 1995a The human bioclimates of western and south Pacific and climate change. *Int. J. Biometeorol.,* 39, 5–12.

McGregor G R 1995b The tropical cyclone hazard over the South China Sea, 1970–1989: annual spatial and temporal characteristics. *Appl. Geog.,* 15, 35–52.

McGregor G R 1995c Theoretical current and future indoor thermal referendum for equatorial western and tropical south Pacific. *Theor. Appl. Climatol.,* 50, 227–33.

McGuffie K, Henderson-Sellers A, Zhang H, Durbridge T, Pitman A 1995 Global climate sensitivity to tropical deforestation. *Global and Planetary Change,* 10, 97–128.

McIlveen R 1992 *Fundamentals of Weather and Climate.* Chapman Hall, London.

McMichael A, Haines A, Sloof R, Kovats A 1996 *Climate Change and Human Health.* World Health Organisation, Geneva.

Meehl G A 1987 The annual cycle and interannual variability in the tropical Pacific and Indian Ocean regions. *Mon. Wea. Rev.,* 115, 27–50.

Meher-Homji V M 1991 Probable impact of deforestation on hydrological processes. *Climatic Change,* 19, 63–73.

Meyers N 1991 Tropical forests: present status and future outlook. *Climatic Change,* 19, 3–32.

Miller A A 1971 *Climatology, 9th Edition.* Methuen, London.

Mink J F 1960 Distribution pattern of rainfall in the leeward Koolau mountains, Ohau, Hawaii. *J. Geophys. Res.,* 65, 2869–76.

Mohr E C J, van Baren F A and Van Schuylenborgh J 1972 *Tropical Soils,* 3rd edition. The Hague, pp. 24–5.

Mooley D A, Parthasarathy B 1984 Fluctuations in all-India summer monsoon rainfall during 1871–1978. *Climatic Change,* 6, 287–301.

Morgan R P C 1974 Estimating regional variations in soil erosion hazard in Peninsula Malaysia. *Malay. Nat. J.,* 28, 94–106.

Morgan R P C, Hatch T, Sulaiman W 1982 A simple procedure for assessing soil erosion risk: a case study for Malaysia. *Z. f. Geomorph. Suppl.,* 44, 69–89.

Moron V, Bigot S, Roucou P 1995 Rainfall variability in subequatorial America and Africa and relationships with the main sea surface temperature modes 1951–1990. *Int. J. Climatol.,* 15, 1297–322.

Murakami M 1983 Analysis of the deep convective activity over the western Pacific and southeast Asia: Part I: Diurnal variation. *J. Met. Soc. Japan,* 61, 60–75.

Murakami T 1979 Winter monsoon surges over East and Southeast Asia. *J. Meteor. Soc. Japan,* 57, 113–58.

Neumann C J 1993 Global overview in *Global Guide to Tropical Cyclone Forecasting.* World Meteorological Organisation, Geneva, WMO/TC-No 560, Report No. TCP-31.

Newell R E 1979 Climate and the ocean. *Amer. Sci.,* 67, 405–16.

Newell R E, Kidson J W 1984 African mean wind changes between Sahelian wet and dry periods. *J. Climatol.,* 4, 27–33.

Nicholls N 1984 The Southern Oscillation, sea-surface temperatures and interannual fluctuations in Australian tropical cyclone activity. *J. Climatol.,* 4, 661–70.

Nicholls N 1992 Recent performance of a method for forecasting Australian seasonal tropical cyclone activity. *Aust. Met. Mag.*, **40**, 105–10.

Nicholls N, Wong K K 1990 Dependence of rainfall variability on mean rainfall, latitude and the southern oscillation. *J. Clim.*, **3**, 163–70.

Nicholls N, Griza G, Jouzel J, Karl T, Ogallo L, Parker D 1996 Observed climate variability and change. In *Climate Change 1995: The Science of Climate Change*. Cambridge University Press, Cambridge, pp. 133–92.

Nicholson S E 1979 Revised rainfall series for West African subtropics. *Mon. Wea. Rev.*, **107**, 620–3.

Nicholson S E 1988 *Atlas of African Rainfall and its Interannual Variability*. Tallahassee.

Nieuwolt S 1966a A comparison of rainfall during 1963 and average conditions in Malaya. *Erdkunde*, **20**, 180.

Nieuwolt S 1966b The urban microclimate of Singapore. *J. Trop. Geog.*, **22**, 31–4.

Nieuwolt S 1968 Uniformity and variation in an equatorial climate. *J. Trop. Geog.*, **27**, 23–39.

Nieuwolt S 1973 Rainfall and evaporation in Tanzania. BRALUP Research Paper No. 24, Dar es Salaam, pp. 7, 14.

Nieuwolt S 1974 Seasonal rainfall distribution in Tanzania. *Geografiska Annaler, Series A*, **56**, 241–50.

Nieuwolt S 1977 *Tropical Climatology*. John Wiley, Chichester.

Nieuwolt S 1981 The climates of continental Southeast Asia. *World Survey of Climatology*, **9**, 1–66.

Nieuwolt S 1982 *Climate and Agricultural Planning in Peninsular Malaysia*. MARDI, Serdang, Malaysia.

Nieuwolt S 1986 Agricultural droughts in the tropics. *Theor. Appl. Climatol.*, **37**, 29–38.

Nieuwolt S 1989 Estimating the agricultural risks of tropical rainfall. *Agric. For. Met.*, **45**, 251–63.

Nieuwolt S 1991 Climatic uniformity and diversity in the Galapagos Islands and the effects on agriculture. *Erdkunde*, **45**, 134–14.

Nullet D, Giambelluca T W 1988 Risk analysis of agricultural drought on low Pacific Islands. *Agr. For. Met.*, **42**, 229–39.

Ogallo L 1979 Rainfall variability in Africa. *Mon. Wea. Rev.*, **107**, 1133–9.

Oguntoyinbo J S 1986 Some aspects of the urban climates of tropical Africa. In *Urban Climatology and Its Application with Special Regard to Tropical Areas*. World Climate Programme, WMO Publication No. 652, World Meteorological Organisation, Geneva, pp. 166–78.

Oguntoyinbo J S, Akintola F O 1983 Rainstorm characteristics affecting water availability for agriculture. In Keller R (ed.) *Hydrology of Humid Tropical Regions*. IAHS Publication No. 140, pp. 63–72.

Oke T R 1986 Urban climatology and the tropical city. In *Urban Climatology and Its Application with Special Regard to Tropical Areas*. World Climate Programme, WMO Publication No. 652, World Meteorological Organisation, Geneva, pp. 1–25.

Oke T R 1987 *Boundary Layer Climates*. Methuen, London.

Olaniran O 1982 The physiological climate of Ilorin, Nigeria. *Arch. Geophys. Bioklimatol.*, **31**, 287–99.

Olaniran O J 1987 A study of the seasonal variation of rain days and rainfall of different categories in Nigeria in relation to the Miller station types for tropical continents. *Theor. Appl. Climatol.*, **38**, 198–209.

Oquist M, Svensson B 1996 Non-tidal wetlands. In *Climate Change 1995: Impacts, Adaptations and Mitigations of Climate Change: Scientific and Technical Analyses*. Cambridge University Press, Cambridge, pp. 215–40.

Padmanabhamurty B, 1986 Some aspects of the urban climates of India. In *Urban Climatology and Its Application with Special Regard to Tropical Areas*. World Climate Programme, WMO Publication No. 652, World Meteorological Organisation, Geneva, pp. 136–65.

Paffen K 1967 Das Verhaeltnis der Tages zu Jahreszeitlichen Temperaturschwankung. *Erdkunde*, **21**, 94–111.

Palmen E, Newton C W 1969 *Atmospheric Circulation Systems*. Academic Press, New York.

Palmer T N 1986 Influence of Atlantic, Pacific and Indian Oceans on Sahel rainfall. *Nature*, **322**, 251–3.

Palmer T N, Anderson D L T 1994 The prospects for seasonal forecasting – a review paper. *Q. J. R. Met. Soc.*, **120**, 744–93.

Palmer W C 1965 *Meteorological Drought*. US Weather Bureau Research Paper No. 45.

Parker D E, Wilson W, Jones P D, Christy J R, Folland C K 1996 The impact of Mt Pinatubo on world-wide temperatures. *Int. J. Climatol.*, **16**, 487–99.

Penman H L 1948 Natural evaporation from open water, bare soil and grass. *Proc. Roy. Soc. Series A*, **193**, 120–45.

Pernetta J C 1992 Impacts of climate change and sea level rise on small island states: national and international responses. *Global Environ. Change*, **2**, 19–31.

Peterson J 1984 Global population projections through the 21st Century: a scenario for this issue. *Ambio*, **13**, 134–41.

Philander S G H 1983 El Nino southern oscillation phenomena. *Nature*, **302**, 295–301.

Piexoto J P and Oort A H 1992 *Physics of Climate*. American Institute of Physics, New York.

Pittock A B, Salinger J M 1982 Towards regional scenarios for CO_2 warmed earth. *Climatic Change*, **4**, 23–40.

Puvaneswaran K M, Smithson P A 1991 Precipitation–elevation relationships over Sri Lanka. *Theor. Appl. Climatol.*, **43**, 113–22.

Quinn W H, Zopf D O, Short K S and Kuo Yang R T W 1978 Historical trends and statistics of the Southern Oscillation, El Nino and Indonesian droughts. *Fishery Bulletin*, **76**, 663–78.

Ramage C S 1955 The cool season tropical disturbances of southeast Asia. *J. Meteorol.*, **12**, 257.

Ramage C S 1964 Diurnal variation of summer rainfall of Malaya. *J. Trop. Geog.* **19**, 66.

Ramage C S 1968 Role of a tropical 'maritime continent' in the atmospheric circulation. *Mon. Wea. Rev.*, **96**, 365.

Ramage C S 1971 *Monsoon Meteorology*. Academic Press, New York.

Rao V B, Hada K 1990 Characteristics of rainfall over Brazil: annual variations and connections with the southern oscillation. *Theor. Appl. Climatol.*, **42**, 81–91.

Rao V B, Satyamurty P, de Brito J I B 1986 On the 1983 drought in northeast Brazil. *J. Climatol.*, **6**, 43–51.

Rao V B, Hada K, Herdies D L 1995 On the severe drought of 1993 in northeast Brazil. *Int. J. Climatol.*, **15**, 697–704.

Rapino M R, Self S 1982 Historic eruptions of Tambora (1815), Krakatoa (1883) and Agung (1963) their stratospheric aerosols and climatic impact. *Quatern. Res.*, **18**, 127–43.

Rasmusson E M, Arkin P 1993 A global view of large-scale precipitation variability. *J. Clim.*, **6**, 1495–522.

Rasmusson E M, Carpenter T H 1982 Variations in tropical sea surface temperature and surface wind field associated with Southern Oscillation/El Nino. *Mon. Wea. Rev.*, **110**, 354–84.

Rasmusson E M, Wallace J M 1983 Meteorological aspects of the El Nino/Southern Oscillation. *Science*, **222**, 1195–202.

Reading A J 1990 Caribbean tropical storm activity over the past four centuries. *Int. J. Climatol.*, **10**, 365–76.

Reed R J, Norquist D C, Recker E E 1977 The structure and properties of African disturbances as observed during phase III of GATE. *Mon. Wea Rev.*, **105**, 317–33.

Reilly J 1996 Agriculture in a changing climate: impacts and adaptations. In *Climate Change 1995: Impacts, Adaptations and Mitigations of Climate Change: Scientific and Technical Analyses*. Cambridge University Press, Cambridge, pp. 427–68.

Restelli G, Angeletti G 1993 *Dimethyl Sulphide: Ocean, Atmosphere and Climate*. Kluwer Academic, London.

Reynolds R 1985 Tropical meteorology. *Prog. Phys. Geog.*, **9**, 157–86.

Riehl H 1954 *Tropical Meteorology*. McGraw-Hill, New York.

Rooy M P 1965 A rainfall anomaly index independent of time and space. *Notos*, **14**, 43.

Ropelewski C F and Halpert M S 1989 Global and regional scale precipitation patterns associated with El Nino/Southern Oscillation. *Mon. Wea. Rev.*, **114**, 2352–62.

Rossby C G 1947 On the general circulation of the atmosphere in the middle latitudes. *Bull. Amer. Met. Soc.*, **28**, 255–80.

Rossow W B 1989 Measuring cloud properties from space: a review. *J. Clim.*, **2**, 201–13.

Rossow W B, Walker A W, Garder L C 1993 Comparison of ISCCP and other cloud amounts. *J. Clim.*, **6**, 2394–418.

Rowell D P, Milford J R 1993 On the generation of African squall lines. *J Clim.*, **6**, 1181–93.

Rowntree K M 1988 Storm rainfall on the Njemps Flats, Baringo District, Nigeria. *Int. J. Climatol.*, **8**, 297–309.

Ruddiman W F, Kutzbach J E 1991 Plateau uplift and climatic change. *Sci. Amer.*, **264**, 42–50.

Sagan C, Toon O B, Pollack J B 1979 Anthropogenic albedo changes and the earth's climate. *Science*, **206**, 1363–8.

Salinger M J, Basher R, Fitzharris B B, Hay J E, Jones P D, MacVeigh J P, Schidely Leleu I 1995 Climate trends in the southwest Pacific. *Int. J. Climatol.*, **15**, 285–302.

Schroeder T A, Kilonsky B J, Ramage C S 1978 Diurnal rainfall variability over the Hawaiian Islands. In *11th Technical Conference on Hurricanes and Tropical Meteorology*. American Meteorological Society, Boston, pp. 72–7.

Sear C B, Kelly P M, Jones P D, Goodess C M 1987 Global surface temperature response to major volcanic eruptions. *Nature*, **346**, 453–6.

Sellers W D 1965 *Physical Climatology*. University of Chicago Press, Chicago.

Shapiro L J 1982 Hurricane climatic fluctuations, part 1: pattern and cycles. *Mon. Wea. Rev.*, **110**, 1007–13.

Shukla J 1987 Long-range forecasting of monsoons. In Fein J S, Stephens P L (eds) *Monsoons*. John Wiley, New York, pp. 523–48.

Shukla J, Misra B M 1977 Relationships between sea surface temperature and wind speed over the central Arabian Sea, and monsoon rainfall over India. *Mon. Wea. Rev.*, **105**, 998–1002.

Shukla J, Paolino D A 1983 The southern oscillation index and long-range forecasting of the summer monsoon rainfall in India. *Mon. Wea. Rev.*, **111**, 1830–7.

Skinner N, Tapper N 1994 Preliminary sea breeze studies over Bathurst and Melville Islands, Northern Australia as part of the Island Thunderstorm Experiment (ITEX). *Met. Atmos. Phys.*, **53**, 77–94.

Stern R D, Coe R 1982 The use of rainfall models in agricultural planning. *Agric. Met.*, **26**, 35–43.

Stern R D, Dennett M D, Garbutt D J 1981 The start of the rains in West Africa. *J. Climatol.*, **1**, 59–70.

Sumner G 1988 *Precipitation Process and Analysis*. John Wiley, Chichester.

Suppiah R 1992 The Australian summer monsoon: a review. *Prog. Phys. Geog.*, **16**, 283–318.

Taylor C M and Lawes E F 1971 *Rainfall Intensity-Duration-Frequency Data for Stations in East Africa*. E.A.M.D., Techn. Memo. No 17, Nairobi.

Thom E C 1958 Cooling degree days. *Air Conditioning, Heating and Ventilating*, **22**, 65–72.

Thompson D 1965 *The Climate of Africa*. Oxford University Press, Nairobi.

Thorncroft C D, Hoskins B J 1994 An idealized study of African easterly waves. I: A linear view. *Q. J. R. Met. Soc.*, **120**, 953–82.

Thornthwaite C W 1943 Problems in the classification of climates. *Geogr. Rev.*, **33**, 233–55.

Thornthwaite C W 1948 An approach toward a rational classification of climate. *Geogr. Rev.*, **38**, 55–94.

Thornthwaite C W, Mather J R 1957 Instruction and tables for computing potential evapotranspiration and the water balance. Publications in Climatology, Drexel Institute of Technology, Vol. X, p. 244.

Trenberth K E 1991 General characteristics of El-Nino Southern Oscillation. In Glantz M H, Katz R W, Nicholls N (eds) *Teleconnections Linking Worldwide Climate Anomalies*. Cambridge University Press, Cambridge.

Trenberth K E (ed.) 1992 *Climate System Modeling*. Cambridge University Press, Cambridge.

Trenberth K E, Paolino D A 1980 The Northern Hemisphere sea level pressure data set: trends, errors and discontinuities. *Mon. Wea. Rev.*, **108**, 855–72.

Trent E M, Gathman S G 1972 Oceanic thunderstorms. *Pure Appl. Geophys.*, **100**, 60–9.

Troll C 1959 Die tropischen Gebirge. Ihre dreidimensionale klimatische und pflanzengeographische Gliederung. *Bonner Geogr. Abhandl.*, **25**.

US Navy Marine Climatic Atlas of the World. Vol. II, North Pacific Ocean 1977; Vol. III, Indian Ocean 1957; Vol. IV, South Atlantic Ocean 1978; Vol. V, South Pacific Ocean 1979; Vol. IX, World Wide Means and Standard Deviations 1981; Washington, DC.

US Weather Bureau 1959 Notes on temperature–humidity index. LS 5922, Washington, DC.

Van Loon H, Shea D J 1985 The Southern Oscillation. Pt IV: The precursors south of 15°S to the extremes of the oscillation. *Mon. Wea. Rev.*, **113**, 2063–74.

Vincent D G 1994 The South Pacific Convergence Zone (SPCZ): a review. *Mon. Wea. Rev.*, **124**, 1949–70.

Waliser D E, Gautier C 1993 A satellite derived climatology of the ITCZ. *J. Clim.*, **6**, 2162–74.

Wallace J M, Hobbs P V 1977 *Atmospheric Science: An Introductory Survey*. Academic Press, London.

Wang B 1994 On the annual cycle in the tropical eastern-central Pacific. *J. Clim.*, **7**, 1926–42.

Wang B 1995 Interdecadal changes in the El Nino onset in the last four decades. *J. Clim.*, **8**, 267–85.

Wang B and Murakami T 1994 Summer monsoon in the eastern North Pacific. In *Proceedings of the International Conference on Monsoon Variability and Predictability*, Trieste, Italy, July 1994. WCRP-84, WMO/TD No. 619, pp. 112–18.

Ward M N 1992 Provisionally corrected surface wind data, worldwide ocean–atmosphere fields and Sahelian rainfall variability. *J. Clim.*, **5**, 454–75.

Warren S G, Hahn C J, London K, Chervin R M, Jenne R L 1986 *Global Distribution of Total Cloud Cover and Cloud Type Amounts Over the Land*. NCAR Technical Note TN-273 + STR, Boulder, Colorado.

Warren S G, Hahn C J, London K, Chervin R M, Jenne R L 1988 *Global Distribution of Total Cloud Cover and Cloud Type Amounts Over the Ocean*. NCAR Technical Note TN-317 + STR, Boulder, Colorado.

Warrick R A, Le Provost C, Meier M F, Oerlemans J, Woodworth P L 1996 Changes in sea level. In *Climate Change 1995: The Science of Climate Change*. Cambridge University Press, Cambridge, pp. 359–406.

Watts I E M 1955 *Equatorial Weather*. University of London Press, London.

Webster P J 1983 Mechanisms of monsoon low frequency variability: surface hydrological effects. *J. Atmos. Sci.*, **40**, 2110–24.

Webster PJ 1987 The elementary monsoon. In Fein J S and Stephens P L (eds) *Monsoons*. John Wiley, New York, pp. 3–32.

Webster P J, Yang, S 1992 Monsoon and ENSO: selectively interactive systems. *Q. J. R. Met. Soc.*, **118**, 877–926.

Webster P J, Clayson C A, Curry J A 1996 Clouds radiation and the diurnal cycle of seas surface temperature in the tropical western Pacific. *J. Clim.* **9**, 1712–30.

Weisner C J 1970 *Hydrometeorology*. Chapman and Hall, London.

Wells N 1986 *The Atmosphere and Ocean: A Physical Introduction*. Taylor and Francis, London.

Wigley T M 1995 Global mean temperatures and sea level consequences of greenhouse gases stabilization. *Geophys. Res. Lett.*, **22**, 45–8.

Wigley T M, Jones P D, Kelly P M 1980 Scenarios for a warm high CO_2 world. *Nature*, **283**, 17–21.

Wilhite D A, Glantz M H 1987 Understanding the drought phenomenon: the role of drought definitions. In Wilhite D A, Easterling W E (eds) *Planning for Drought, Toward a Reduction of Societal Vulnerability*. Westview Press, Boulder Colorado, pp. 11–27.

Williamson T J, Coldicutt S, Penny R E C 1991 Aspects of thermal preferences in housing in a hot humid climate, with particular reference to Darwin. *Australia. Int. J Biometeorol.*, **34**, 251–8.

WMO 1995 *The Global Climate System Review*. World Meteorological Publication No. 819.

World Bank Atlas 1992 25th edition. Washington, DC.

World Market Atlas 1992 Bus. Intern. Corp., New York.

World Survey of Climatology, Vol. 9, S and W Asia 1981; Vol. 10, Africa 1972; Vol. 12, Central and South America 1976; Vol. 15, The Oceans 1984. Amsterdam and New York.

Wright I R, Gash J H C, Rocha H R, Shuttleworth W J, Nobre C A, Carvallo P R A, Leitao M V B R, Maitelli G T, Zamparoni C A G P 1992 Dry season micrometeorology of Amazonian ranchland. *Q. J. R. Met. Soc.*, **118**, 1083–99.

Wright W J 1993 Tropical–extratropical cloudbands and Australian rainfall. Preprint for the *4th Int. Conf. on Southern Hemisphere Meteorology and Oceanography*, Hobart, Australia, March 29–April 2 1993. American Meteorological Society, pp. 283–4.

Wycherley P R 1963 Variation in the performance of hevea in Malaya. *J. Trop. Geog.*, **17**, 155–6.

Yang S 1996 Snow–monsoon associations and seasonal–interannual predictions. *Int. J. Climatol.*, **16**, 125–34.

Yasunari T 1980 A quasi-stationary appearance of the 30–40 day period in the cloudiness fluctuations during the summer monsoon over India. *J. Met. Soc. Japan*, **58**, 225–9.

Yasunari T 1990 The impact of the Indian monsoon on the coupled atmosphere/ocean system in the tropical Pacific. *Meteorol. Atmos. Phys.*, **44**, 29–41.

Zangvil A 1975 Temporal and spatial behaviour of large scale disturbances in tropical cloudiness deduced from satellite brightness data. *Mon. Wea. Rev.*, **103**, 904–20.

Zhang G J, Ramanathan V, McPhaden M J 1995 Convection–evaporation feedback in the equatorial Pacific. *J. Clim.*, **8**, 3040–51.

Zhang H A, Henderson-Sellers A, McGuffie K 1996a Impacts of tropical deforestation. Part I: Process analysis of local climatic change. *J. Clim.*, **9**, 1497–517.

Zhang H A, McGuffie K, Henderson-Sellers A 1996b Impacts of tropical deforestation. Part II: The role of large-scale dynamics. *J. Clim.*, **9**, 2498–521.

Author index

Geographical index

Subject Index